Recent Progress of Battery Design, Modeling and Testing in Electric Vehicles

Recent Progress of Battery Design, Modeling and Testing in Electric Vehicles

Guest Editors

Jie Deng

Chulheung Bae

MDPI

Basel • Beijing • Wuhan • Barcelona • Belgrade • Novi Sad • Cluj • Manchester

Guest Editors

Jie Deng
Energy Systems and Applied Sciences
Ford Motor Company
Dearborn
USA

Chulheung Bae
Electrification and Energy Solutions
Caterpillar Inc.
Peoria
USA

Editorial Office
MDPI AG
Grosspeteranlage 5
4052 Basel, Switzerland

This is a reprint of the Special Issue, published open access by the journal *Batteries* (ISSN 2313-0105), freely accessible at: https://www.mdpi.com/journal/batteries/special_issues/39X1EPZF68.

For citation purposes, cite each article independently as indicated on the article page online and as indicated below:

Lastname, A.A.; Lastname, B.B. Article Title. *Journal Name* **Year**, *Volume Number*, Page Range.

ISBN 978-3-7258-6600-7 (Hbk)
ISBN 978-3-7258-6601-4 (PDF)
https://doi.org/10.3390/books978-3-7258-6601-4

Contents

About the Editors

Jie Deng

Jie Deng is a technical expert in high-voltage battery systems and analysis in the Department of Energy Systems and Applied Science at Ford Motor Company. He has extensive experience in battery array/system development and validation, including structural design, thermal management system, and thermal runaway mitigation. He also leads computer-aided engineering (CAE) analysis and optimization in various automotive applications, such as safety, durability, thermal, and packaging. He obtained his PhD degree in Mechanical Engineering from Florida State University and has published 40+ papers with 2000+ citations. In his current role, he works on improving battery performance across several key attributes to support next-generation electrified vehicles.

Chulheung Bae

Chulheung Bae is a Chief Engineer at Caterpillar Inc., where his research activities focus on various chemistry cell technologies and validations for machine applications. Dr. Bae has over 24 years of experience in advanced battery materials and various energy storage devices including lithium ion, NiZn, lead–acid, and redox flow batteries and ultra-capacitors. Dr. Bae has a Doctorate in Chemical Engineering from the University of Manchester in the UK.

Preface

Driven by advances in battery technologies, electrified vehicles (EVs) have become more and more acceptable, and their market share has increased rapidly in recent years. Nevertheless, to compete with internal combustion engine vehicles, EV batteries still need to overcome some barriers, such as energy density, fast charging capacity, safety, and durability. This reprint gathers contributions from leading researchers in battery technology and covers a diverse range of EV battery development that aims to improve one or multiple key attributes of EV batteries.

We thank all authors for their high-quality contributions and reviewers for their thoughtful feedback that elevated the quality of this reprint. We are also grateful to the editorial team at *Batteries* for their professional support. We hope that this reprint will serve as a valuable resource for readers, shed light on the recent progress in improving EV battery performance, and inspire more thoughts in advancing battery technologies that pave the road for future EV adoption.

Jie Deng and Chulheung Bae
Guest Editors

batteries

MDPI

Article

Influence of Cell Selection and Orientation within the Traction Battery on the Crash Safety of Electric-Powered Two-Wheelers

Alessio Sevarin, Markus Fasching *, Marco Raffler and Christian Ellersdorfer *

Vehicle Safety Institute, Graz University of Technology, Inffeldgasse 23, 8010 Graz, Austria
* Correspondence: fasching@tugraz.at (M.F.); christian.ellersdorfer@tugraz.at (C.E.)

Abstract: The crash safety of lithium-ion traction batteries is a relevant concern for electric vehicles. Current passive safety strategies of traction batteries usually come at the cost of their volumetric or gravimetric energy density. This work analyses the influence of the variables cell selection and orientation within the traction battery on the crash safety of an electric-powered two-wheeler. These two variables do not negatively influence the traction battery's volumetric or gravimetric energy density in the design process. Metamodels and numerical simulations are used to evaluate the crash safety of an electric-powered two-wheeler's traction battery in a potentially dangerous crash scenario. The influence of the variable's cell selection and orientation is evaluated through the internal short circuit risk of the integrated cells. The comparison of the metamodels shows that the cell orientation reduces the internal short circuit risk by up to 51% on average in the analysed crash scenario. The cell selection reduces it only up to 21% on average. The results show that crash safety can be increased in the design process, and a combination with the current protection strategies can increase crash safety further.

Keywords: Li-ion batteries; battery safety; integration of battery system; design of battery system; crash safety of battery system

1. Introduction

Lithium-ion (Li-ion) represents nowadays the most common technology for the cells of the traction batteries of electric vehicles (EVs), including electric-powered two-wheelers (E-PTWs), thanks to its high volumetric and gravimetric energy density compared to other technologies [1]. Li-ion cells are available in three main shapes: cylindrical, prismatic, and pouch, which differ in construction [2]. Different formats (as an example for cylindrical cells in 18650, 21700, 4680, etc.) [3] and different chemistries (e.g., $LiCoO_2$, $LiFePO_4$, $LiMn_2O_4$, etc.) [4,5] are available for each shape.

Independent of the cell shape, relevant hazards, such as the venting of toxic gases, the release of harmful liquids, or the combustion of the entire cell due to the uncontrolled temperature rise called thermal runaway, can arise in case of abuse conditions due to thermal, electrical, and mechanical loads, such as those acting on the cells in the case of a crash [6,7].

The safety of Li-ion cells is commonly evaluated through experimental tests defined in various norms and standards [8–10] to identify safe cells for vehicle integration. The cells are subjected to abuse conditions representative of different load scenarios, inter alia mechanical loading [11–13]. The experimental tests evaluate the safety of a cell based on the cell reactions achieved during the test and categorise it through the EUCAR hazard level table [14]. A maximal level of four is usually aimed for vehicle integration [15]. This approach allows us to identify cells with potentially dangerous behaviour. However, it does not allow us to forecast the safety performance of the cell on the crash safety of the traction battery after integration into the vehicle. The effect of the cell is superimposed with other variables (e.g., load introduction points or energy dissipation components) introduced in the integration process.

The design and integration of the traction battery in EVs play a relevant role in its crash safety in case of a collision [16,17].

With the aim to reduce the risk of an internal short circuit (ISC) due to mechanical loads, commonly the trigger of hazardous reactions under crash loads, several passive safety strategies are present in the current state of the art. These strategies can be categorised into three working principles: (1) increase the energy dissipation performance of the traction battery, (2) reduce the intrusion into the traction battery, and 3) manipulate the load path In case of a crash.

Energy dissipation components to improve the crashworthiness of the traction battery can be applied to the traction battery housing [18,19], its connection to the frame [20], or inside the modules [11]. In all approaches, the authors of the studies achieved a reduction in the number of cells experiencing an ISC. However, these approaches decrease the traction battery's volumetric energy density or increase the traction battery's mass due to the volume and mass needed to integrate the energy dissipation components [19,21].

The intrusion in the traction battery, which can potentially damage the inner cells, can be reduced using stiff traction battery housing. By stiffening the traction battery housing or parts of it, a higher mechanical load is necessary to produce an intrusion and potentially damage the traction battery cells [19]. Various approaches can be found in the current state of the art using this principle. One research direction consists of optimising the geometry of the traction battery housing by determining the optimal thickness of each part of the traction battery enclosure [22] or optimising the geometry of the case itself [17].

In particular, Shui et al. [22] showed, by applying a multi-step optimisation process, the possibility of reducing the deformation of the traction battery in case of crash with similar loads by circa 22%, while also reducing the mass of the enclosure by circa 11%.

Further studies focus on the increase in the stiffness of the traction battery housing using different materials, such as composite material [23,24] or high-strength alloys [16,25], for the traction battery housing or part of it. Pan et al. [16] achieved a mass reduction of circa 10% using high-strength steels for the traction battery housing while maintaining the crash safety requisites for the integration of the traction battery into the vehicle.

Nevertheless, optimising the traction battery's geometry or using high-strength materials does not always lead to a traction battery housing mass reduction, as shown by Szabo et al. [17]. Furthermore, the modification of the stiffness of the structure [26], by means of geometry or material modification can influence the response of the structure to external vibrations and the dynamic response of the vehicle itself under driving conditions. Therefore, the stiffness increase process is commonly conducted in combination with vibrational analyses to ensure the achievement of a feasible structure for the vehicle integration, as in the work of Shui et al. [22].

Faßbender et al. [27] proposed manipulating the traction battery load path in case of a crash by adopting a damage-tolerant traction battery through a module design based on a triangular shape. This allows a relative movement of the modules in the event of traction battery deformation, thus increasing the intrusion into the traction battery needed to damage the cells. Faßbender et al. [27] provide no information on the effects in terms of cells' protection or changes in the traction battery's gravimetric and volumetric energy density of the traction battery. However, such an effect can be expected due to the extra volume of the module housings needed to allow their relative displacement.

The strategies presented in the aforementioned studies have been successfully verified in a numerical environment. However, they can have a negative impact on the volumetric or gravimetric energy density of the traction battery due to added components or stiffening structures.

This study aims to investigate the use of mechanical capabilities of cells to absorb deformation energy until ISC occurs in the case of a crash. Therefore, the influence of the two variables, cell selection and orientation, on the crash safety of an E-PTW traction battery is investigated. The cell selection and orientation within the traction battery are two variables in the design process of a traction battery, which come at no volumetric or gravimetric energy density cost. The approach presented in this work represents an innovation in the current state of the art as it uses the cells' mechanical properties to improve the traction battery's crash safety without integrating further components into the traction battery that reduce its gravimetric or volumetric energy density.

2. Materials and Methods

To achieve the goal of this study, a finite element (FE) model of the entire E-PTW is built. The EPTW FE model is subjected to a potentially dangerous crash load condition through FE simulations, analysing its crash safety with the traction battery cells' internal short circuit risk (ISCR).

With the FE model, the influence of the two variables is evaluated through:

1. The cell selection, by considering two 18650 cells from different manufacturers;
2. The cells orientation within the traction battery, by considering two different cell orientations.

Additionally, a thickness range of two components of the traction battery housing defining its mass and crashworthiness is analysed to evaluate the influence of cell selection and orientation independent from the crashworthiness of a specific traction battery housing. This measure ensures the overall validity of the study.

Metamodels of the ISCR as a function of the FE model variables are built to analyse the influence of the cell selection and orientation on the crash safety of the traction battery beyond the simulated cases, reducing the computational effort of the study.

The comparison of the ISCR trends of the traction battery of the E-PTW with different integrated cells and cell orientation permits us to observe their influence on the crash safety of the traction battery.

2.1. Reference E-PTW, Cells, and Crash Scenario

This work is based on an E-PTW concept of category L3e-A2 [28] for urban and commuting purposes. The E-PTW concept can integrate two structural (i.e., substituting the vehicle frame) traction battery concepts. Both integrated traction battery concepts (Concept Y and Concept Z) are composed of three identical modules (216 cells per module) connected in their series (see Figure 1) but differing in their orientation. Each module consists of 18650 cells with a 24s9p configuration. The modules are oriented with the cells' longitudinal axis in the y_M direction in Concept Y (see Figure 1a) and with an angle of 25.5° relative to the z_M direction in Concept Z (see Figure 1b). The traction battery concepts integrate the connection to the steering head, motor, and swingarm. The volume of Concept Y and Concept Z is the same; therefore, integrating the same cells results in an identical volumetric energy density.

This study considers two commercial Li-ion cylindrical cells of the 18650 format with the chemical composition Lithium-Nickel-Manganese-Cobalt-Oxide (commonly called NMC). The two cells, referred to in this study as cell A and B, have the same format and do have the same dimensions according to the datasheets, namely a nominal diameter of 18 mm and an axial length of 65 mm. The cells are from different manufacturers and differ minimally in their electrical specifications. However, the electrical performance of both cells, based on the data sheets, is comparable. Additionally, both cells have their protection systems integrated (such as the current interrupt device, positive thermal coefficient switch, or safety vent).

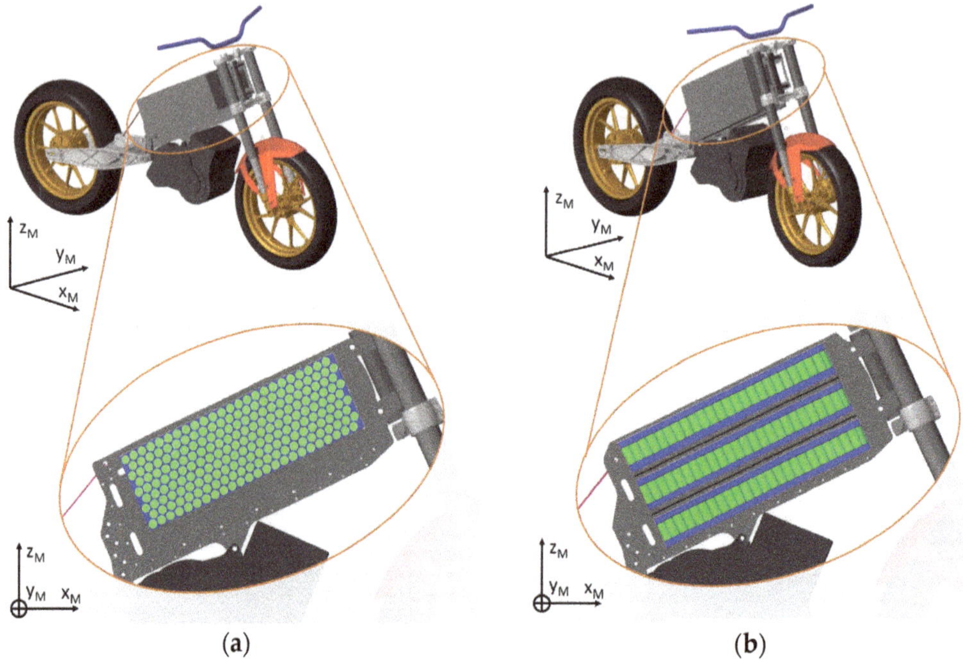

Figure 1. CAD models of the E-PTW with traction battery: (**a**) Concepts Y and (**b**) Concept Z.

The current literature [6,29,30] refers to the collision of an EPTW and a cylindrical object as a worst-case crash scenario for E-PTWs; therefore, this scenario is selected in this study. The cylindrical object has a diameter of 150 mm, as proposed in the SAE J2464 [31]. The impact speed is set to 8.9 m/s to represent a collision in urban areas, following the collision speed defined by Matsuda [20] and Ellersdorfer [6]. The collision area is set in the middle of the traction battery side to maximise the potential deformation of the traction battery.

2.2. E-PTW, Traction Battery, Cells, and Crash Scenario FE Modelling

Data on the traction battery's crash safety are obtained using the FE simulations of the considered crash scenario. The cylindrical collision object is modelled with two-dimensional (2D) elements and a rigid material model.

The E-PTW concept is simplified in the FE modelling to reduce the computational effort of the analysis due to the high total number of cases that needed to be analysed numerically for the metamodel building (see Figure 2). The components of the E-PTW not part of the traction battery are modelled using 112 one-dimensional (1D) rigid elements. The mass of the 1D elements is chosen to obtain a mass and a centre of gravity position for the entire vehicle in the simplified model, as in the concept E-PTW.

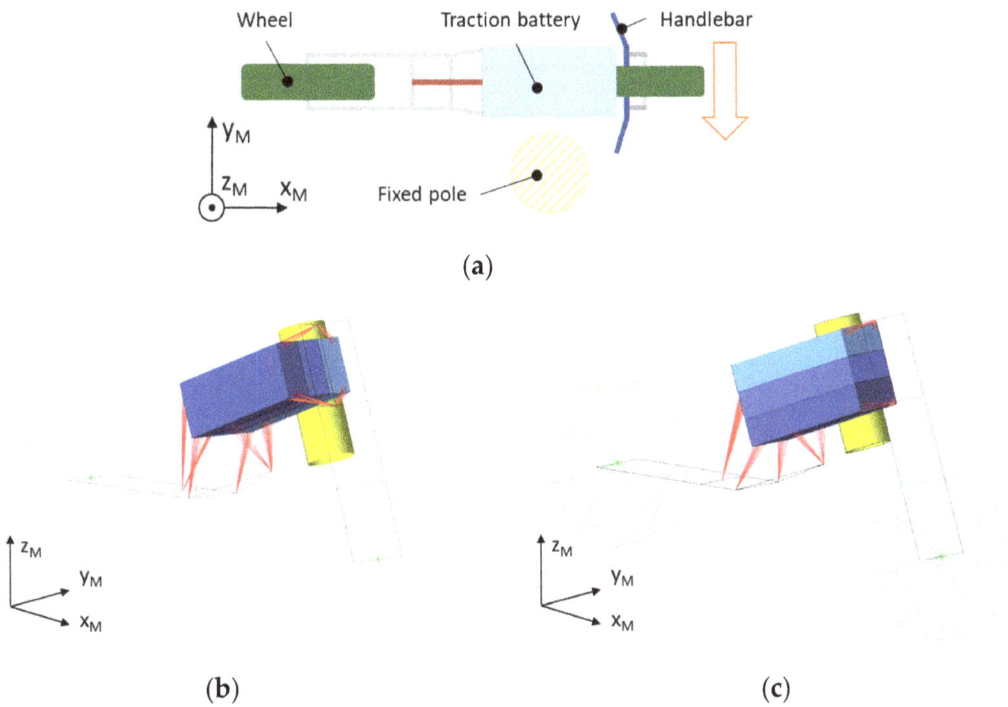

Figure 2. Schematic top view representation of the collision configuration (**a**) and simplified FE E-PTW model with traction battery: (**b**) Concepts Y and (**c**) Concept Z.

More details are present in the FE model of the traction battery to achieve a higher accuracy in approximating its mechanical behaviour in the crash scenario. The FE model of the traction battery consists of the following components (presented in Figure 3 using Concept Y as reference):

- The traction battery housing is composed of two components: the external plates (EPs), which define the external contour of the housing, and the longitudinal plates (LPs), which are placed between the modules for their fixation and thermal management. Both housing components are modelled with a total of 27,286 2D quadratic elements (average side length of 5 mm) with Belytschko-Tsay formulation with 5 integration points and using an elastoplastic aluminium material model without strain-rate hardening behaviour and with an effective plastic strain-based failure criterion in tension. The material model's effective plastic strain at failure was calibrated using data retrieved from tensile tests executed with specimens of the same material. The thickness range of the EPs and LPs is defined through variable parameters to be used as the input in the analysis.
- The modules consist of the following subcomponents:
    - The cell holders are responsible for holding the cells together and are modelled with a total of 86,910 2D elements, combining quadratic and triangular elements (average side length 5 mm), with Belytschko-Tsay formulation with 5 integration points using an elastic material model without strain-rate hardening and with an effective plastic strain-based failure criterion resembling the material behaviour of an acrylonitrile butadiene styrene thermoplastic.
    - The cells are modelled using a combination of 544 2D and 1881 1D elements, according to Raffler et al. [32]. Two-dimensional elements are used to geomet-

rically model the outer surface of the cell, and one-dimensional elements are used to represent the cell's mechanical behaviour under load.

- The connection between the modules and traction battery housing (i.e., connection screws) is modelled with 60 1D elements (average length 10 mm) with Hughes-Liu formulation with cross-section integration formulation and a steel elastic material model to simulate the screw connection needed to fix the modules within the traction battery housing.

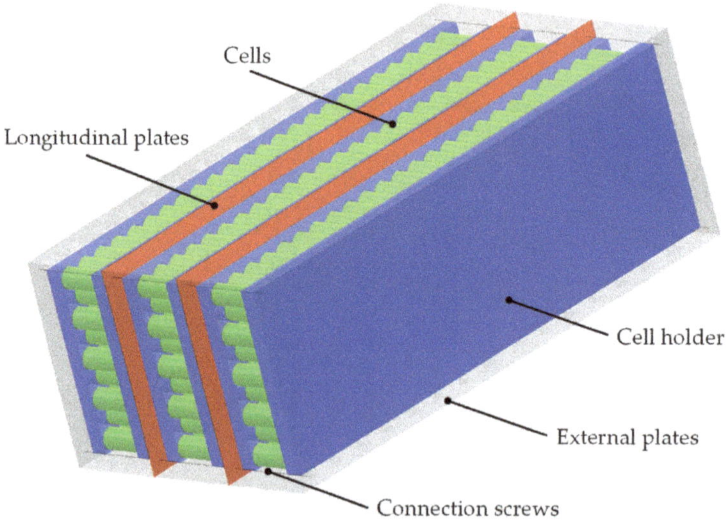

Figure 3. Schematic sketch of the traction battery of Concept Y with its components. The external plates are semi-transparent to show the inner components of the traction battery [33].

The smallest time step of the undeformed model is 6×10^{-4} ms. Due to the deformation in the load case, the smallest time step can drop up to 1×10^{-4} ms. The smallest time step is limited to 1×10^{-4} ms by adding artificial mass to the elements to ensure the simulation's stability and reasonable calculation time. Nevertheless, the maximal mass added was equal to 1.45 kg (less than 1% of the vehicle model mass).

A contact between all of the components of the FE model of the traction battery, except the 1D elements of the FE model of the cells, and the impactor using an automatic formulation are defined using the penalty method. The penalty method is used due to its good convergence in different contact conditions [34] and its easiness of implementation [35].

The FE simulations are performed with the single precision solver version R 9.20 of LS-DYNA® on the High-Performance Computing cluster of the Graz University of Technology, running on a Red Hat Enterprise Linux 5.4 platform.

The experimental tests, subjecting the cells with a 100% state of charge to quasistatic mechanical loads under three load configurations, as performed by Raffler et al. [33] (see Figure 4), provided the data on the cells' mechanical behaviour. Five repetitions are conducted for every load configuration and cell.

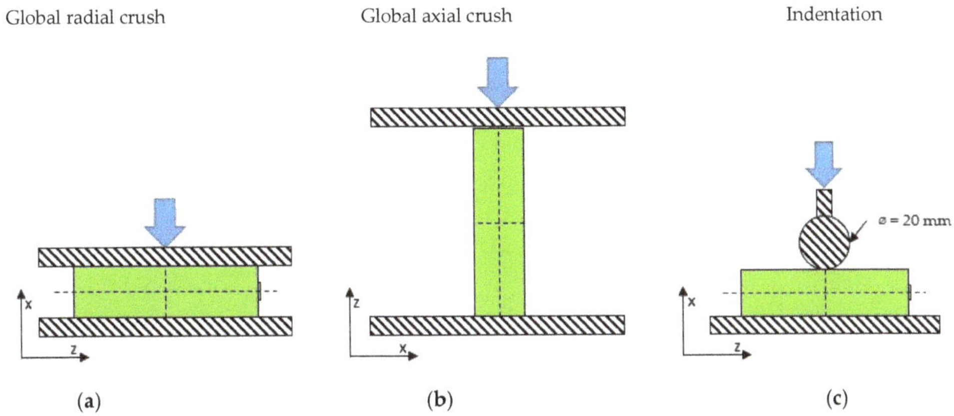

Figure 4. Mechanical load configurations for data acquisition: (**a**) global radial crush, (**b**) global axial crush and (**c**) indentation. The blue arrow indicates the acting force.

The experiment data (Tables A1 and A2) are used to calibrate the mechanical behaviour of the cell FE model and to determine the short circuit criterion using the method by Raffler et al. [33]. Previous studies [33,36,37] identified different ISC behaviours due to a deformation in the radial and axial cell direction; therefore, an internal short circuit risk at the cell level is defined for radial deformation ($ISCR_{r\ cell}$) and axial deformation ($ISCR_{a\ cell}$) (Equations (1) and (2)) with the description in Table 1.

$$ISCR_{r\ cell} = \frac{\varepsilon_r}{\varepsilon_{r\ SC}} \quad (1)$$

$$ISCR_{a\ cell} = \frac{\varepsilon_p}{\varepsilon_{p\ SC}} \quad (2)$$

Table 1. Description of the acronyms for the ISCR at the cell level.

Acronym	Description
ε_r	Radial cell deformation. Calculated as proposed by Raffler et al. [32]
$\varepsilon_{r\ SC}$	Radial cell deformation at ISC onset.
ε_p	Pole cell deformation. Calculated as proposed by Raffler et al. [32]
$\varepsilon_{p\ SC}$	Pole cell deformation at ISC onset.

The maximum between $ISCR_{r\ cell}$ and $ISCR_{a\ cell}$ defines the ISCR of a single cell ($ISCR_{cell}$), while the maximum between the $ISCR_{cell}$ of all of the cells of the traction battery defines the ISCR at the traction battery level.

2.3. Assessment of FE Models

The correlation and analysis (CORA) method is used to assess the validity of the mechanical behaviour of the FE models. Gehre et al. [38] first proposed the CORA method as an objective comparison method to assess the quality of compliance of the two curves. The method is used to compare different curves, including time-displacement [38], time-force [38,39], and displacement-force curves [40]. The CORA method consists of four independent procedures to separately evaluate the agreement for the size, phase, shape, and corridor of the two curves [38]. Every procedure provides a score (R) between zero and one, where zero defines no agreement between the signals and one is a perfect agreement. The four procedures are grouped into the "Corridor method", including the corridor agreement comparison; and the "Correlation method", consisting of the size, phase, and shape agreement comparison. The ISO/TR-9790 procedures [38,41], followed in this work, suggest calculating an overall rating divided into five grades (see Table 2) related to the agreement of the two curves by calculating the weighted average of the score of the four procedures.

Table 2. Grades of the CORA method.

Grade	Score Range
Excellent	$0.86 \leq R < 1.00$
Good	$0.65 \leq R < 0.86$
Fair	$0.44 \leq R < 0.65$
Marginal	$0.26 \leq R < 0.44$
Unacceptable	$0.00 \leq R < 0.26$

The CORA method differs from other methods that also assess magnitude, shape, and phase, such as the "Enhanced Error Assessment of Response Time Histories" (EEARTHs) [42] or "The Normalized Integral Square Error" (NISE) [43] by the feature of corridor assessment. Additionally, there are no disadvantages that CORA has to the mentioned methods. The "Correlation of signals" method (CoSi) [44] assesses the same metrics as the CORA but has the disadvantages of individual factors for each metric and not supplying an overall rating. The minimum area discrepancy method (MADM) [45] is developed especially to compare force-displacement curves and has the advantage of also working for non-monotonic functions. However, the MADM does not directly evaluate the size and shape. In the literature, there are also numerous other evaluation methods, such as the "Sprague and Geers method" or "Cumulative standard deviation (CSD)", but these methods cannot identify the key features such as phase, magnitude, and slope [39,44,46]. Therefore, the CORA is extensively used and implemented in norms and standards and is well-suited for use in this article because of the displacement-based data acquisition in the experiments.

The used parameters are chosen according to recommended values from the literature [47]. The FE model is considered valid when the overall score of the three load configurations achieved at least the grade of "Excellent".

The CORA method is used to validate the FE models of the components for which hardware specimens are available to generate experimental data, i.e., the cells integrated into the traction battery and material models. Therefore, all components directly involved in the area of contact between the vehicle and pole in the crash scenario are valid. The remaining parts of the vehicle (wheels, swingarm, handlebar, ...) are only responsible for reproducing the vehicle's kinematics (kinetic energy, centre of gravity) and the load introduction between the traction battery and connected components. For this reason, only the centre of gravity and the vehicle mass are validated based on the CAD model of the baseline vehicle.

The curves that are compared through the CORA method for the validation of the FE cell model are the average force-displacement curve of the conducted experimental tests and the force-displacement curve of the FE simulation subjected to the same load configuration. All force-displacement curves of a load configuration are considered and resampled with a displacement resolution of 0.05 mm. The force values of every curve for the same displacement are averaged to obtain an average curve. The CORAplus tool [48] is used to develop the CORA analysis.

2.4. Metamodel Building and Assessment

Within this study, a metamodel-based approach [49,50] is used to investigate the influence of cell selection and orientation on the crash safety of the traction battery.

The metamodels are built using the response surface method (RSM) [51]. The method is called the "response surface" as the response (i.e., output) values of the phenomena investigated are interpolated using regression analysis to create a continuous surface [52] with a reduced number of experiments needed and a high computational efficiency [53]. The response surface is calculated using a multiquadric (MQ) radial basis function network (RBFN). These functions are commonly used to easily and accurately interpolate scattered data [54].

The variables of the study are the selected cell and the cell orientation. Additionally, the design parameters of the thickness of the LPs and thickness of the EPs are introduced. The thickness of the traction battery housing plates defines the crashworthiness and mass of the traction battery. The output responses are the ISCR of the traction battery in the crash scenario and the mass of the traction battery housing, both retrieved from the FE simulations.

To compare the ISCR of traction batteries with different cells integrated and different cell orientations within the traction battery, a metamodel is built for each combination of the selected cell and cell orientation. For each of these metamodels, only two design parameters, the thickness of the EPs and LPs, are used to reduce the complexity of the metamodel. The design space of the design parameters is limited between a minimal thickness of 3 mm and a maximal thickness of 20 mm. The limits of the two parameters are defined based on thickness manufacturing constraints for an aluminium moulding process.

The design points, i.e., combinations of the two design parameters of each metamodel in the design space, are generated using a space-filling approach based on a "maximin" method, i.e., maximising the pairwise minimum distance across the design points [49]. Sixty design points are used to build each response surface.

The results of each metamodel are validated by controlling that the ISCR of further ten design points is in the tolerance range (TR) of the ISCR response surface with a confidence interval of 95%. The design points are selected through simple random sampling and are not used to build the response surface. After proving that the ISCR error distribution between the observed ISCR in the FE simulations of the design points and the metamodel's one is normal through the chi-squared goodness-of-fit test, the TR of the ISCR response surface is calculated, as in Equation (3), where:

- μ is the mean ISCR error between the FE simulations and the metamodel;
- σ is the standard deviation of the ISCR error.

$$TR = \mu \pm 1.96\sigma \quad (3)$$

The metamodels are built with LS-OPT® 6.0.

2.5. Evaluation of the Crash Safety

The ISCR at the traction battery level of the metamodels is used in this study as the evaluation criteria for crash safety. Multiple ISCRs of the same mass point are possible because of the combination of EPs and LPs with different thicknesses. To ensure an objective comparison and evaluate the maximal influence of the cell selection and orientation, the lowest ISCR of every mass value point is considered. The ISCR, as a function of the traction battery housing mass, is further calculated for every combination of the investigated cell and cell orientation within the traction battery design parameters.

Comparing these functions permits the calculation of the maximal difference in the ISCR by substituting the integrated cell and changing the cell orientation for traction battery housing with the same mass.

3. Results

3.1. Influence of Cell Selection on ISCR—Concept Y

Due to the simple geometric construction, the response surface of the traction battery housing mass of Concept Y for cell A and B can be approximated to the function presented in Equation (4). The used variables are described in Table 3.

$$TBH_m = \rho(TH_{LP} * S_{LP} + TH_{EP} * S_{EP}) \quad (4)$$

Table 3. Description of the acronyms for the approximation of the traction battery housing mass.

Acronym	Description and Units	Value
TBH_m	Mass of the traction battery housing [kg]	Output
ρ	Density of the housing material [kg/mm^3]	2.7×10^{-6}
TH_{LP}	Thickness of the longitudinal plates [mm]	Variable
S_{LP}	Middle surface of the longitudinal plates [mm^2]	171,7831
TH_{EP}	Thickness of the external plates [mm]	Variable
S_{EP}	Middle surface of the external plates [mm^2]	515,192

The response surface of the ISCR of Concept Y with cell A achieves an ISCR of 100% in the entire design space, i.e., due to the cell mechanical properties and their orientation, every combination in the design space of EPs' and LPs' thicknesses cannot avoid the onset of an ISC in the analysed crash scenario. The response surface of the ISCR of Concept Y with cell A is, therefore, not presented graphically. The response surface of the ISCR as a function of the EPs and LPs thickness of Concept Y with cell B is presented in Figure 5.

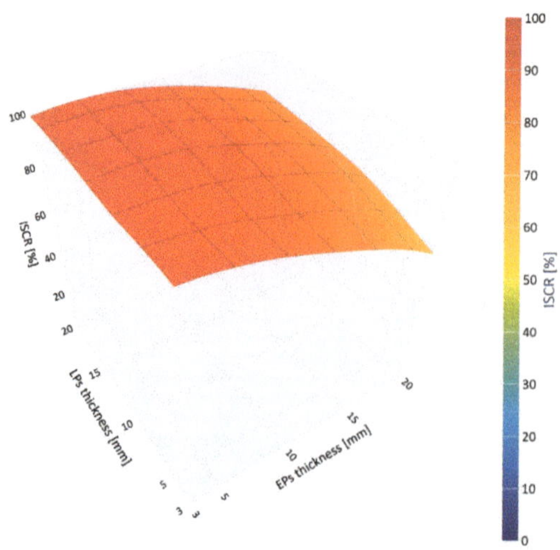

Figure 5. Response surfaces of the maximal ISCR in relation to the two traction battery housing variables for Concept Y with cell B.

The global ISCR minimum in the response surface of Concept Y with cell B is 63% at a thickness of 20 mm and 3 mm for the EPs and LPs, respectively, while the maximum is 100% in the design area near a thickness of the LPs equal to 20 mm. Every point of the design space with a thickness of the EPs less than circa 10 mm achieves an ISCR higher than 90%.

As the thickness of the LPs increases, the ISCR increases. Analysing the load path during impact can explain this phenomenon. The LPs of Concept Y are not directly deformed during impact but limit the displacement of the cell in the y_M direction. Therefore, increasing the thickness of the LPs and, thus, their stiffness reduces their bending under the load transmitted by the cells and increases the maximal deformation of the cells in the impact zone (see Figure 6a,b). An increase in the thickness of the EPs produces a decrease in the ISCR. The EPs deform directly upon contact. Therefore, increasing their thickness increases the stiffness, and reduces the deformation of the cells in the impact zone with a consequent reduction of the ISCR (see Figure 6a,c).

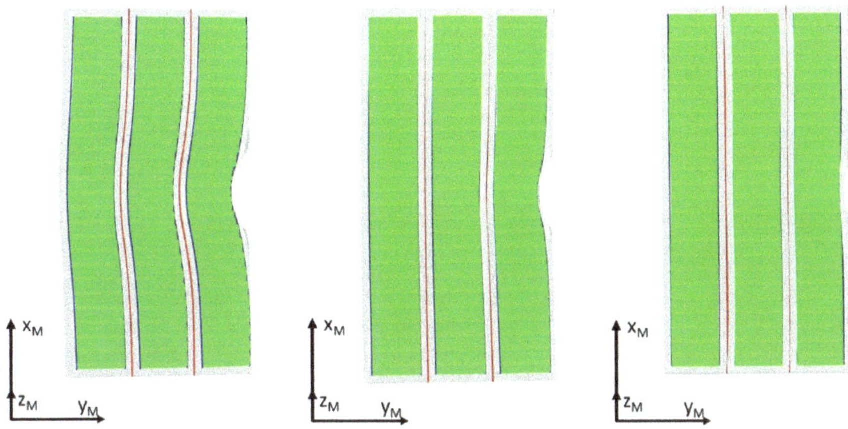

Figure 6. Traction battery deformation for Concept Y with cell B at 6.5 ms after impact for: (**a**) EPs' thickness of 3 mm and LPs' thickness of 3 mm, (**b**) EPs' thickness of 3 mm and LPs' thickness of 20 mm, and (**c**) EPs' thickness of 20 mm and LPs' thickness of 3 mm. Note that the EPs are semi-transparent.

The error distributions in ISCR between the FE model of Concept Y with cells A and B are proven to be normal.

The evaluation of error distributions in ISCR between the FE model of Concept Y with cells A and B and their respective metamodel is presented in Table 4, where the used units are the percentage points of the relative ISCR.

Table 4. Statistical analysis of the error distribution for Concept Y.

Cell Integrated	Mean	Standard Deviation	Upper Tolerance Range	Lower Tolerance Range
A	0	0	0	0
B	−0.2	1.2	2.2	−2.6

3.2. Influence of Cell Selection on ISCR—Concept Z

The response surface of the traction battery housing mass for Concept Z, due to the same geometry of Concept Y but a different orientation of the modules, follows the same approximation presented in Equation (4).

Figure 7 shows the response surface of the ISCR as a function of the EPs' and LPs' thickness for Concept Z with cell A and B.

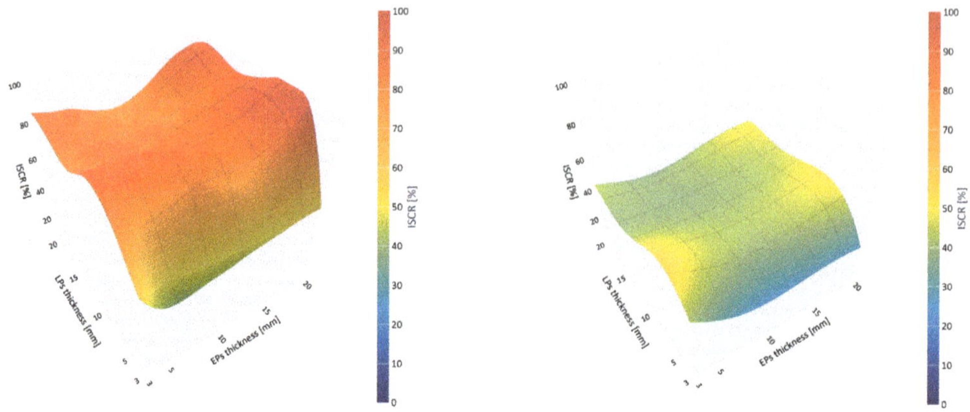

(a) (b)

Figure 7. Response surfaces of the maximal ISCR in relation to the thickness of EPs and LPs for (**a**) Concept Z with cell A and (**b**) Concept Z with cell B.

Based on the metamodel, the minimum and maximum ISCRs for Concept Z with cell A are 39% for an EP thickness of circa 6.5 mm and an LP thickness of 3.0 mm, and are 100% in the area for an EP thickness of 20 mm and an LP thickness of circa 10 mm, respectively.

In the case of Concept Z with cell B, the maximum and minimum ISCRs based on the metamodel are 19% for an EP thickness of circa 8.7 mm and an LP thickness of 3.0 mm, and are 46% for an EP thickness of 20.0 mm and LP thickness of 10.0 mm, respectively.

Increasing and decreasing the EP thickness of concept Z with cell A, starting from the point of minimum ISCR, increases the ISCR. In particular, decreasing the EPs' thickness produces more intrusion in the traction battery, consequently increasing the ISCR. Increasing the Eps' thickness reduces the intrusion in the traction battery. Nevertheless, the acceleration loads to which the cells are subjected to increase and lead to their deformation due to the impact on the LPs, thus increasing the ISCR. Increasing the LPs' thickness and maintaining the EPs' thickness constant produces the same effect. The effect is even more pronounced because the LPs have an important stiffening function for the entire traction battery housing due to the geometry of Concept Z.

The ISCR response surface for concept Z with cell B shows a striking resemblance to that of Concept Z with cell A, despite the former exhibiting a lower ISCR throughout the entire design space. This finding indicates that similar deformation modes of the traction battery and the integrated cells upon impact occur. However, a higher impactor displacement needed to trigger an ISC for cell B compared to cell A in all three tested configurations (see also Table 5) leads to a reduction of the ISCR of the metamodel of Concept Z with cell B.

Table 5. Comparison of the impactor displacement at ISCR onset for cell A and cell B in the three tested configurations at cell level.

Cell	Impactor Displacement at ISC Onset [mm]		
	Global Radial Crush	Global Axial Crush	Indentation
A	5.5	3.5	4.5
B	6.7	5.6	5.6

The error distributions in ISCR between the FE model of Concept Z with cells A and B were proven to be normal.

The evaluation of error distributions in ISCR between the FE model of Concept Z with cell A and B and their respective metamodel is presented in Table 6, where the used units are the percentage points of the relative ISCR.

Table 6. Statistical analysis of the error distribution for Concept Z.

Cell Integrated	Mean	Standard Deviation	Upper Tolerance Range	Lower Tolerance Range
A	0.3	2.3	4.9	−4.2
B	0.0	0.9	1.8	−1.9

3.3. Influence of Cell Orientation on ISCR

The trends of the ISCR, inclusive tolerance range, as a function of the traction battery housing mas for Concept Y and Z with cell A and B are presented in Figure 8.

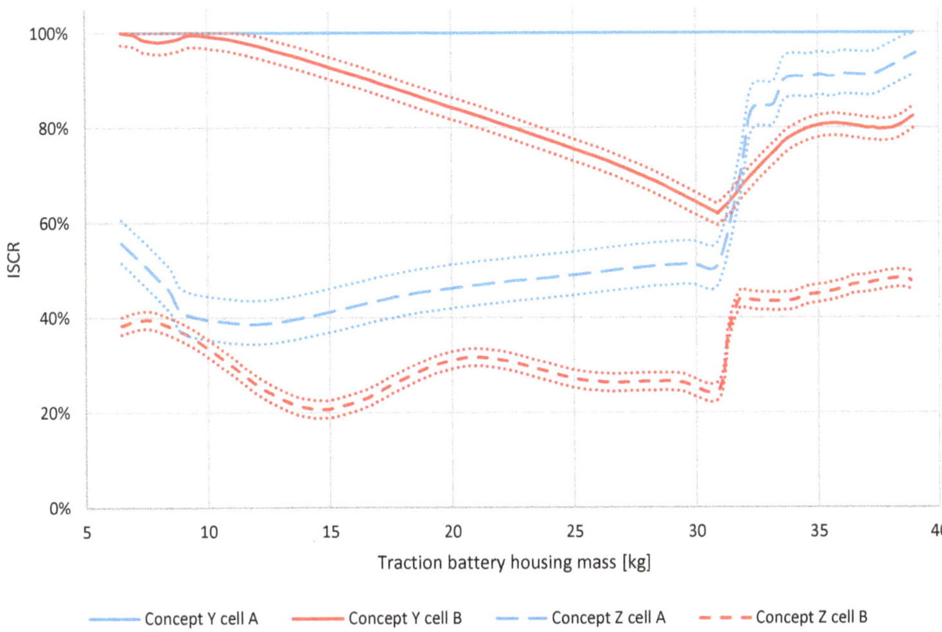

Figure 8. Comparison of ISCR and mass trends for Concepts Y and Z with cells A and B. The dotted lines represent the upper and lower limit of the TR for each metamodel.

The ISCR trend of Concept Y with cell A is constant at 100%, as every design point in the design area achieved an ISCR of 100%.

The same concept but with cell B shows an ISCR equal to or higher than 95%, up to a traction battery housing mass of circa 13.5 kg. The ISCR monotonically decreases from a traction battery housing mass of 9.4 kg up to circa 31 kg when the EPs achieve the upper thickness limit of 20 mm. The ISCR then increases non-monotonically until reaching the maximal traction battery housing mass of 39 kg due to the thickness increase in the LPs in this traction battery housing mass range, due to the behaviour described in Section 3.2.

The ISCR for Concept Z with cell A ranges from a minimum of 39% to a maximum of 96%. The ISCR decreases monotonically in the traction battery housing mass range of 5 to 11.3 kg. In this range, the thickness of the LPs is constant at the lower limit and the thickness of the EPs increases. A further mass increase produces a non-monotonical rise in the ISCR. In particular, up to a traction battery housing mass of circa 31 kg, the thickness of the EPs increases up to the upper limit, while the LPs' thickness is constant at the lower limit. For a traction battery housing mass of more than 31 kg, the combination of the variables that minimise the ISCR for a fixed mass, due to the form of the response surface, lays on a different area of the ISCR response surface, producing a leap in the curve trend.

The ISCR as a function of the mass for Concept Z with cell B shows a minimal ISCR of circa 21% and a maximal ISCR of circa 48%. Starting from the minimal mass, the ISCR increases to a traction battery housing mass of 7.9 kg and then decreases to the global minimum at a traction battery housing mass of circa 15 kg. If the mass further increases, a non-monotonical ISCR increase is produced. The same effect for Concept Z with cell A produces the sudden rise of the ISCR at a traction battery housing mass of circa 31 kg.

4. Discussion

The results of this study show a relevant influence of both the cell selection and orientation in the crash safety of an E-PTW traction battery.

In particular, the switch from cell A to cell B reduces the ISCR on average, in absolute values, by 17% in the case of Concept Y and by 24% in the case of Concept Z, with a maximal reduction of 38% for Concept Y and 48% for Concept Z. The differences in the cell deformation of the ISC onset for the two analysed cells explain these results. Cell B achieves an ISC at circa 60% higher deformation in global axial loading and circa 25% higher deformation in global radial loading than cell A. Therefore, more deformation of the traction battery housing is possible before achieving an ISCR of 100%. Consequently, the crashworthiness of the traction battery housing structure can be better exploited with cell B compared to cell A.

The cell orientation has an even stronger influence on the ISCR. Concept Z achieves a reduction in the ISCR on average, in absolute value, by 44% with cell A and 51% with cell B compared to Concept Y. Maximal ISCR reductions, in absolute values, up to 61% and 73%, respectively, for cell A and B are observed in Concept Z compared to Concept Y. The first reason for this effect is based on the cell's properties; the deformation that triggers the ISC onset is up to 37% smaller in the case of global axial cell loading than in global radial cell loading. Secondly, it can be explained by the influence of the LPs in the load path of the traction battery housing; the main load path of Concept Z is oriented in the cells´ radial direction, while for Concept Y, it is in the cell´s axial direction. Therefore, Concept Z leverages the cells' anisotropic ISC behaviour, decreasing the ISCR compared to Concept Y.

Considering the influence of the LPs, they are directly loaded upon impact in Concept Z, which is not the case in Concept Y, where only the EPs are mechanically directly loaded in the crash phase. Therefore, an LP's thickness increase does not reduce the ISCR in Concept Y. In the case of Concept Z, a counterintuitive effect is present: an LP's thickness increase raises the ISCR. In Concept Z, an LP's thickness increase decreases the intrusion in the traction battery housing, but the stiffness increase produces higher acceleration pulses upon impact. These dynamic loads oscillate the inner modules and their collision with the traction battery housing. This provokes a local deformation of the cells, which increases the ISCR.

5. Limitations

The results provided in this study are assessed for one vehicle type only. Due to different boundary conditions and load scenarios, they can differ if applied to different EPTW categories such as off-road motorcycles or different vehicle types such as four-wheelers.

Furthermore, this work analysed two cells of the same shape and format, namely 18650 cells. Comparing cells of the same shape but different formats (e.g., 18650 and 21700 cells) and different shapes (e.g., cylindrical and pouch cells) could identify new insights to guide the selection of the optimal cell for the integration into EVs.

Finally, it should be considered that the cells' FE models are validated under quasistatic mechanical loads. Current studies observed differences in the mechanical behaviour of

Li-ion cells between quasistatic and dynamic loads, as the ones that cells are subjected to in the case of a crash [55,56]. These differences in mechanical behaviour can lead to different ISCRs of the metamodels.

6. Conclusions

In this work, the influence of the cell selection and orientation within the traction battery on the crash safety of the traction battery of an electric-powered two-wheeler under a potentially dangerous crash scenario for the traction battery is analysed through metamodels fitted to the numerical results provided by finite element simulations of the crash scenario. The integrated traction battery cells evaluate crash safety by the internal short circuit risk (ISCR). The following conclusions are obtained:

1. The cell orientation within the traction battery has the biggest influence on the crash safety of the traction battery. In particular, changing only the cell orientation can reduce the ISCR by 73%. This knowledge can be leveraged in the design phase of the traction battery of an electric vehicle if a potentially dangerous crash scenario is known by orienting the cell to obtain a loading in the cell radial direction during the collision. Therefore, the gravimetric energy density of traction batteries can be increased because fewer additional reinforcements are needed.
2. The cell selection can influence the crash safety of the traction battery of an electric-powered two-wheeler but with a smaller contribution than the cell orientation. In particular, switching from cell A to cell B can reduce the ISCR by 48%. Therefore, the crash safety of a traction battery can be improved without structural reinforcements or design changes in the traction battery housing or vehicle frame. Knowing this aspect, the crash safety of already existing traction batteries can be improved with small effort.
3. In a crash load condition, the longitudinal plates of the traction battery housing can improve the crashworthiness of the entire traction battery housing, acting as energy-absorbing structures while fulfilling their functional role for the traction battery assembly. Therefore, their design should be subjected to a crashworthiness assessment to avoid landing in an area of the design space with a high internal short circuit risk.
4. The investigation approach, using metamodels, has been proven as both computationally inexpensive and adequately precise (maximal absolute internal short circuit risk error range of 9.1%) for investigating the internal short circuit risk of traction battery concepts. The same approach can be used in a vehicle development phase to identify optimal combinations of the thickness of traction battery housing components, cell orientations, and selections to accelerate the development of crashworthy traction batteries.

Finally, it should be considered that the analysed variables, cell selection and orientation, can be combined with state-of-the -art passive protection strategies of the traction battery, further improving the crash safety of electric-powered two-wheelers and electric vehicles in general.

Author Contributions: Conceptualization, C.E.; Data curation, A.S.; Formal analysis, M.F.; Funding acquisition, C.E.; Investigation, A.S. and M.R.; Methodology, A.S. and C.E.; Project administration, C.E.; Resources, A.S.; Supervision, C.E.; Validation, M.F.; Writing—original draft, A.S.; Writing—review and editing, M.F. and C.E. All authors have read and agreed to the published version of the manuscript.

Funding: This research was funded by the Austrian Research and Promotion Agency (FFG), grant number 845337. Open Access Funding by the Graz University of Technology.

Institutional Review Board Statement: Not applicable.

Informed Consent Statement: Not applicable.

Data Availability Statement: Not applicable.

Acknowledgments: The authors acknowledge KTM AG, represented by Silvio Marcon and Martin Truss, for helpful discussions and the Zentral Informatikdienst (ZID) of the Graz University of Technology for the usage of the High-Performance Computing cluster. The Austrian Research and Promotion Agency (FFG), grant number 845337. Open Access Funding was provided by the Graz University of Technology.

Conflicts of Interest: The authors declare no conflict of interest.

Appendix A

The results of the FE cell model for cells A and B are presented in the following tables, including the CORA analysis outputs. The curves are trimmed at the displacement at which the first ISC onset in the experimental tests was observed.

Table A1. Comparison of the force-displacement curves of the FE model and experimental tests for cell A and results of the CORA evaluation.

	Graphic	Method	Rating	Weight
Radial global crush	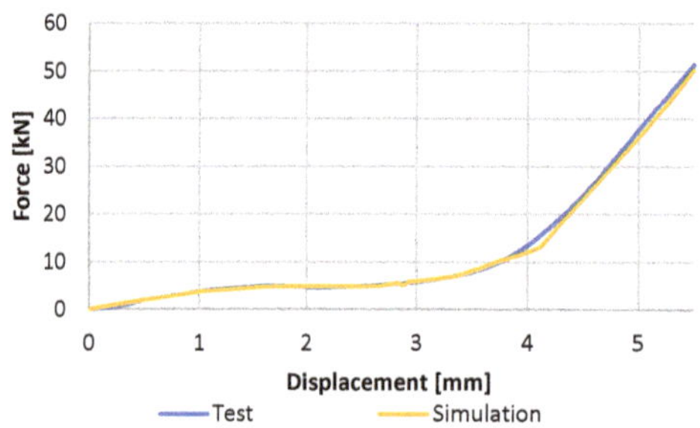	Corridor method	1.0	0.5
		Correlation method	0.974	0.5
		Total	0.990	
Axial global crush	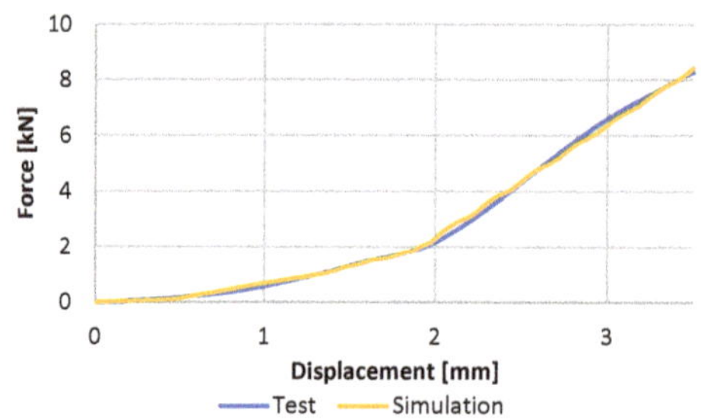	Corridor method	1.0	0.5
		Correlation method	0.995	0.5
		Total	0.997	
Indentation	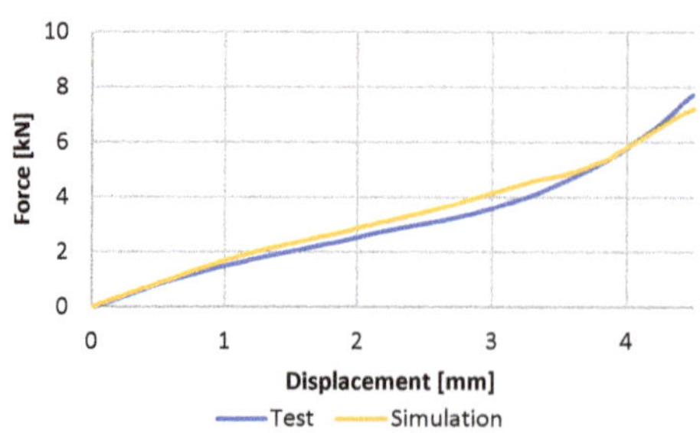	Corridor method	0.972	0.5
		Correlation method	0.971	0.5
		Total	0.972	

The FE model of cell A achieves a total rating of 0.986 and a grade of "Excellent" and is, therefore, assessed positively to be integrated into the FE model of the E-PTW concept for the analysis of its crash safety.

Table A2. Comparison of the force-displacement curves of the FE model and experimental tests for cell B and results of the CORA evaluation.

<table>
<tr><th></th><th>Graphic</th><th>Method</th><th>Rating</th><th>Weight</th></tr>
<tr><td rowspan="3">Radial global crush</td><td rowspan="3">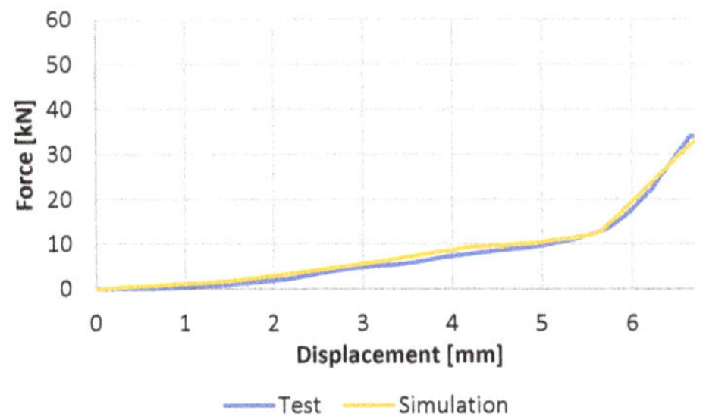
</td><td>Corridor method</td><td>1.0</td><td>0.5</td></tr>
<tr><td>Correlation method</td><td>0.971</td><td>0.5</td></tr>
<tr><td>Total</td><td>0.985</td><td></td></tr>
<tr><td rowspan="3">Axial global crush</td><td rowspan="3">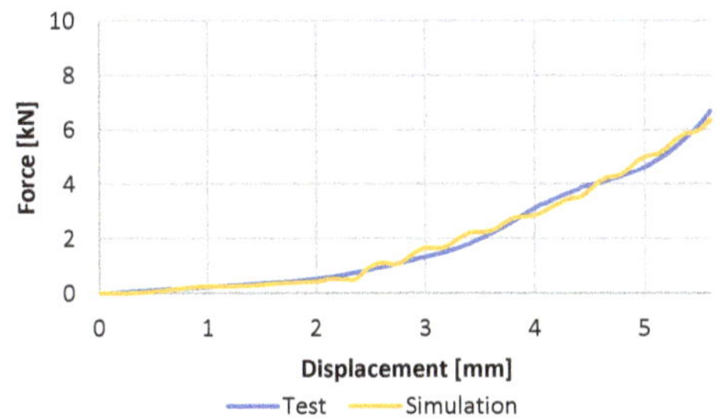
</td><td>Corridor method</td><td>0.998</td><td>0.5</td></tr>
<tr><td>Correlation method</td><td>0.974</td><td>0.5</td></tr>
<tr><td>Total</td><td>0.986</td><td></td></tr>
<tr><td rowspan="3">Indentation</td><td rowspan="3">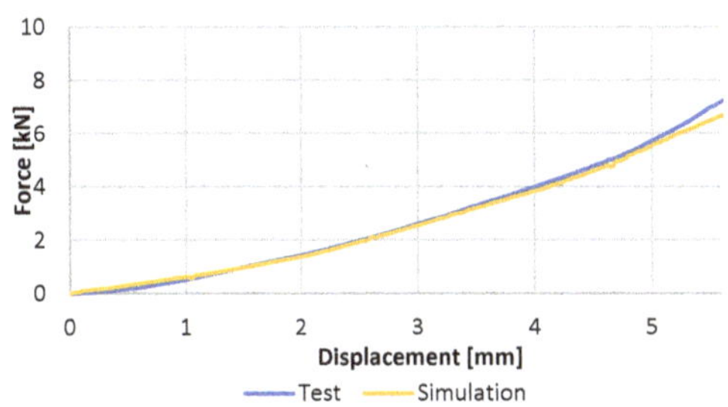
</td><td>Corridor method</td><td>0.992</td><td>0.5</td></tr>
<tr><td>Correlation method</td><td>0.901</td><td>0.5</td></tr>
<tr><td>Total</td><td>0.946</td><td></td></tr>
</table>

The FE model of cell B achieves a total rating of 0.973 and a grade of "Excellent" and is, therefore, assessed positively to be integrated into the FE model of the E-PTW concept for the analysis of its crash safety.

References

1. Weiss, M.; Dekker, P.; Moro, A.; Scholz, H.; Patel, M.K. On the electrification of road transportation—A review of the environmental, economic, and social performance of electric two-wheelers. *Transp. Res. D Transp. Environ.* **2015**, *41*, 348–366. [CrossRef]
2. Birke, P. Lithium-Ionen-Akkumulatoren Diskussion der verschiedenen Zellformate. *ATZ Elektron* **2014**, *9*, 38–43. [CrossRef]
3. Schröder, R.; Aydemir, M.; Seliger, G. Comparatively Assessing different Shapes of Lithium-ion Battery Cells. *Proc. Manuf.* **2017**, *8*, 104–111. [CrossRef]

4. Etacheri, V.; Marom, R.; Elazari, R.; Salitra, G.; Aurbach, D. Challenges in the development of advanced Li-ion batteries: A review. *Energy Environ. Sci.* **2011**, *4*, 3243. [CrossRef]
5. Nitta, N.; Wu, F.; Lee, J.T.; Yushin, G. Li-ion battery materials: Present and future. *Mater. Today* **2015**, *18*, 252–264. [CrossRef]
6. Ellersdorfer, C. Abbildung und Bewertung des Crashverhaltens von lithiumbasierten Batterien für elektrisch betriebene Motorräder. Ph.D. Thesis, Technische Universität Graz, Graz, Austria, 2016.
7. Bisschop, R.; Willstrand, O.; Amon, F.; Rosengren, M. *Fire Safety of Lithium-Ion Batteries in Road Vehicles;* RISE Research Institute of Sweden: Boras, Sweden, 2019; ISBN 978-91-88907-78-3.
8. *IEC 62660-2:2010;* Secondary Lithium-Ion Cells for the Propulsion of Electric Road Vehicles: Part 2: Reliability and Abuse Testing, Ed. 1.0. International Electrotechnical Commission: Geneva, Switzerland, 2010; ISBN 978-2-88912-302-5.
9. Battery Safety Standards Committee. *Electric and Hybrid Vehicle Propulsion Battery System Safety Standard—Lithium-based Rechargeable Cells;* SAE International: Warrendale, PA, USA, 2011.
10. United Nations. Recommendations on the Transport of Dangerous Goods—Manual of Tests and Criteria: UN 38.3 Lithium Metal and Lithium Ion Batteries. ST/SG/AC 10 ST/SG/AC.10/11/Rev.6; New York, NY, USA,, 2015.
11. United Nations Economic Comission for Europe. *Regulation No 100 of the Economic Commission for Europe of the United Nations (UNECE)—Uniform Provisions Concerning the Approval of Vehicles with Regard to Specific Requirements for the Electric Power Train, Revision 2;* Economic Commission for Europe of the United Nations: Geneva, Switzerland, 2015.
12. *UL 1642;* UL Standard for Safety: Lithium Batteries, Version 5. Underwriters Laboratories Inc.: Northbrook, IL, USA, 2012.
13. *UL 2580;* Batteries for Use in Electric Vehicles, Edition 2.0. Underwriters Laboratories Inc.: Northbrook, IL, USA, 2016.
14. Doughty, D.H.; Crafts, C.C.; Doughty, D.; Crafts, C. *FreedomCAR: Electrical Energy Storage System Abuse Test Manual for Electric and Hybrid Electric Vehicle Applications;* Sandia Report SAND2005-3123; FreedomCAR: Albuquerque, NM, USA; Livermore, CA, USA, 2006.
15. EUCAR. Battery requirements for future automotive applications: Battery Requirements 2030, 2019.
16. Pan, Y.; Xiong, Y.; Wu, L.; Diao, K.; Guo, W. Lightweight Design of an Automotive Battery-Pack Enclosure via Advanced High-Strength Steels and Size Optimization. *Int. J. Automot. Technol.* **2021**, *22*, 1279–1290. [CrossRef]
17. Szabo, I.; Scurtu, L.I.; Raboca, H.; Mariasiu, F. Topographical Optimization of a Battery Module Case That Equips an Electric Vehicle. *Batteries* **2023**, *9*, 77. [CrossRef]
18. Rawlinson, P.D.; Clarke, A.; Gadhiya, H.L.; Sumpf, R.D.; Edwards, B.P. System for absorbing and distributing side impact energy utilizing an integrated battery pack. U.S. Patent US20140182958A1, 30 January 2014.
19. Funcke, M.; Schäfer, S.; Sturk, D.; Dufaut, D. Simulation and Active Protection of Li-Ion Traction Batteries. *ATZ Worldw.* **2015**, *117*, 10–15. [CrossRef]
20. Matsuda, Y. *A Study of Electric Motorcycle;* SAE Technical Paper 2014-32-0012; SAE International: Warrendale, PA, USA, 2014. [CrossRef]
21. Kukreja, J.; Nguyen, T.; Siegmund, T.; Chen, W.; Tsutsui, W.; Balakrishnan, K.; Liao, H.; Parab, N. Crash analysis of a conceptual electric vehicle with a damage tolerant battery pack. *Extrem. Mech. Lett.* **2016**, *9*, 371–378. [CrossRef]
22. Shui, L.; Chen, F.; Garg, A.; Peng, X.; Bao, N.; Zhang, J. Design optimization of battery pack enclosure for electric vehicle. *Struct. Multidisc. Optim.* **2018**, *58*, 331–347. [CrossRef]
23. Zhao, Y.; Shi, J.; Wang, K.; Wang, B.; He, C.; Deng, X. Mechanical Properties and Optimization Analysis on Battery Box with Honeycomb Sandwich Composite Structure. *Int. J. Automot. Technol.* **2023**, *24*, 1–14. [CrossRef]
24. Um, H.-J.; Hwang, Y.-T.; Bae, I.-J.; Kim, H.-S. Design and manufacture of thermoplastic carbon fiber/polyethylene terephthalate composites underbody shield to protect the lithium-ion batteries for electric mobility from ground impact. *Compos. Part B Eng.* **2022**, *238*, 109892. [CrossRef]
25. Biharta, M.A.S.; Santosa, S.P.; Widagdo, D.; Gunawan, L. Design and Optimization of Lightweight Lithium-Ion Battery Protector with 3D Auxetic Meta Structures. *WEVJ* **2022**, *13*, 118. [CrossRef]
26. Xia, Y.; Wierzbicki, T.; Sahraei, E.; Zhang, X. Damage of cells and battery packs due to ground impact. *J. Power Sources* **2014**, *267*, 78–97. [CrossRef]
27. Faßbender, S.; Eckstein, L.; Hören, B.; Stein, J.; Hesse, L.; Urban, P. Prospects of Holistic Purpose Design by the Example of the Electric Vehicle Concept "SpeedE". In Proceedings of the 21st Aachen Colloquium Automobile and Engine Technology 2012, Aachen, Germany, 8–10 October 2012.
28. European Commission. Regulation (EU) No 168/2013 of the European Parliament and of the Council of 15 January 2013 on the approval and market surveillance of two- or three-wheel vehicles and quadricycles (Text with EEA relevance). *Off. J. Eur. Union* **2013**, *L60*, 52–128.
29. Ellersdorfer, C.; Sevarin, A.; Tomasch, E.; Sinz, W.; Ebner, A.; Deitermann, N.; Forster, M. Evaluation method of the crash safety of traction batteries for electric driven motorcycles. In *Tagungsband der 11. Internationalen Motorradkonferenz;* Institut für Zweiradsicherheit e.V: Essen, Germany, 2016; ISBN 3923994222.
30. Ellersdorfer, C.; Sevarin, A.; Tomasch, E.; Sinz, W.; Ebner, A.; Deitermann, N.; Forster, M. Battery Safety Evaluation of Electric Driven Motorcycles from the Perspective of Accident Research. In Proceedings of the 30th International Electric Vehicle Symposium (EVS30), Stuttgart, Germany, 9–11 October 2017.
31. Doughty, D.H. SAE J2464 "EV & HEV Rechargeable Energy Storage System (RESS) Safety and Abuse Testing Procedure". In Proceedings of the SAE 2010 World Congress & Exhibition, Detroit, MI, USA, 12–15 April 2010.

32. Raffler, M.; Sevarin, A.; Ellersdorfer, C.; Heindl, S.F.; Breitfuß, C.; Sinz, W. Finite element model approach of a cylindrical lithium ion battery cell with a focus on minimization of the computational effort and short circuit prediction. *J. Power Sources* **2017**, *360*, 605–617. [CrossRef]
33. Sevarin, A.; Fasching, M.; Ellersdorfer, C. Crash safety optimisation method for the integration of the traction batteries into electric powered-two-wheelers. In *Tagungsband der 13. Internationalen Motorradkonferenz;* Institut für Zweiradsicherheit e.V: Cologne, Germany, 2020.
34. Nasdala, L. *FEM-Formelsammlung Statik und Dynamik;* Vieweg+Teubner: Wiesbaden, Germany, 2010; ISBN 978-3-8348-0980-3.
35. Hallquist, J.O. *LS-DYNA Theory Manual;* Livermore, CA, USA, 2006.
36. Plaimer, M.; Breitfuß, C.; Sinz, W.; Heindl, S.F.; Ellersdorfer, C.; Steffan, H.; Wilkening, M.; Hennige, V.; Tatschl, R.; Geier, A.; et al. Evaluating the trade-off between mechanical and electrochemical performance of separators for lithium-ion batteries: Methodology and application: Methodology and application. *J. Power Sources* **2016**, *306*, 702–710. [CrossRef]
37. Jantscher, K.; Breitfuß, C.; Miklau, M.; Ismail, K.; Dobusch, P. Virtual Detection of Mechanically Induced Short Circuits in a Cylindrical Lithium-Ion Battery Cell Based on Finite Element Simulation. *Batteries* **2021**, *7*, 79. [CrossRef]
38. Gehre, C.; Gades, H.; Wernicke, P. Objective Rating of Signals Using Test and Simulation Responses. In *The 21st ESV Conference Proceedings. International Technical Conference on the Enhanced Safety of Vehicles, Stuttgart, Germany, 15–18 June 2009;* NHTSA: Washington, DC, USA, 2009; pp. 1–8.
39. Barbat, S.; Fu, Y.; Zhan, Z.; Yang, R.-J.; Gehre, C. Objective Rating Metric for Dynamic Systems. In *The 23rd ESV Conference Proceedings. International Technical Conference on the Enhanced Safety of Vehicles, Seoul, Republic of Korea, 27–30 May 2013;* NHTSA: Washington, DC, USA, 2013; pp. 1–10.
40. Fasching, M.; Sevarin, A.; Ellersdorfer, C. Investigate the elastoplastic deformation behaviour of a motorcycle frame under different mechanical laod configurations. *J. Crashworthiness*, 2023; in press.
41. Giordano, C.; Kleiven, S. Development of an Unbiased Validation Protocol to Assess the Biofidelity of Finite Element Head Models used in Prediction of Traumatic Brain Injury. *Stapp Car Crash J.* **2016**, *60*, 363–471.
42. Zhan, Z.; Fu, Y.; Yang, R.-J. Enhanced Error Assessment of Response Time Histories (EEARTH) Metric and Calibration Process. In *SAE Technical Paper Series;* SAE International: Warrendale, PA, USA, 2011.
43. Donnelly, B.R.; Morgan, R.M.; Eppinger, R.H. Durability, Repeatability and Reproducibility of the NHTSA Side Impact Dummy. In *SAE Technical Paper Series;* SAE International: Warrendale, PA, USA, 1983.
44. Murmann, R.; Harzheim, L.; Dominico, S.; Immel, R. CoSi: Correlation of signals—A new measure to assess the correlation of history response curves. *Mech. Syst. Signal Process.* **2016**, *80*, 482–502. [CrossRef]
45. Peres, J.; Bastien, C.; Christensen, J.; Asgharpour, Z. A minimum area discrepancy method (MADM) for force displacement response correlation. *Comput. Methods Biomech. Biomed. Engin.* **2019**, *22*, 981–996. [CrossRef] [PubMed]
46. Vavalle, N.A.; Jelen, B.C.; Moreno, D.P.; Stitzel, J.D.; Gayzik, F.S. An evaluation of objective rating methods for full-body finite element model comparison to PMHS tests. *Traffic Inj. Prev.* **2013**, *14*, S87–S94. [CrossRef] [PubMed]
47. Carsten Thunert. *CORAplus Release 4.0.4: User's Manual;* Braunschweig, Germany, 2017.
48. Gehre, C. *CORAplus Release 4.0.4;* pdb—Partnership for Dummy Technology and Biomechanics: Gaimersheim, Germany, 2017.
49. Stander, N.; Basudhar, A.; Roux, W.; Witowski, K.; Eggleston, T.; Goel, T.; Craig, K. *LS-OPT User ´s Manual: A Design Optimization and Probabilistic Analysis Tool for the Engineering Analyst, Version 6.0;* Livermore, CA, USA, 2019.
50. Wang, G.G.; Shan, S. Review of Metamodeling Techniques in Support of Engineering Design Optimization. *Struct. Multidisc. Optim.* **2007**, *129*, 370–380. [CrossRef]
51. Kleijnen, J.P.C. Response Surface Methodology. *SSRN J.* **2014**, *60*, 735. [CrossRef]
52. Christensen, J.; Bastien, C. *Nonlinear Optimization of Vehicle Safety Structures: Modeling of Structures Subjected to Large Deformations;* Elsevier; Oxford, UK, 2016; ISBN 9780124173095.
53. Aydar, A.Y. Utilization of Response Surface Methodology in Optimization of Extraction of Plant Materials. In *Statistical Approaches With Emphasis on Design of Experiments Applied to Chemical Processes;* Silva, V., Ed.; InTech: London, UK, 2018; ISBN 978-953-51-3877-8.
54. Misra, R.K.; Kumar, S. Multiquadric Radial Basis Function Method for Boundary Value and Free Vibration Problems. *Indian J. Industr. Appl. Math.* **2013**, *4*, 138. [CrossRef]
55. Zhou, M.; Hu, L.; Chen, S.; Zhao, X. Different mechanical-electrochemical coupled failure mechanism and safety evaluation of lithium-ion pouch cells under dynamic and quasi-static mechanical abuse. *J. Power Sources* **2021**, *497*, 229897. [CrossRef]
56. Bulla, M.; Schmandt, C.; Kolling, S.; Kisters, T.; Sahraei, E. An Experimental and Numerical Study on Charged 21700 Lithium-Ion Battery Cells under Dynamic and High Mechanical Loads. *Energies* **2023**, *16*, 211. [CrossRef]

batteries

MDPI

Article

Modeling of Lithium Plating and Stripping Dynamics during Fast Charging

Polina Brodsky Ringler [1,†], Matthew Wise [2,†], Prashanth Ramesh [3,†], Jung Hyun Kim [3,†], Marcello Canova [3,*,†], Chulheung Bae [4,†], Jie Deng [4,†] and Heechan Park [5]

1 Department of Mechanical Engineering, Colorado School of Mines, Golden, CO 80401, USA; pbringler@mines.edu
2 Schaeffler Group USA, Wooster, OH 44691, USA; wisemtt@schaeffler.com
3 Center for Automotive Research, The Ohio State University, Columbus, OH 43210, USA; ramesh.47@osu.edu (P.R.); kim.6776@osu.edu (J.H.K.)
4 Ford Motor Company, Dearborn, MI 48126, USA; cbae@ford.com (C.B.); jdeng9@ford.com (J.D.)
5 SK On, Daejeon 34124, Republic of Korea; parkhc@sk.com
* Correspondence: canova.1@osu.edu; Tel.: +1-614-247-2336
† These authors contributed equally to this work.

Abstract: This paper proposes a new model that predicts the cell voltage dynamics and capacity degradation induced by lithium plating and stripping. The proposed model uses a single equilibrium reaction to describe the deposition and dissolution of metallic lithium, predicting the partial reversibility of the plating/stripping reaction, the characteristic voltage plateau during relaxation, and the capacity loss due to the Loss of Cyclable Lithium (LCL). The model is integrated with a Doyle–Fuller–Newman (DFN) electrochemical model, calibrated and validated with experimental data. The model has the potential to improve the accuracy of predicting the effects of lithium plating in Li-ion cells and aid in the development of Extreme Fast Charging (XFC) technology for BEVs.

Keywords: fast charging; lithium plating; modeling

1. Introduction

Enabling Extreme Fast Charging (XFC) is a key priority for increasing the adoption of Battery Electric Vehicles (BEVs) and decarbonizing the transportation sector. On the other hand, lithium ion (Li-ion) cell technology is currently limited in terms of charge rate capability by a side reaction known as lithium plating, in which solid metallic lithium forms in the anode, where its local potential falls below the potential of solid lithium [1]. Lithium plating has been shown to be favored at low temperatures, high State of Charge (SOC) and high rates of charging, which leads to lower intercalation rates, larger overpotentials, values of the anode Open Circuit Potential (OCP) close to 0V (vs. Li/Li^+), and slower solid-phase diffusion [2–5]. While the plating reaction can be partially reversible, irreversibly plated lithium can form solid dendrites, leading to a loss in the accessible surface area of the anode active material and a loss of cyclable lithium through formation of a Solid Electrolyte Interface (SEI) [6–9].

Due to the practical challenges of measuring the onset of plating in a commercial cell in situ during fast charging, modeling and simulation resources have received considerable attention [10–12]. Methods to predict the lithium plating and its effects on capacity loss have been created by postulating different mechanisms, rates, and locations of these phenomena [1,13–18]. On the other hand, capturing the interactions between the side reactions and the lithium intercalation in the anode during the cell charging and subsequent relaxation is rather complex due to the dynamics and couplings inherent to this process. A lack of fidelity in modeling these effects ultimately limits the ability of models to predict the characteristic voltage response that can be observed from experimental tests when plating occurs.

This paper proposes a new model that predicts the cell voltage dynamics and capacity degradation induced by lithium plating and stripping via a single side reaction. Most plating models assume that the plating and stripping reaction kinetics are described by two separate equations. However, this assumption contrasts with the principles of equilibrium reactions, in that a cathodic and an anodic reaction are simultaneously present at the same electrode/electrolyte interface. According to this assumption, the lithium plating and stripping kinetics can be described using a single equation. In this study, a single equilibrium reaction is used to describe the deposition and dissolution of metallic lithium. The resulting model predicts the partial reversibility of the plating/stripping reaction, the characteristic voltage plateau during relaxation, and the capacity loss due to the Loss of Cyclable Lithium (LCL). The model is validated with experimental data to show the ability to correctly represent the process.

The paper is organized as follows: Section 2 presents the experimental tests and data analysis for lithium plating detection and capacity loss analysis. Section 3 illustrates the plating/stripping model. Section 4 describes the integration of the model with a Doyle–Fuller–Newman (DFN) electrochemical model, the calibration process, and verification.

2. Experimental Tests and Results

A small format, 1.6 Ah pouch cell with commercial-grade electrodes was selected for the experimental study. The 50 × 50 mm cell contains a blended NMC811 and NMC111 cathode and a blended artificial and natural graphite anode and utilizes a 1M $LiPF_6$ electrolyte with a ceramic coated separator.

The pouch cells were retrofitted by inserting a reference electrode, providing an in-situ, non-destructive method to measure anode potential. After conducting initial tests for baseline capacity and internal resistance, the cells were pierced in one of the lower corners, and a 22 AWG copper wire was inserted into the cell. Prior to the insertion, the copper wire was coated in solid lithium using electroplating to ensure proper electrical contact and minimize noise in the measurements. After the insertion, electrolyte solvent was injected into the cell, and the hole was sealed with an epoxy resin [19]. Figure 1 shows a schematic of the reference electrode insertion.

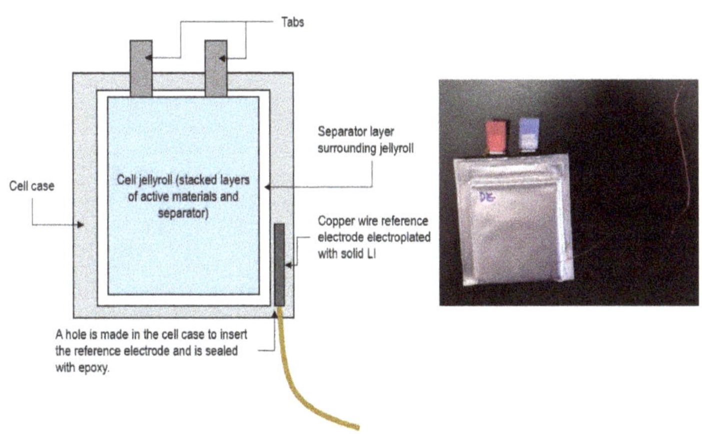

Figure 1. Li-ion pouch cell setup with lateral reference electrode.

The retrofitted cells were then placed in an aluminum fixture to apply pressure (34 kPa) and temperature control via Peltier junctions. The cells were cycled with a 0–5 V Arbin Instruments LBT21044 eight-channel battery cycler, capable of a current range of 0-100/10/1/0.1A with 100 ppm precision.

After a baseline characterization, a series of fast charge tests was performed at three different temperatures (0 °C, 10 °C, 25 °C). In this protocol, a cell is charged starting from a low SOC at a constant current (CC) until the terminal voltage of the cell reaches the maximum cell voltage cutoff (4.2V), and then the charge mode is switched to constant voltage (CV). As discussed in [18,20–27], one of the most accessible detection methods in-situ for the presence of Li-plating in a charge cycle consists of a plateau in cell voltage during relaxation. Figure 2 shows the difference in terminal voltage during cell relaxation between typical settling behavior and a voltage plateau indicative of the Li-stripping side reaction, which indirectly indicates that Li-plating has occurred during the preceding cell charging.

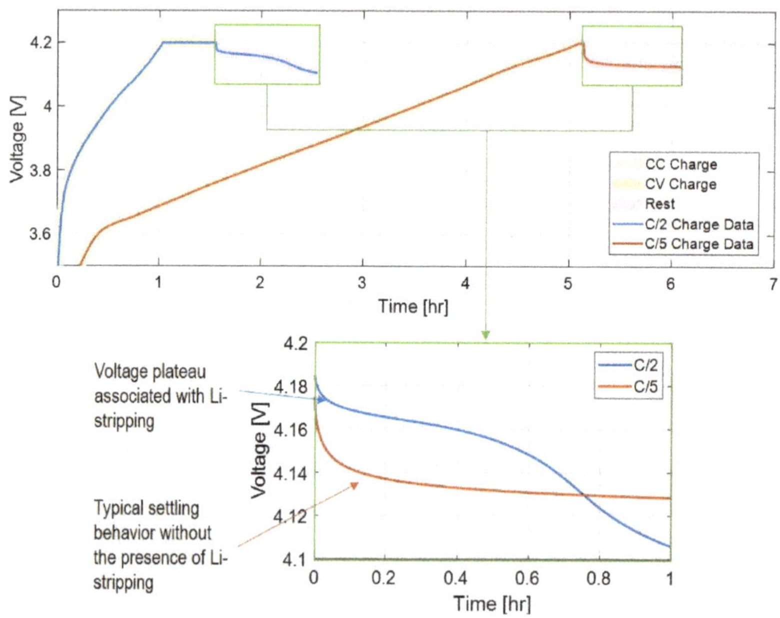

Figure 2. Experiment cell terminal voltage during relaxation for a typical charge cycle vs. a cycle indicating the presence of Li-stripping.

According to literature, plating can occur when the local anode potential falls below 0 V vs. Li/Li^+ [1,18,23,26,28,29]. If charging ends while Li-plating is still occurring, or if the overpotential is sufficiently small to cause the anode potential to rise above 0 V vs. Li/Li^+, the plating reaction will reverse (stripping), leading to a characteristic plateau in the voltage [15,18,21,23,26,30,31].

It is noteworthy to observe that plating may extend through the early phase of the CV portion during a CC-CV charging event. If charging ends while lithium is still plating, the stripping reaction will begin as soon as the anode potential rises above 0 V. This appears as a voltage plateau as the cell relaxes, as shown experimentally in Figure 3. This figure (relative to a 5C charge with a 1C CV cutoff at 25 °C) indicates the occurrence of Li-stripping by the presence of a peak in the rate change of the cell voltage profile (dV/dt) during relaxation, which is equivalent to the characteristic voltage plateau.

On the other hand, if the anode potential rises above 0V vs Li/Li^+ during the CV phase, the stripping process will begin while the cell is still charged. Because the voltage is held constant, the transition to stripping during the CV phase can be detected by monitoring the rate of change in current (dI/dt), which displays a characteristic peak when the stripping side reaction begins. Figure 4 illustrates this behavior for a CC-CV charge at 4.2 V with a C/10 CV cutoff at 25 °C).

It should be noted that while the plateau in the voltage profile during relaxation is commonly assessed in the literature as an indicator of Li-stripping [22,28,32–34], the peak in the applied current during the CV phase under very low currents is a novel indicator of stripping in this study. The presence of a plating-to-stripping transition during CV charging is strongly dependent on the parameters of the CC-CV charge profiles. For the cells tested in this study, the transition occurred during the CV phase, when the target cell voltage was 4.2 V and the cutoff current was set lower than C/2 at 25 °C.

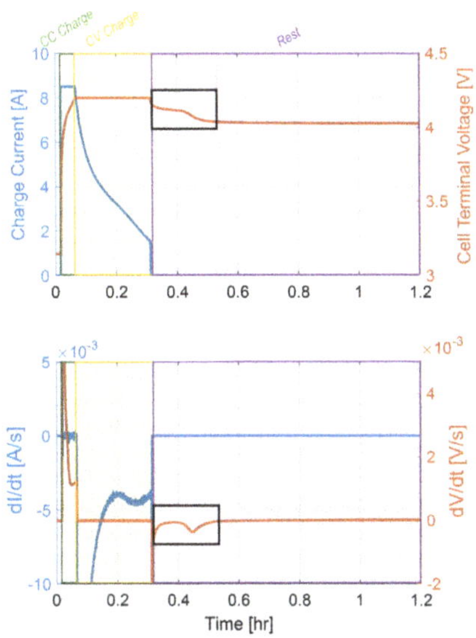

Figure 3. Example of lithium stripping detection for CC−CV charging profile during rest.

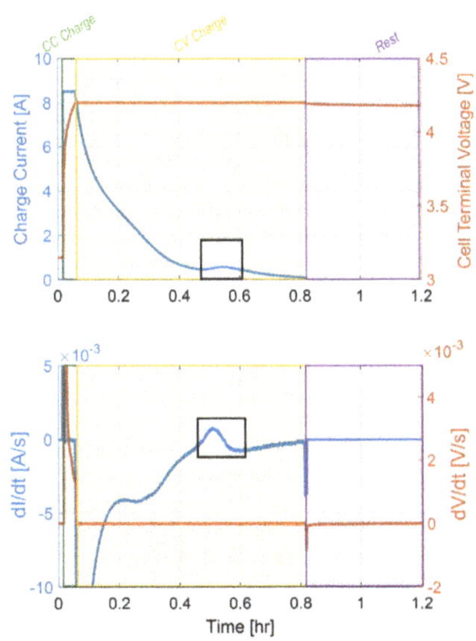

Figure 4. Example of lithium stripping detection for CC−CV charging profile during CV phase.

The magnitude of the CC current and the CV cutoff voltage were varied to explore their effects on the onset of Li-plating in the cell. For each selected CC charge rate and CV current cutoff, the cell was charged to its maximum cutoff voltage, charged at CV until the set cutoff current, allowed to rest for 4 h, and then discharged at C/5 to its minimum cutoff voltage to perform a capacity test.

For each tested condition, the rate change of the voltage profile was analyzed for the indicative peak in the dV/dt profile. If a peak is present, this suggests that a voltage plateau has occurred during the relaxation after the charge, which indicates the presence of Li-stripping and, by consequence, of Li-plating. The current profile during the CV phase was also analyzed to detect a localized increase in the current decay profile (or the peak in the dI/dt profile), which may be indicative of an Li-stripping side reaction occurring during CV charge. If no voltage plateau was present in the relaxation voltage data or peaks present in the CV current profile, it was determined that there was no indicative sign of Li-stripping in the reaction. This investigation was repeated for multiple C-rates and cutoff currents for CV, and the presence of an Li-stripping plateau in each test is tabulated in Table 1 for tests performed at 25 °C.

Table 1. Li-plating test conditions and results of Li-stripping plateau detection.

CC Charge Rate	**CV Current Cutoff**		
	None (CC Only)	**1.5 A**	**0.1 A**
4C	Stripping during relaxation	Stripping during relaxation	Stripping during CV
5C	Stripping during relaxation	Stripping during relaxation	Stripping during CV
6C	No Stripping	Stripping during relaxation	Stripping during CV

A summary of the outcomes of all tests performed at the three chosen temperatures is shown in Figure 5. Here, it can be seen that varying the C-rate for CC charging, the CV cutoff current, and the cell temperature changes the lithium stripping behavior in the cell, and that this behavior is sensitive to the charge conditions which include temperature, charge current, and CV current cutoff.

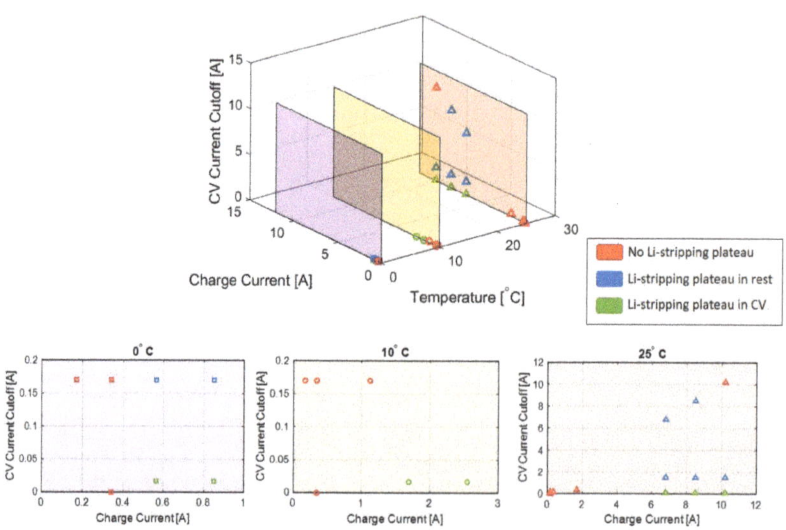

Figure 5. Summary of all experimental charge results and occurrence of stripping plateau.

As documented in the literature and verified in these experiments, Li-plating increases with increased charge current and decreased temperature. At 0 °C, there were signs of Li-stripping (and thus Li-plating) at charge rates as low as C/3, while Li-plating did not occur in the same charge cycles at 10 °C until rates of 1C and at 25 °C until rates of 3C.

The CV cutoff, which has an impact when the Li-stripping plateau occurs (during CV or during relaxation), can provide additional insight into the behavior of the anode during the plating/stripping portion of the charge cycle. Figure 6 highlights how studying the effects of the CV cutoff at one fixed CC charge rate can give insight into the behavior of the anode potential in situ.

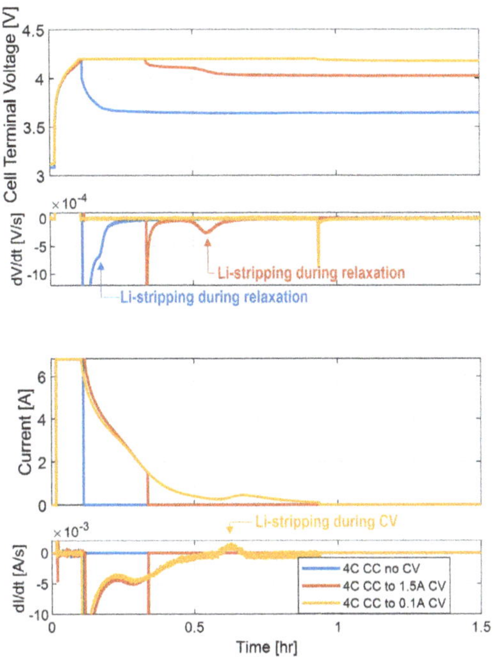

Figure 6. Example of various 4C charge profiles at 25°C with various CV cutoffs and implications on plating.

In the 4C charge test with no CV portion (CC only), there is a very small plateau associated with the occurrence of Li-stripping during the rest phase. In the 4C charge with a 1.5 A CV cutoff current, there is a more noticeable and distinct voltage plateau, which lasts longer after charge is cut off in the cell. When the cell is charged at 4C with a 0.1 A CV cutoff current, Li-stripping occurs during the CV phase.

From these conditions, it can be deduced that in all three charge conditions, the anode reaches a potential of less than 0 V vs Li/Li^+ during the CC charge portion. In fact, even with the 4C charge with no CV, there are signs of Li-stripping, indicating that Li-plating has occurred. The 4C charge with a 1.5A CV cutoff current exhibits a longer and more distinct voltage plateau, indicating that more Li-stripping is occurring after this charge and implying that this is caused by more Li-plating. Because the voltage plateau happens after charge is stopped, the anode must be plating (at $<$ 0V vs. Li/Li^+) during the entire CV mode in this charge cycle. This indicates that the Li-plating starts in the CC region (as seen in the 4C charge with no CV) and continues through the CV region until the 1.5 A charge current cutoff. In the longer CV test (at which the CV current is cut off at 0.1 A), the plateau occurs around the point at which the cell is being charged at 0.5 A. Because the current peak indicates the occurrence of an Li-stripping side reaction, this represents the point at which the anode potential rises above 0 V to transition from the Li-plating to Li-stripping reaction.

These sequential charges at different rates, exemplified in Figure 6, can shed further light on the anode behavior during different portions of the charge phase and can lend understanding to both the onset and stop of the Li-plating and allow the charge region to be broken into the regions shown in Figure 7.

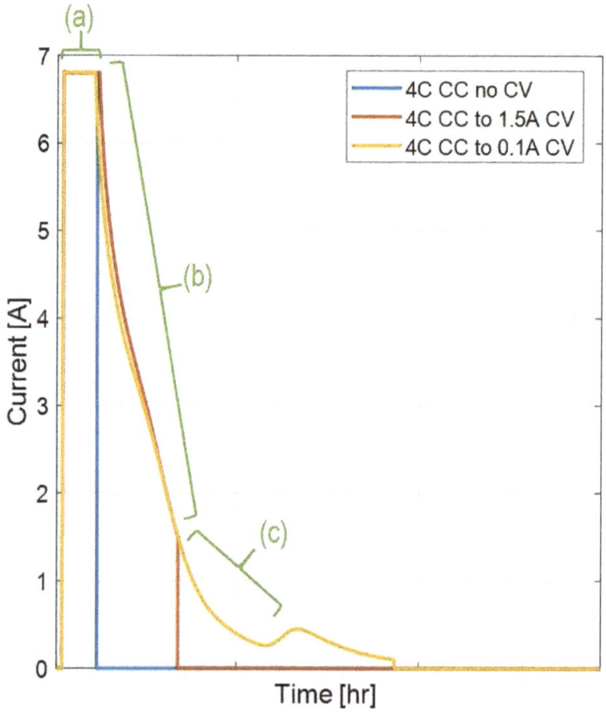

Figure 7. Understanding of charge regions from 4C charge profiles at 25 °C with various CV cutoffs and implications on plating.

In region (a), the anode reaches the Li-plating favored potential lower than 0 V vs Li/Li^+ and maintains this condition through region (b). Region (c) indicates the final portions of the Li-plating condition being favored before the cell transitions into conditions of Li-stripping, when the anode rises to a potential greater than 0 V vs Li/Li^+. This methodology of mapping various charge profiles to indicators of anode behavior can help to give insight into the underlying mechanisms and conditions favoring Li-plating. Additionally, this provides a useful data set for the model calibration explored in this work.

3. Modeling

The lithium plating side reaction is a local phenomenon that can occur in parallel with the intercalation and other degradation mechanisms. As pointed out in the literature—see for instance [28]—plating often occurs in conjunction with other side reactions, such as SEI layer growth. On the other hand, for a low number of consecutive charge cycles at high C-rate conditions, as performed in the experiments described in this paper, it is reasonable to assume that the amount of capacity loss due to SEI layer growth is minimal compared to plating. This allows us to simplify the integration of the plating/stripping model into an electrochemical model.

If the local potential of the anode drops below the potential of solid Li (<0V vs. Li/Li^{+}), the lithium plating side reaction is favored [1,18,23,28]. The process of deposition and dissolution of the solid lithium can be implemented as a side reaction occurring at each location within the anode thickness. The reference reaction can be written as

$$Li^{+} + e^{-} \rightleftharpoons Li^{0} \tag{1}$$

The deposition and dissolution of metallic lithium is described by a single Butler–Volmer equation:

$$j_{Li} = i_{0,Li}\left[\frac{c_{Li}}{c_{Li}^{ep}}\exp\left(\frac{\alpha_{a,Li}F}{RT}\eta_{Li}\right) - \frac{c_e}{c_e^{ep}}\exp\left(\frac{-\alpha_{c,Li}F}{RT}\eta_{Li}\right)\right] \tag{2}$$

where j_{Li} is the side reaction current density, $i_{0,Li}$ is the exchange current density, $\alpha_{a,Li}$ and $\alpha_{c,Li}$ are the respective anodic and cathodic transfer coefficients for the plating/stripping reaction, c_{Li} is the concentration of Li metal, c_e is the concentration of Li^{+} in the electrolyte, c_{Li}^{eq} and c_e^{eq} are the bulk concentrations of Li metal and ions in the electrolyte respectively, R is the universal gas constant, T is the temperature, and F is Faraday's constant.

The lithium overpotential η_{Li} is defined as

$$\eta_{Li} = \phi_s - \phi_e - U_{Li} \tag{3}$$

where ϕ_s is the local solid phase potential, ϕ_e is the local electrolyte potential, and U_{Li} is the equilibrium potential of metallic lithium.

The exchange current density for the side reaction is calculated by

$$i_{0,Li} = Fk_{0,Li}(c_{Li}^{eq})^{\alpha_{c,Li}}(c_e^{eq})^{\alpha_{a,Li}} \tag{4}$$

where $k_{0,Li}$ is the reaction rate constant of the plating/stripping reaction. Since the deposition/dissolution reaction is modeled as homogeneous along the surface, the lithium concentration is set to 1 when plated lithium is present, causing the amount of plated lithium to not affect the reaction rate. When plated lithium is not present, the concentration is set to 0 so that stripping does not occur. A correction term $\left(\frac{\beta n_{Li,rev}}{1+\beta n_{Li,rev}}\right)$ on the anodic branch replaces this plated lithium concentration to bulk concentration ratio $(\frac{c_{Li}}{c_{Li}^{eq}})$ to compute the stripping side reaction within the Butler–Volmer equation [28].

$$j_{Li} = i_{0,Li}\left[\left(\frac{\beta n_{Li,rev}}{1+\beta n_{Li,rev}}\right)\exp\left(\frac{\alpha_{a,Li}F}{RT}\eta_{Li}\right) - \frac{c_e}{c_e^{ep}}\exp\left(\frac{-\alpha_{c,Li}F}{RT}\eta_{Li}\right)\right] \tag{5}$$

where $n_{Li,rev}$ is the quantity of reversibly plated lithium, and β is a large value to effectively make the $\left(\frac{\beta n_{Li,rev}}{1+\beta n_{Li,rev}}\right)$ term equal 1 when $n_{Li,rev} >> 1$, and 0 when $n_{Li,rev} = 0$. The inclusion of this term in Equation (5) allows us to compute the current density for the stripping reaction while reversibly plated lithium is present and restricts stripping when the reversibly plated lithium is not present.

At each node, the amount of plated lithium is calculated by integrating the side reaction current:

$$\frac{dn_{Li}}{dt} = -\frac{a_n}{F}\cdot j_{Li} \tag{6}$$

where n_{Li} is the quantity of plated lithium, and a_n is the surface area of the anode. It is assumed that a portion of the plated lithium becomes electrically inactive and no longer participates in the reaction. The reversibility of the plating reaction is governed by

$$\frac{dn_{Li,rev}}{dt} = -\eta_r\frac{a_n}{F}\cdot j_{Li} \tag{7}$$

$$\frac{dn_{Li,irr}}{dt} = -(1-\eta_r)\frac{a_n}{F} \cdot j_{Li} \tag{8}$$

which introduces a "reversibility factor" η_r as a function of current sign:

$$\eta_r = \alpha + 0.5(1-\alpha)\left(1 + \frac{\arctan(kj_{Li})}{90}\right) \tag{9}$$

where α is the reversible fraction of plated lithium (tunable parameter), and k is a large number.

During plating ($j_{Li} < 0$), the side reaction current leads to the accumulation of both irreversible and reversible plated lithium. When stripping is favored ($j_{Li} \geq 0$), the side reaction current contributes to removing only the reversibly plated lithium. The total quantity of plated lithium at each location within the anode thickness is the sum of the reversible and irreversible parts:

$$n_{Li,tot}(x,t) = n_{Li,irr}(x,t) + n_{Li,rev}(x,t) \tag{10}$$

4. Integration with P2D Model

The plating and stripping model can be integrated into any electrochemical Li-ion cell model as a side reaction. In this work, the Doyle–Fuller–Newman (or Pseudo Two Dimensional, P2D) model was utilized as the source model due to its ability to predict the distribution of lithium concentration and potential through the thickness of the electrodes [35,36].

The P2D model assumes uniform use of the electrodes in the planar direction (i.e., uniform intercalation across anode surface). For this reason, this work assumes the same uniform Li plating/stripping on the anode surface.

The side reaction current density j_{Li} represents both the forward lithium plating and reverse Li-stripping reactions. The lithium plating side reaction occurs in parallel with the intercalation/deintercalation of graphite. At each location within the anode, the net current density (j_n) is split between the side reaction current density (j_{Li}) and the intercalation/deintercalation current density (j_{int}):

$$j_n(x,t) = j_{Li}(x,t) + j_{int}(x,t) \tag{11}$$

Figure 8 illustrates the concept. When any location in the anode reaches a potential favoring the plating side reaction, the amount of plated lithium n_{Li} grows due to Equation (6). This process occurs in parallel with the intercalation reaction, described by j_{int}. This net current density is the same magnitude of the current density that is subtracted from the electrolyte solution at each location through the electrode thickness.

When the local potential no longer favors the Li-plating side reaction and there is reversible plated lithium present on the particle ($n_{Li,rev} \geq 0$), the amount of plated lithium decreases as described by Equations (5) and (7). The side reaction current density that dissolves the reversible lithium results in an increased local concentration of ionic lithium in the electrolyte solution, with successive intercalation into the graphite.

Since the magnitude of the plating reaction is dependent on the local value of the anode potential in both location and time, there can be different amounts of plated lithium at different locations within the anode domain as the cell is charged.

The model was integrated within PyBaMM, an open-source Python implementation of the P2D model [36]. Figure 9 illustrates the essential equations of the P2D model and its interaction with the plating and stripping model, via block diagram representation.

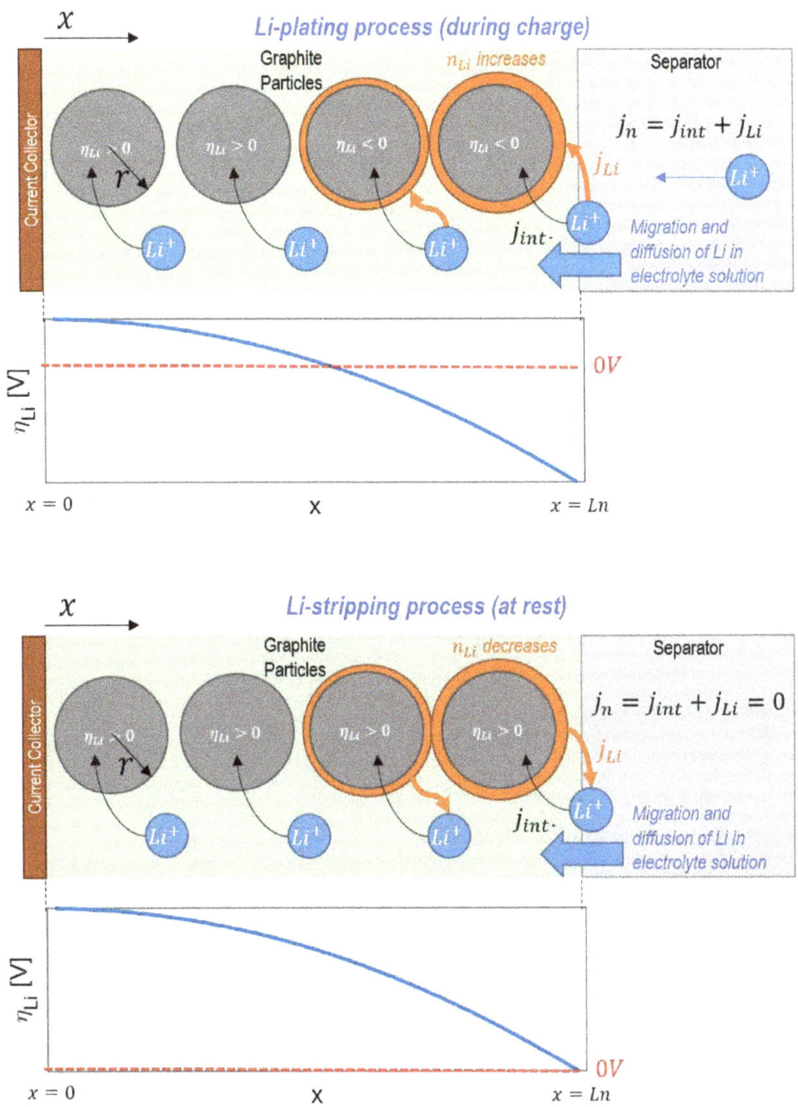

Figure 8. Mechanism of plating and stripping reactions integrated in P2D model.

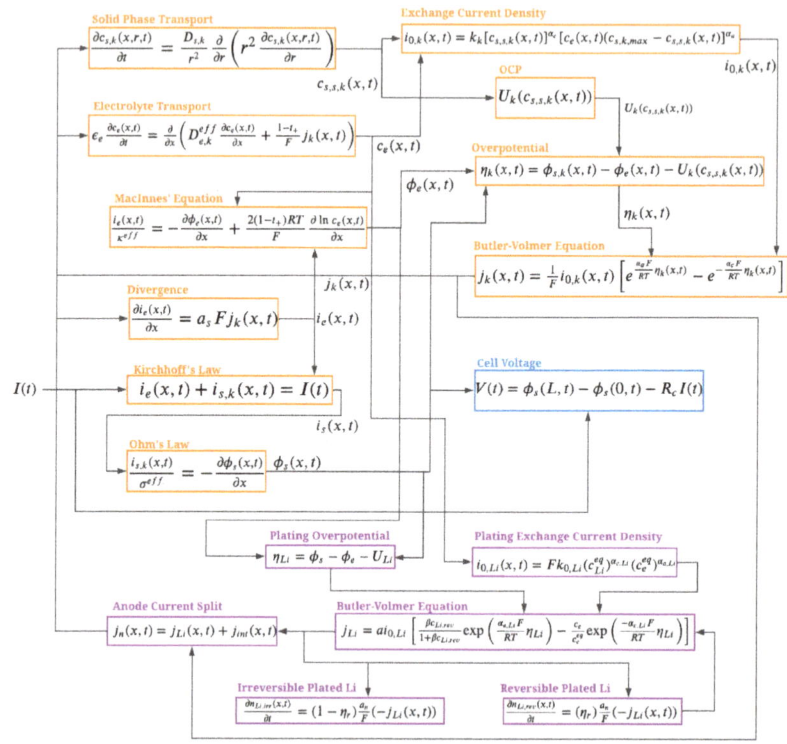

Figure 9. Block Diagram of P2D Model with Integrated Plating/Stripping Model.

5. Calibration and Verification

The plating and stripping model has two parameters for calibration, namely the reaction rate constant $k_{0,Li}$ and the degree of reversibility α. The reaction rate parameter affects the exchange current density for the plating and stripping reaction and is the main fitting parameter. The model behaviors for different $k_{0,Li}$ values are presented in Figure 10.

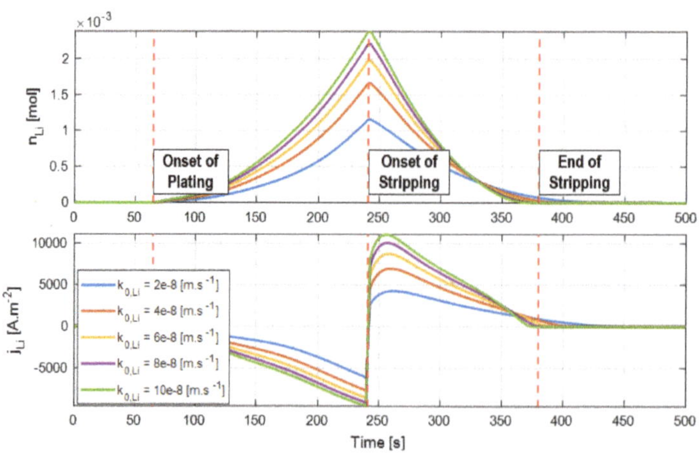

Figure 10. Plating/stripping behavior for different values of $k_{0,Li}$ (5C CC fast charge).

While the parameter $k_{0,Li}$ dictates the amount of lithium plating that occurs, the duration of the plating/stripping process is mainly dependent on the overpotential η_{Li}. For this reason, accurate prediction of the anode potential (which requires use of reference electrode data) during fast charging is critical to correctly capture the intercalation and plating reactions.

The voltage plateau associated with the stripping reaction during cell relaxation is commonly used to calibrate the reaction rate constant. When stripping induces a voltage plateau, the magnitude of the peak in the differential voltage is an indication of the quantity of lithium being stripped away, so the reaction rate constant can be tuned to match this peak. Figure 11 shows how the voltage and differential voltage profiles during relaxation are affected by changes in the reaction rates for the formation/dissolution of metallic lithium.

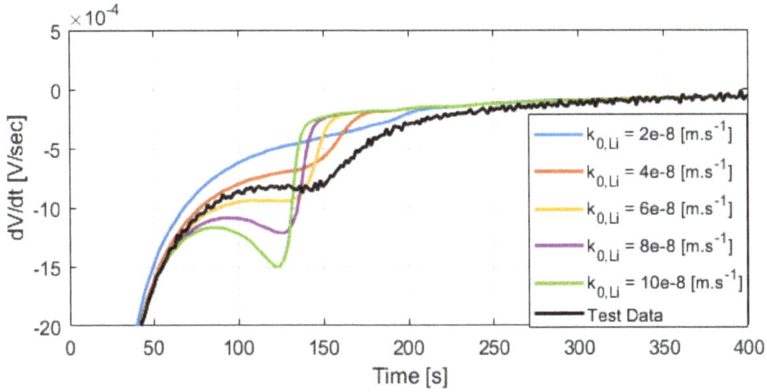

Figure 11. Effects of $k_{0,Li}$ parameter on voltage derivative during relaxation (5C CC fast charge).

One single fast charging test (5C CC fast charge) was used to calibrate the parameter $k_{0,Li}$. The differential voltage profiles in Figure 12 and 13 show good agreement between the model and the data for other test conditions.

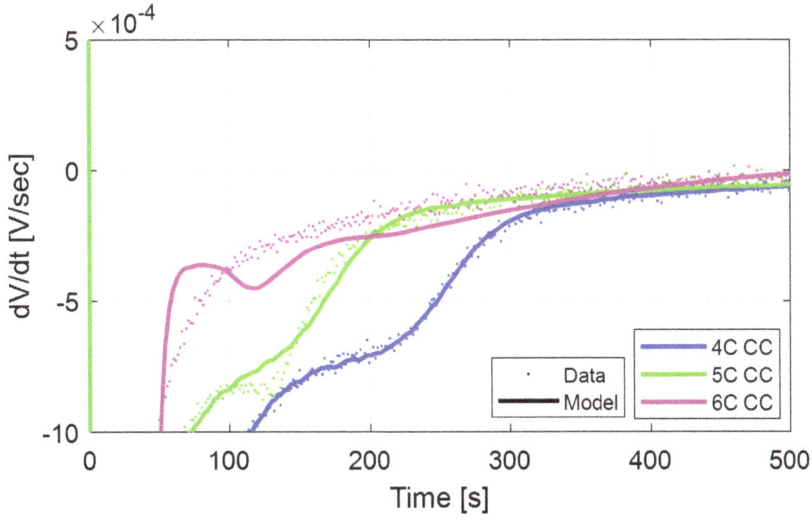

Figure 12. Verification of calibrated $k_{0,Li}$ parameter for CC fast charge tests.

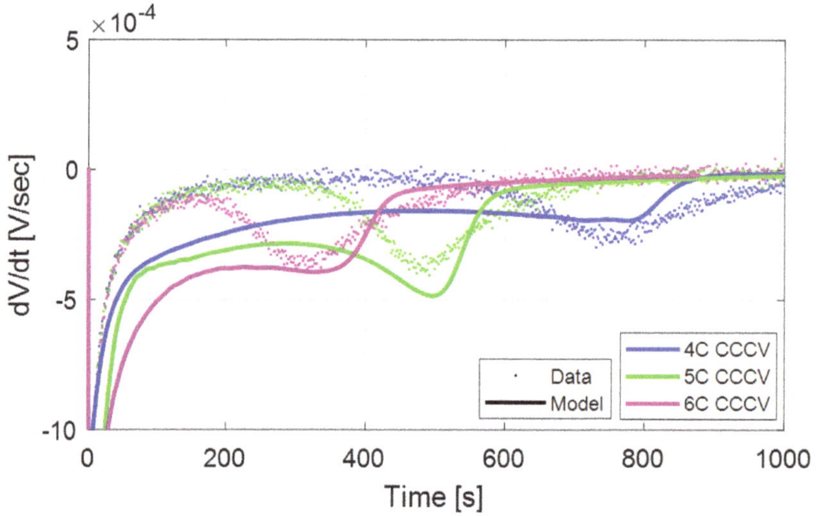

Figure 13. Verification of calibrated $k_{0,Li}$ parameter for CC−CV fast charge tests.

In Figure 12, the modeled 4C CC relaxation dV/dt has an early peak occurring at 90 s that is not observed in the data. This peak is caused by a characteristic cathode OCP slope change and is not related to the lithium stripping side reaction.

In Figure 13, the model presents errors in capturing the relaxation dV/dt curves at 4C and 6C for the CC-CV test. The errors might be the result of an imperfect model calibration at Beginning of Life (BOL), rather than issues with the stripping model. In fact, although the predicted dV/dt profile does not exactly match the data, the time of occurrence of the stripping peak matches quite well, which indirectly proves that the model is correctly predicting the onset of the stripping behavior.

The reversibility parameter α governs the fraction of reversible and irreversible plated lithium and can be calibrated to match capacity test data following a fast charging cycle. To this extent, the C/20 discharge test data were used to compute the capacity reduction after each fast charge cycle. Since severe capacity fade is observed within just a few charge cycles, it can be assumed that the main aging mechanism causing this degradation is lithium plating.

Simulations of consecutive fast charge cycles (4C, 5C, then 6C) were conducted to predict the quantity of irreversibly plated lithium, from which the capacity loss was computed via Faraday's law:

$$Q_{pl} = \frac{n_{Li}F}{3600} \quad (12)$$

The reversibility parameter α was tuned to match the capacity fade experienced due to CC fast charging (Figure 14).

Note that changing the reversibility of the side reaction impacts the accuracy of the voltage plateau prediction, requiring further tuning after capacity loss data are matched. The final calibrated lithium plating parameters are provided in Table 2. As previously mentioned, once the reaction rate parameter $k_{0,Li}$ is tuned to match the relaxation profile, the reversibility parameter α can be calibrated to match the capacity loss, which is measured in the C/20 discharge experiments. Figure 14 shows how the model can match the capacity loss from lithium plating by calibrating the α parameter.

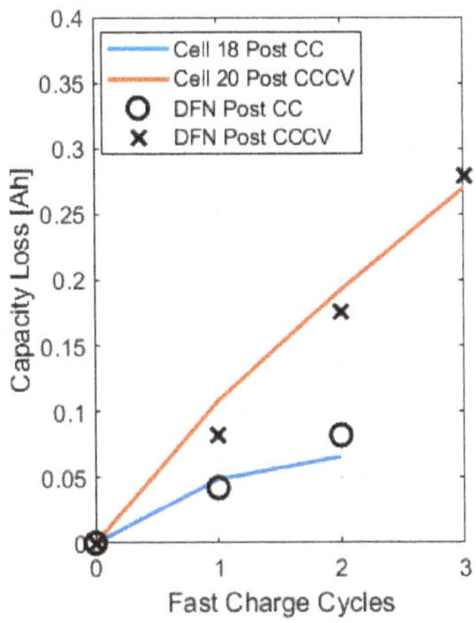

Figure 14. Modeled capacity agreement with RPT measured capacity loss.

Table 2. Calibrated plating model parameters.

Plating Parameters	Value
Rate Constant $k_{0,Li}$	5.5×10^{-8} [m/s]
Reversibility α	0.65 [-]

6. Simulation and Analysis

The model predictions can be further analyzed to understand the behavior of the cell during Li-plating and Li-stripping conditions. Figure 15 illustrates the phenomenon of Li-stripping during relaxation, namely how the profile of plated lithium through the anode thickness changes as it is oxidized and returns to the electrolyte solution during the rest period.

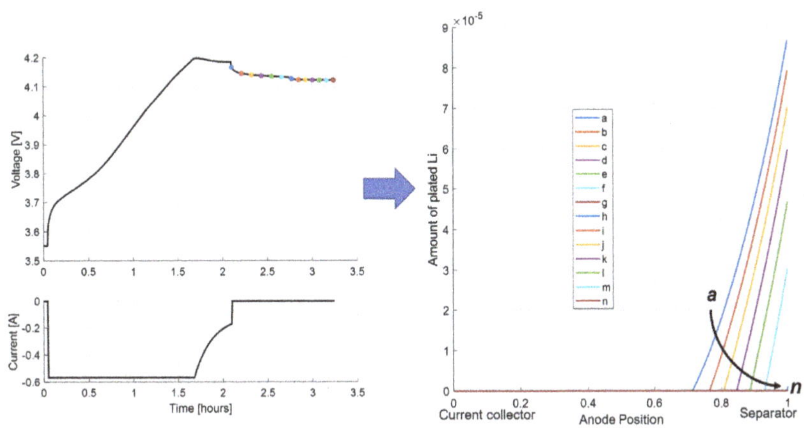

Figure 15. Investigation of plated lithium quantity through the anode thickness after relaxation with the DFN model.

The individual points in the Li-stripping plateau in time show that the amount of plated lithium decreases at each time step, with the peak of plated lithium located at the interface with the separator. As the stripping process continues, each location in the cell where plated lithium exists becomes progressively depleted.

Investigation of the net anodic current density ($j_n = j_{int} + j_{Li}$) in Figure 16 during Li-stripping shows a location-dependent net current. At the interface between the anode and separator, the net anodic current is positive, meaning the oxidation reaction is favored. This means that there is a positive contribution of lithium from the stripping reaction, since there is a larger quantity of accumulated lithium in these regions as confirmed in Figure 15. Closer to the current collector, the net anodic current density is negative, meaning that the reduction reaction is favored. Here, the net reaction is the charge of the graphite from stripped Li intercalating back into the anode. Because lithium stripping is favored near the separator side of the anode but lithium intercalation is favored on the current collector side, this implies the net movement of lithium ions from the separator to current collector side of the anode as the stripping reaction occurs.

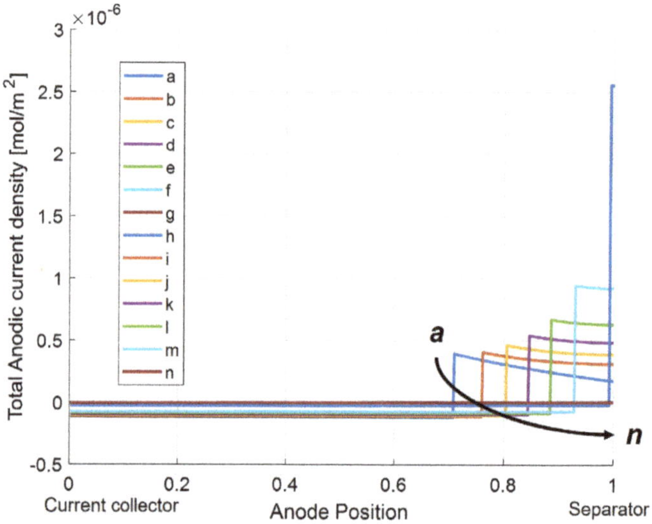

Figure 16. Investigation of anodic current density through the anode thickness after relaxation with the DFN model.

Figure 17 confirms that as lithium strips from the surface of the graphite particles, it moves towards the current collector side of the anode to re-intercalate back into the graphite material. As the concentration of lithium in the electrolyte increases, ionic diffusion drives the movement of lithium towards the current collector, where less lithium is in solution due to the lack of stripping in this region of the anode. Additionally, the potential gradient in the electrolyte solution after charge favors the migration of Lithium towards the current collector, as there is a higher potential in the liquid phase on the separator side of the anode.

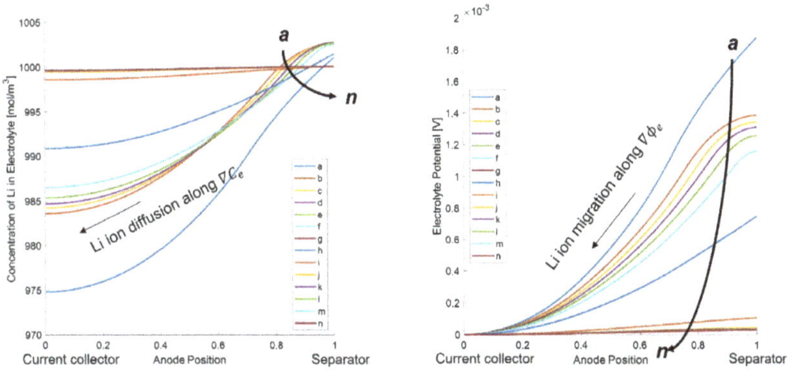

(a) Concentration gradient in electrolyte solution.

(b) Potential gradient in electrolyte solution

Figure 17. Concentration and potential gradients in the electrolyte solution.

7. Conclusions

This paper proposes a new model that predicts the cell voltage dynamics and capacity degradation induced by lithium plating and stripping. Unlike most models that utilize multiple reactions to describe the process, a single equilibrium reaction is here derived to describe the deposition and dissolution of metallic lithium. The resulting model predicts the partial reversibility of the plating/stripping reaction, the characteristic voltage plateau during relaxation, and the capacity loss due to the Loss of Cyclable Lithium (LCL).

The plating and stripping model was integrated with a Doyle–Fuller–Newman (DFN) electrochemical model, calibrated and validated with experimental data to show the ability to correctly represent the process. The model described in this paper can be utilized to analyze the physical phenomena occurring in the cell during fast charging, giving better insight into the cell's behavior during this undesirable side reaction. If the behavior can be better predicted and understood and the the root causes and dynamics of the Li-plating and Li-stripping reactions can be quantified, these models can be used for design studies for cell construction and applied current profiles to avoid Li-plating and enable fast charging in BEVs.

Ongoing work focuses on extending this modeling effort to predict the effects of dendrite formation on impedance growth and loss of active material. Furthermore, teardown and post-mortem analysis will be utilized to prove the occurrence of plating and stripping.

Author Contributions: Conceptualization: P.B.R., P.R., J.H.K., M.C.; Methodology: P.B.R., P.R.; Model Development and Validation: P.B.R., M.W., M.C.; Investigation: P.B.R., M.W., P.R., J.H.K., M.C., C.B., J.D.: Resources, C.B., J.D., H.P.; Writing—Original Draft Preparation: P.B.R., M.C.; Writing—Review and Editing: P.B.R., M.W., P.R., J.H.K., M.C., C.B., J.D.; Supervision: C.B., J.D., H.P.; Project Administration: P.R., M.C., J.H.K. All authors have read and agreed to the published version of the manuscript.

Funding: The research activity that led to this work was funded by Ford Motor Company.

Data Availability Statement: The data presented in this study are available on request from the corresponding author. The data are not publicly available due to confidentiality.

Conflicts of Interest: The authors declare no conflict of interest.

References

1. Arora, P. Mathematical Modeling of the Lithium Deposition Overcharge Reaction in Lithium-Ion Batteries Using Carbon-Based Negative Electrodes. *J. Electrochem. Soc.* **1999**, *146*, 3543. [CrossRef]
2. Smart, M.C.; Ratnakumar, B.V. Effects of Electrolyte Composition on Lithium Plating in Lithium-Ion Cells. *J. Electrochem. Soc.* **2011**, *158*, A379. [CrossRef]

3. Fan, J.; Tan, S. Studies on Charging Lithium-Ion Cells at Low Temperatures. *J. Electrochem. Soc.* **2006**, *153*, A1081. [CrossRef]
4. Jansen, A.N.; Dees, D.W.; Abraham, D.P.; Amine, K.; Henriksen, G.L. Low-temperature study of lithium-ion cells using a LiySn micro-reference electrode. *J. Power Sources* **2007**, *174*, 373–379. [CrossRef]
5. Huang, C.K.; Sakamoto, J.S.; Wolfenstine, J.; Surampudi, S. The Limits of Low-Temperature Performance of Li-Ion Cells. *J. Electrochem. Soc.* **2000**, *147*, 2893. [CrossRef]
6. Janakiraman, U.; Garrick, T.R.; Fortier, M.E. Lithium Plating Detection Methods in Lithium-ion Batteries. In *Electrochemical Society Meeting Abstracts Prime2020*; The Electrochemical Society, Inc.: Pennington, NJ, USA, 2020; p. 47.
7. Zier, M.; Scheiba, F.; Oswald, S.; Thomas, J.; Goers, D.; Scherer, T.; Klose, M.; Ehrenberg, H.; Eckert, J. Lithium dendrite and solid electrolyte interphase investigation using OsO_4. *J. Power Sources* **2014**, *266*, 198–207. [CrossRef]
8. Lu, W.; López, C.M.; Liu, N.; Vaughey, J.T.; Jansen, A.; Dennis W.D. Overcharge Effect on Morphology and Structure of Carbon Electrodes for Lithium-Ion Batteries. *J. Electrochem. Soc.* **2012**, *159*, A566–A570. [CrossRef]
9. Lin, H.p.; Chua, D.; Salomon, M.; Shiao, H.C.; Hendrickson, M.; Plichta, E.; Slane, S. Low-Temperature Behavior of Li-Ion Cells. *Electrochem.-Solid-State Lett.* **2001**, *4*, A71. [CrossRef]
10. Subramanian, V.R.; Diwakar, V.D.; Tapriyal, D. Efficient Macro-Micro Scale Coupled Modeling of Batteries. *J. Electrochem. Soc.* **2005**, *152*, A2002–A2008. [CrossRef]
11. Whittingham, M.S. History, Evolution, and Future Status of Energy Storage. *Proc. IEEE* **2012**, *100*, 1518–1534. [CrossRef]
12. Rahn, C.D.; Wang, C.Y. *Battery Systems Engineering*; John Wiley & Sons: Hoboken, NJ, USA, 2013.
13. Lin, N.; Röder, F.; Krewer, U. Multiphysics Modeling for Detailed Analysis of Multi-Layer Lithium-Ion Pouch Cells. *Energies* **2018**, *11*, 2998. [CrossRef]
14. Smith, K. *Micro/Macro-Scale Modeling for Battery Fast Charge Applications*; National Renewable Energy Lab. (NREL): Golden, CO, USA, 2018; p. 28.
15. Zhao, X.; Yin, Y.; Hu, Y.; Choe, S.Y. Electrochemical-thermal modeling of lithium plating/stripping of Li(Ni 0.6 Mn 0.2 Co 0.2) O2/ Carbon lithium-ion batteries at subzero ambient temperatures. *J. Power Sources* **2019**, *418*, 61–73. [CrossRef]
16. Yang, X.G.; Leng, Y.; Zhang, G.; Ge, S.; Wang, C.Y. Modeling of lithium plating induced aging of lithium-ion batteries: Transition from linear to nonlinear aging. *J. Power Sources* **2017**, *360*, 28–40. [CrossRef]
17. Zhang, Z.; Jin, Z.; Wyatt, P. Electrochemical Modeling of Lithium Plating of Lithium Ion Battery for Hybrid Application. *Sae. Int. J. Altern. Powertrains* **2017**, *6*. [CrossRef]
18. von Lüders, C.; Keil, J.; Webersberger, M.; Jossen, A. Modeling of lithium plating and lithium stripping in lithium-ion batteries. *J. Power Sources* **2019**, *414*, 41–47. [CrossRef]
19. Brodsky, P.; Canova, M.; Kim, J.H.; Ramesh, P.; Bae, C.; Deng, J.; Park, H. *Calibration of Electrochemical Models for Li-Ion Battery Cells Using Three-Electrode Testing*; SAE Technical Paper: Warrendale, PA, USA, 2020; p. 14. [CrossRef]
20. Petzl, M.; Danzer, M.A. Nondestructive detection, characterization, and quantification of lithium plating in commercial lithium-ion batteries. *J. Power Sources* **2014**, *254*, 80–87. [CrossRef]
21. Uhlmann, C.; Illig, J.; Ender, M.; Schuster, R.; Ivers-Tiffée, E. In situ detection of lithium metal plating on graphite in experimental cells. *J. Power Sources* **2015**, *279*, 428–438. [CrossRef]
22. Schindler, S.; Bauer, M.; Petzl, M.; Danzer, M.A. Voltage relaxation and impedance spectroscopy as in-operando methods for the detection of lithium plating on graphitic anodes in commercial lithium-ion cells. *J. Power Sources* **2016**, *304*, 170–180. [CrossRef]
23. Yang, X.G.; Ge, S.; Liu, T.; Leng, Y.; Wang, C.Y. A look into the voltage plateau signal for detection and quantification of lithium plating in lithium-ion cells. *J. Power Sources* **2018**, *395*, 251–261. [CrossRef]
24. Bugga, R.V.; Smart, M.C. Lithium Plating Behavior in Lithium-Ion Cells. *Ecs. Trans.* **2010**, *25*, 241. . [CrossRef]
25. Bauer, M.; Rieger, B.; Schindler, S.; Keil, P.; Wachtler, M.; Danzer, M.A.; Jossen, A. Multi-phase formation induced by kinetic limitations in graphite-based lithium-ion cells: Analyzing the effects on dilation and voltage response. *J. Energy Storage* **2017**, *10*, 1–10. [CrossRef]
26. von Lüders, C.; Zinth, V.; Erhard, S.V.; Osswald, P.J.; Hofmann, M.; Gilles, R.; Jossen, A. Lithium plating in lithium-ion batteries investigated by voltage relaxation and in situ neutron diffraction. *J. Power Sources* **2017**, *342*, 17–23. [CrossRef]
27. Waldmann, T.; Wohlfahrt-Mehrens, M. Effects of rest time after Li plating on safety behavior—ARC tests with commercial high-energy 18650 Li-ion cells. *Electrochim. Acta* **2017**, *230*, 454–460. [CrossRef]
28. Ren, D.; Smith, K.; Guo, D.; Han, X.; Feng, X.; Lu, L.; Ouyang, M.; Li, J. Investigation of Lithium Plating-Stripping Process in Li-Ion Batteries at Low Temperature Using an Electrochemical Model. *J. Electrochem. Soc.* **2018**, *165*, A2167–A2178. [CrossRef]
29. Yang, X.; Miller, T. *Fast Charging Lithium-Ion Batteries*; SAE Technical Paper: Warrendale, PA, USA, 2017; p. 12. [CrossRef]
30. Zinth, V.; von Lüders, C.; Hofmann, M.; Hattendorff, J.; Buchberger, I.; Erhard, S.; Rebelo-Kornmeier, J.; Jossen, A.; Gilles, R. Lithium plating in lithium-ion batteries at sub-ambient temperatures investigated by in situ neutron diffraction. *J. Power Sources* **2014**, *271*, 152–159. [CrossRef]
31. Hein, S.; Danner, T.; Latz, A. An Electrochemical Model of Lithium Plating and Stripping in Lithium Ion Batteries. *Acs Appl. Energy Mater.* **2020**, *3*, 8519–8531. [CrossRef]
32. Konz, Z.M.; McShane, E.J.; McCloskey, B.D. Detecting the Onset of Lithium Plating and Monitoring Fast Charging Performance with Voltage Relaxation. *Acs Energy Lett.* **2020**, *5*, 1750–1757. [CrossRef]

33. Konz, Z.M.; McShane, E.J.; McCloskey, B.D. *Voltage Relaxation to Detect the Onset of Lithium Plating on Graphite for Fast Charging*; p. 19. Available online: https://chemrxiv.org/engage/chemrxiv/article-details/60c74a07bdbb898b2ea392d6 (accessed on 8 May 2023).
34. Carter, R.; Klein, E.J.; Kingston, T.A.; Love, C.T. Detection of Lithium Plating During Thermally Transient Charging of Li-Ion Batteries. *Front. Energy Res.* **2019**, *7*, 144. [CrossRef]
35. Doyle, M. Modeling of Galvanostatic Charge and Discharge of the Lithium/Polymer/Insertion Cell. *J. Electrochem. Soc.* **1993**, *140*, 1526. [CrossRef]
36. Sulzer, V.; Marquis, S.G.; Timms, R.; Robinson, M.; Chapman, S.J. Python Battery Mathematical Modelling (PyBaMM). *J. Open Res. Softw.* **2021**, *9*. [CrossRef]

batteries

MDPI

Article

Experimental Application of the Global Technical Regulation on In-Vehicle Battery Durability

Gian Luca Patrone * and Elena Paffumi

Joint Research Centre, European Commission, 21027 Ispra, Italy
* Correspondence: gian-luca.patrone@ec.europa.eu

Abstract: Battery aging of electrified vehicles is a key parameter to be controlled in order to ensure sufficient energy efficiency and driving range across the whole vehicle lifespan. The United Nations Economic Commission for Europe has recently adopted a new regulatory framework, the Global Technical Regulation No. 22, prescribing minimum performance requirements for in-vehicle battery durability. With the implementation of this new GTR, monitors of the battery state of certified energy and range will be available in every production vehicle, the accuracy of which will be tested statistically by applying an in-use verification procedure (Part A). Once the monitors' correctness is checked, the battery durability performances are controlled in Part B against the defined limit values by a fleet monitoring procedure. This work presents the results of a testing campaign executed at the Joint Research Centre testing facilities on an aged pure electric vehicle to measure its capacity and range fade. The aim is to explore the applicability of GTR No. 22, assessing the in-vehicle battery performance fade of an aged electric vehicle, illustrating the several steps of the developed regulation and experimental methodology.

Keywords: electric vehicles; battery durability; battery aging; state of health; vehicle testing; UN GTR No. 22

1. Introduction

The transport sector is contributing significantly to greenhouse gas (GHG) emissions and global warming and has the highest dependence on fossil fuels of any sector, globally accounting for 37% of CO2 emissions from end-use sectors [1,2]. Despite being one of the sectors heavily impacted by the COVID-19 pandemic, emissions are on the rise due to growing demands, and the adoption of alternative fuels remains partial. That growth is most prominent in developing and emerging economies. European country policies and vehicle manufacturers need to put concrete measures into practice to respect commitments of the Paris Agreement during COP21 [3] and the European Green Deal [4], targeting a 90% cut in GHG production in the European Union by 2050 [5,6].

The electrification of the powertrain seems to be the most attractive solution, especially if cleaner assortments of energy production are employed in the future, and customer acceptance is growing significantly [7]. In the last decade, huge improvements have been reached, ensuring affordable batteries with higher and higher specific energy and power. Extensive research and development activities are pushing battery cathode, anode, and electrolyte chemistries to increase the specific energy content of batteries [8,9], and the adoption of modern battery generations is contributing to increased electro-mobility, reducing range anxiety, allowing shorter charging times with high power chargers and making battery electric vehicles (BEVs) an economically feasible choice.

An important aspect of BEVs is the topic of aging; in fact, battery performance characteristics are worsening over the battery's lifetime, influenced by the storage and usage conditions and depending on complex physical and chemical processes occurring in the cells. Battery aging is due to several secondary reactions happening in the battery components and manifests mainly in capacity and power fading [10,11]. Battery aging is a complex phenomenon influenced by many operating variables, such as calendar time and Ah throughput during cycling, but also Depth of Discharge (DoD), C-rate and temperature [12,13].

The results of this work contributed to inform the discussion within the United Nations Economic Commission for Europe (UN ECE) Electric Vehicles and Environment Informal Working Group (EVE IWG). This group is among the IWGs of the Working Party on Pollution and Energy (GRPE) [14] subsidiary body of the World Forum for Harmonization of Vehicle Regulations (WP.29) preparing regulatory proposals on vehicle emission and energy consumption. One recent development of the EVE IWG has been the GTR No. 22 [15] regulating in-vehicle battery aging within electrified vehicles. The discussion group addressed both the electric range and energy efficiency decrease during the vehicle lifetime due to battery aging. A loss of driving range might lead to diminished utility, resulting in decreased electric vehicle usage and a corresponding decrease in displaced travel distance that might then be covered with a different means of transport; it can also influence electrification and electric vehicle sales. A decrease in vehicle efficiency could impact the upstream emissions by increasing the amount of energy needed per unit of vehicle distance covered. Both efficiency and range have the potential to influence not only the utility but also the environmental performance of the vehicle. Furthermore, alongside changes in range and energy consumption, hybrid electric vehicles frequently incorporate both a conventional and electric powertrain. In the case of these vehicles, the criteria pollutant emissions from the conventional powertrain could potentially be influenced by the degradation of the battery over time [16].

GTR No. 22 [15] prescribes that modern off-vehicle charging hybrid electric vehicles (OVC-HEVs), i.e., plug-in hybrid vehicles and pure electric vehicles (PEVs), should have vehicle-specific monitors on the actual state of health (SOH) of the battery pack, expressed in terms of state of certified energy (SOCE) and state of certified range (SOCR), available for the customers. These values will then be checked against minimum performance requirements, as set in GTR No. 22 for the SOCE for categories 1–1 and 1–2 [17]. The mentioned limits on energy capacity fading foresee a minimum SOCE of 80% over 5 years or 100,000 km, whichever comes first, and 70% up to 8 years or 160,000 km, whichever comes first, for the light-duty vehicles.

This work is presenting the results of a test campaign carried out in the European Commission Joint Research Centre (JRC) Vehicle Emission Laboratory (VELA) on an aged mid-sized BEV. The vehicle was tested with the applicable test cycle [18] to derive the measured SOH. Comparing it with the SOH indication retrieved from the CAN bus of the vehicle, we are approximating a first application of what would happen in Part A of GTR No. 22.

Details on GTR No. 22 and the results of the test campaign, considerations about the aging of the vehicle, and an application of pass/fail statistics foreseen in Part A of the GTR [15] are reported in the following chapters.

2. Materials and Methods

2.1. *In-Vehicle Battery Durability GTR Overview*

In Part A of GTR No. 22 [15], the accuracy of the monitors on the actual state of health (SOH) of the battery pack, expressed in terms of SOCE and SOCR, is verified for each vehicle family by experimentally measuring the usable battery energy (UBE) and the electric driving range by applying the Worldwide Harmonized Light-duty Test Procedure (WLTP) on some vehicle samples [18]. By dividing the measured UBE and range by the respective values from the certification, it is possible to calculate the measured SOCE/SOCR values to be compared with the onboard metrics read from the monitors of the vehicles to verify that their accuracy falls within a predetermined tolerance range.

A pass-or-fail decision about the correctness of the monitors will be reached through a statistical method on a sample of a minimum of 3 up to a maximum of 16 vehicles, evaluating the deviation of the monitor from the measured value according to a statistical formula reported in the GTR No. 22 [15]. More precisely, the initial foreseen sample size is 3 vehicles. If the average deviation of the read and measured values falls below a defined limit, the monitor is accepted; if it is over the higher boundary limit, it is rejected; if the average deviation falls in between the acceptance and rejection limit, the sample size is increased progressively by 1 vehicle up to a maximum of 16 (see Appendix A for the tabulated parameters of Part A of GTR No. 22 [15]).

In Part B of GTR No. 22 [15], the battery durability is checked against the MPRs: since the accuracy of the monitors has been statistically proofed in the previous phase, it is possible to verify the battery durability of aged vehicles through a remote collection of the onboard SOCE/SOCR values for a statistically adequate sample of vehicles within the same battery durability family, as defined in the same GTR No. 22 [15], together with additional information such as the age of the vehicle, the distance traveled, and eventual V2X applications. A battery durability family shall pass if equal to or more than 90 percent of the monitor values read from the vehicle sample is above the MPR. Figure 1 shows a graphical representation of the GTR No. 22 [15] steps for the case of SOCE.

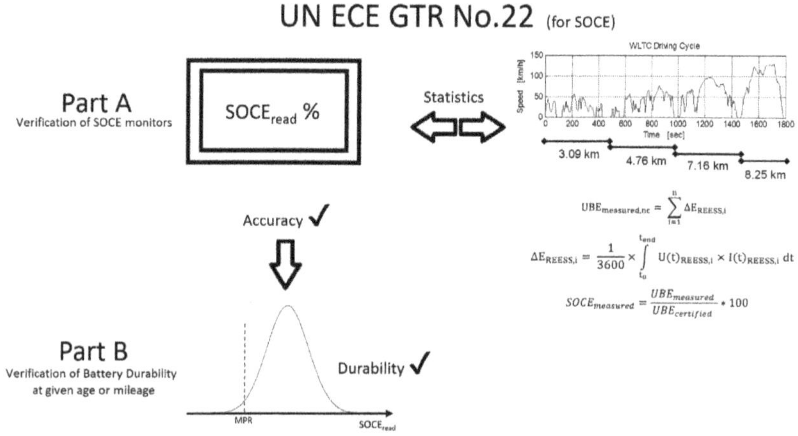

Figure 1. Graphical representation of the GTR No. 22 [15] steps for the case of SOCE.

2.2. Tested Vehicle and the Laboratory

The tested BEV is a JRC property vehicle used for service purposes. It is a mid-size 5-seat vehicle, having an empty mass of 1520 kg and powered with an 80 kW/280 Nm synchronous electric motor at the front axle. The vehicle's main characteristics are reported in Table 1.

Table 1. Main characteristics of the tested vehicle.

Architecture	Battery Electric Vehicle
Propulsion	Synchronous electric motor
Drive Type	FWD
Max. Power (kW)	80
Max. Torque (Nm)	280
Empty mass (kg)	1520
Battery	24 kWh 192 Li-ion cells (96S-2P)
F/R tire and wheel size	205/55 R16
Length (mm)	4440
Width (mm)	1770
Height (mm)	1549
Wheelbase (mm)	2700

The battery is composed of 192 Lithium-Ion cells (96S-2P architecture) for a 24 kWh nominal capacity and circa 360 V nominal voltage.

The tested vehicle was registered on 29 April 2015. The experimental tests reported here were performed in December 2021, with an odometer reading of 9050 km. As evident, this is a special-purpose vehicle, only used inside the JRC premises, and the total distance covered by the vehicle is quite low if compared to its age (6 years and 8 months). The vehicle has been used for normal driving in urban environments and exceptionally in extra urban and highway conditions; it has been used for interoperability tests with charging stations, including high-power ones, and kept parked both externally exposed to weather conditions and internally in temperature-controlled environments. Despite being a company vehicle, its usage seems similar to a low mileage user operation with both cycle and calendar aging effects.

The experimental tests were performed at the JRC VELA in Ispra (Italy) [19,20], precisely in the VeLA-8 test cell, equipped with a 4 $\times$ 4 independent roller benches chassis dynamometer that has a nominal power per axle of 300 kW for full-road simulations with a maximum speed and maximum acceleration, respectively, of 260 km/h and 10 m/s^2 and an inertia range of 250–4500 kg. The chassis dyno wheelbase can be adjusted depending on the tested vehicle from 1800 mm up to 4600 mm. The laboratory test cell is designed for testing light-duty and small commercial vehicles with internal combustion engines (ICEs) or full-electric and hybrid vehicles.

Environmental conditions can be established using a robust climatization system, allowing the control of the ambient temperature from -30 °C to 50 °C and humidity. The VeLA-8 emission measurement system is also customized to allow hybrid vehicles to be tested properly during the phases when the combustion engine is switched off. A description of the testing facility is reported in [21,22].

2.3. Driving Cycles

The tests were performed in a chassis dyno test cell conditioned at 23 °C and with controlled humidity, applying the WLTP testing procedures and calculations for PEVs [18,23,24].

In detail, the two following driving cycles were applied [18,23,24] as shown in Figure 2:

- The Worldwide Harmonized Light-duty Test Cycle (WLTC);
- The Worldwide Harmonized Light-duty Shorten Test Procedure (WLTP STP).

The WLTC is the European certification driving cycle for the light-duty vehicles (LDVs) [18,23,24]. It was designed to reproduce the real-world operating conditions of LDVs more closely than the earlier NEDC cycle. The cycle is comprised of four phases, created to reproduce the urban, the rural, the extra-urban and the highway conditions, respectively, with a duration in time and driven distance of (Figure 2a): low speed (589 s and 3.09 km), medium speed (433 s and 4.76 km), high speed (455 s and 7.16 km), and extra high speed (323 s and 8.25 km).

According to the WLTP Consecutive Cycle Test (CCT), the driving range is derived by driving continually during the WLTC starting with a fully charged battery until the break-off criterion is achieved. This occurs when the driver can no longer adhere to the driving trace because of the vehicle power reduction; the vehicle shall be brought to a standstill, and the driving shall be interrupted.

The WLTP STP is foreseen for high energy capacity, pure electric vehicles in order to shorten the testing duration for the driving range determination [18,23,24]. The cycle consists of two dynamic segments (DS1 and DS2) and two constant speed segments selected to be at 100 km/h (CSS_M and CSS_E) (Figure 2b). The dynamic segments DS1 and DS2 are needed to calculate the energy consumption of the specific phase. The duration of the constant speed segments CSS_M and CSS_E are calculated for each specific tested vehicle characteristic and are designed to cut the test duration by discharging the battery quicker than with the CCT test. The calculated length of the 100 km/h sequences depends on the vehicle's available battery capacity.

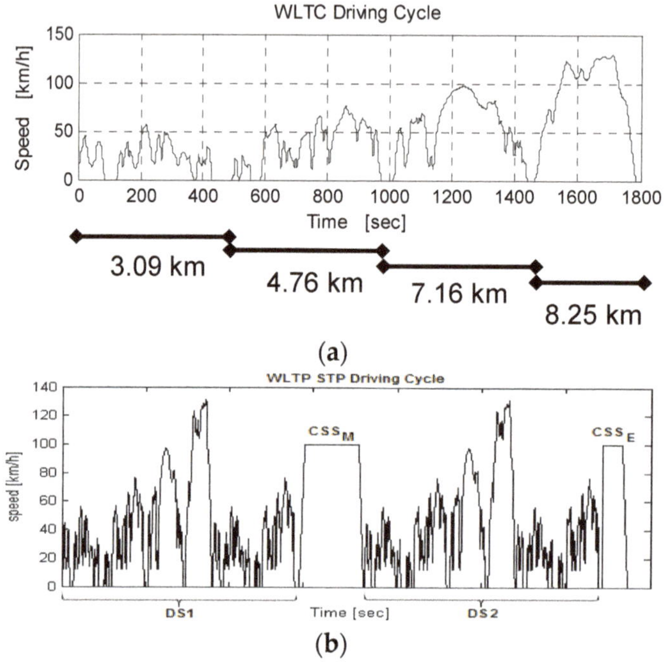

Figure 2. Driving cycles adopted: (**a**) WLTC; (**b**) WLTP STP [18,19,23,24].

After each driving range test, the fully depleted battery is recharged with a 6.6 kW AC charger.

2.4. Measurement Points

During the driving cycle tests, the voltages and the currents were measured at different component levels in the vehicle to calculate the electric power consumption and efficiencies. A complete depiction of the measurement points is presented in Figure 3 and Table 2.

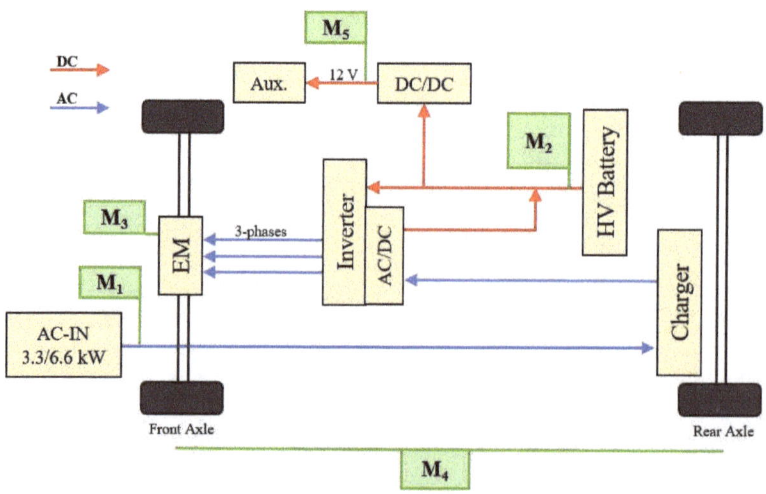

Figure 3. Schematic representation of the vehicle systems and measurement points [19].

Table 2. Detail of measurement locations [21,22] (see Figure 3).

Measurement Point Marker	Explanation
M_1	Electrical energy from the mains to the high-voltage battery (Wh) (acquired directly at the recharging station)
M_2	Current (A) and Voltage (V) from the high-voltage battery to the inverter, the low-voltage auxiliary systems and the HVAC systems; (acquired both by CAN bus and current clamp measurements)
M_3	Rotational speed (rpm) and torque (N·m) of the electric motor; (acquired by CAN bus)
M_4	Mechanical energy at the wheel (Wh); (acquired by the dyno)
M_5	12V battery electrical energy measurement (acquired both by CAN bus and current clamp measurements)

3. Results

The results of the testing campaign on the aged vehicle are discussed in this chapter. Section 3.1. presents the energy and driving range measurement and the derived energy consumption. Section 3.2. compares the aged vehicle performances with the initial figures obtained during certification and an estimate of the SOH is derived. Section 3.3. compares the obtained measured SOH value with the value read from the CAN, and finally, the GTR No. 22 Part A statistics is applied as an example, assuming the tested vehicle to be one of the vehicles sampled for the SOCE/SOCR monitor validation.

3.1. Usable Battery Energy, Driving Range and Energy Consumption

The WLTP CCT and STP procedures [18,23,24] have been applied at 23 °C to determine the aged vehicle UBE, energy consumption, and driving range. During the CCT procedure, the break-off is reached in the fourth phase of the fifth repeated WLTC; instead, during the STP, the break-off occurs during the second CSS, as prescribed, corresponding to approximately the same driven distance and lower test time with respect to the standard CCT procedure.

The UBE measured according to WLTP calculations [18,23,24] is approximately 17,607 Wh for the CCT test and 17,385 Wh for the STP, while the driving range is, respectively, 113.06 km and 114.00 km. The resulting WLTP energy consumption values are 155.74 Wh/km for the CCT test and 152.51 Wh/km for the STP procedure.

Table 3 is reporting the details of the specific energy consumption for each cycle repetition; a higher value is measured in the first cycle both for CCT and STP.

Table 3. Details of specific energy consumption expressed in Wh/km for the different parts of the cycles: CCT procedure on the left and STP procedure on the right.

WLTC Cycle n.	WLTP CCT	○	WLTP STP
1	158.16	WLTC cycle 1	155.58
2	153.66	WLTC cycle 2	151.71
3	152.50	DS1	147.43
4	159.15	DS2	144.06
Tot. up to break-off	158.31	Tot. up to break-off	156.19
WLTP post processed energy consumption	155.74	WLTP post processed energy consumption	152.51

Table 4 below is a further split of the results, reporting energy consumption values for the CCT and STP procedures phase by phase. The energy consumption decreases while driving the cycles, and it is also evident how the specific energy consumption is higher for the fourth phase of each repeated WLTC cycle, reproducing highway driving conditions up to over 130 km/h, with respect to other phases corresponding to a lower speed.

Table 4. Details of specific energy consumption expressed in Wh/km for each phase of the cycles: CCT procedure on the left and STP procedure on the right.

WLTC Cycle n.	Phase n.	WLTP CCT	WLTC Cycle n.	Phase n.	WLTP STP
1	1	145.57	DS1	1	138.16
	2	142.99		2	137.69
	3	145.20		3	143.12
	4	191.06		4	190.22
2	1	138.02		1	126.69
	2	131.58		2	127.80
	3	143.54	CSS1		167.02
	4	190.73	DS2	1	124.61
3	1	136.10		2	129.20
	2	129.69		3	140.85
	3	142.44		4	191.31
	4	190.50		1	123.49
4	1	138.10		2	127.46
	2	129.65	CSS2		210.14
	3	158.12	WLTP post-processed energy consumption		152.51
	4	194.84			
5	1	144.54			
	2	155.82			
	3	152.02			
	4	234.80			
WLTP post-processed energy consumption		155.74			

3.2. Calculate the SOCE/SOCR Monitor

The measured UBE and ranges during the tests have been divided by the values of certification to obtain the measured SOCE and SOCR, an indication of the in-vehicle battery aging at this specific point in its lifetime. The values are reported in Table 5 for both the CCT and the STP procedure.

Table 5. Measured SOCE and SOCR values for the aged vehicle being tested.

	$SOCE_{meas} = \frac{UBE_{meas}}{UBE_{cert}}$ (%)	$SOCR_{meas} = \frac{Range_{meas}}{Range_{cert}}$ (%)
CCT	73.4	70.7
STP	72.4	71.2

The aged vehicle (6 years and 8 months), despite the low accumulated mileage, is quite close to the MPR enforced for 8 years or 160,000 km (30%). The company service vehicle has been used for testing fast charging stations and has been kept parked outside the JRC buildings both in summer and in winter conditions, with a significant contribution expected from calendar aging. The battery SOH from the CAN bus is estimated to have a value of 78%, underestimating the aging with respect to the measured value, but it has to be noted that this value does not reflect the new global technical regulation provisions since the vehicle was registered in 2015.

3.3. Applying the GTR No. 22 Part A Statistics

The accuracy of the SOCE monitors will be verified by GTR No. 22 Part A, according to which the acceptance or rejection of the monitor value will be checked with a statistical analysis based on confidence intervals. An initial sample of three aged vehicles will be taken from the market and tested with the same procedure applied in this work; the difference between the value read from the SOCE/SOCR monitor and the one measured during tests will be evaluated, and the average and standard deviation will be calculated for the sample together with the limit values, as tabulated in GTR No. 22, to verify the correctness of the monitors even if statistically dispersed. If the average value of the difference between the read value and the measured value is sufficiently low, the monitor is accepted. If the value is too high, the monitor value is rejected. If the difference is in the middle, an additional vehicle should be tested. Limit values are set to certainly obtain a pass-or-fail decision with a maximum sample size of 16 values. Details about the calculation and limit values tabulated in GTR No. 22 are reported in Appendix A.

For simplification, it is assumed that the three sampled vehicles are identical to the vehicles tested at the JRC. We are assuming as reference the measured state of certified energy over the CCT procedure, $SOCE_{measured}$ = 73.4%, and the read value from the CAN variable $SOCE_{read}$ = 78%. By applying the formulas in GTR No. 22 [15] (see Appendix A), the difference x_i between the read and measured values is 4.6%, and in this particular case, the standard deviation becomes zero. According to the statistical criterion prescribed in the GTR No. 22, this would result in a pass decision, and the monitor reading would be considered correct and within the tolerance defined in the GTR No. 22. (A = 5%). Table 6 resumes the calculation steps and the obtained result.

Table 6. GTO No. 22 calculation steps and the obtained result.

Vehicle Tested	$SOCE_{read}$ (%)	$SOCE_{meas}$ (%)	x_i (%)	Pass Boundary (%)	Fail Boundary (%)	X_{test} (%)	Decision
1	78	73.4	4.6			4.6	
2	78	73.4	4.6			4.6	
3	78	73.4	4.6	5	5	4.6	PASS

This is just an example case that oversimplifies the situation since the vehicles sampled from the field that will be encountered during the verification of the SOCE/SOCR monitor will be statistically dispersed. In order to also take into consideration this aspect, another example is reported here where two normal distributions are created around the $SOCE_{meas}$ and $SOCE_{read}$ values with a standard deviation of sd = 1.56. Figure 4 shows the normalized probability histograms of the generated distributions, where it is noted that they are only partially overlapping (shaded area) and the average of the $SOCE_{meas}$ distribution is below the $SOCE_{read}$ one. In Table 7, additional information on the distribution quantiles is reported. Iterations of a random sampling over the two distributions are then performed to obtain $SOCE_{meas}$ and $SOCE_{read}$ couples. The GTR statistic was applied for each iteration, and the pass-or-fail decision was recorded for each given sample size.

Table 7. Additional information on generated $SOCE_{meas}$ and $SOCE_{read}$ distributions.

	Minimum (%)	1st Quarter (%)	Median (%)	Mean (%)	3rd Quarter (%)	Maximum (%)	Std. Deviation (%)
$SOCE_{meas}$	67.086	72.356	73.428	73.415	74.488	79.005	1.56
$SOCE_{read}$	72.061	76.937	77.992	77.986	79.048	83.448	1.56

In Figure 5, the cumulative pass curve obtained as a function of the sample size is plotted for this sampling. Table 8 shows the corresponding numerical values for completeness. About 5% of the samples reached a pass decision with an accepted monitor value with a sample size of three. The cumulative percentage of acceptance grew to about 65%, which is the acceptance probability with a sample size of up to 16 vehicles. Since the two generated distributions are not completely overlapping, not all the iterations gave a pass decision; this is due to the characteristic of the developed method aiming at avoiding false passes.

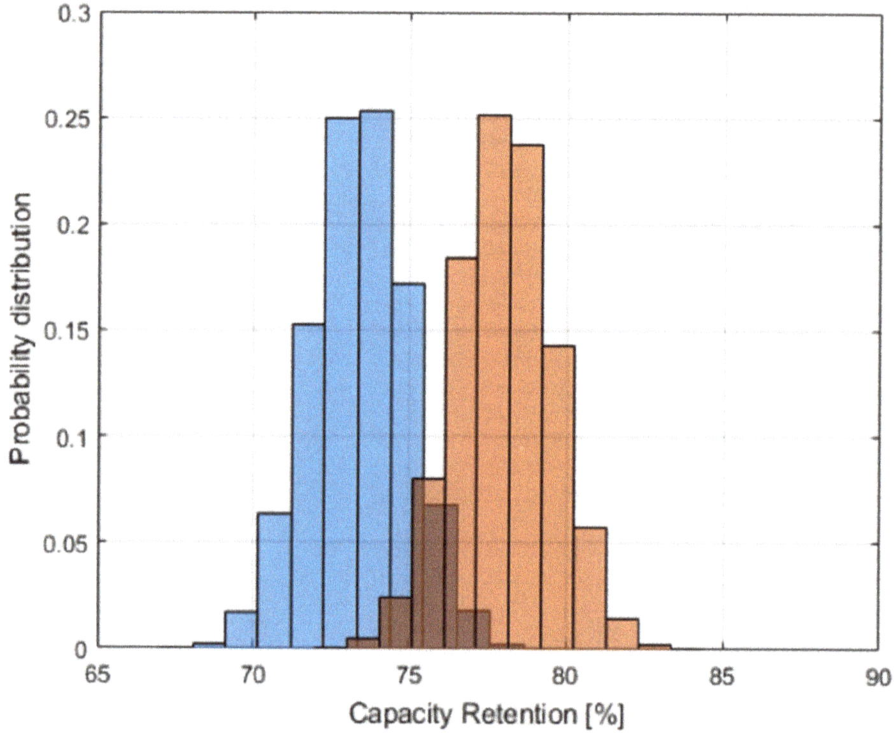

Figure 4. Normalized probability histograms of the generated $SOCE_{meas}$ (blue) and $SOCE_{read}$ (orange) normal distributions.

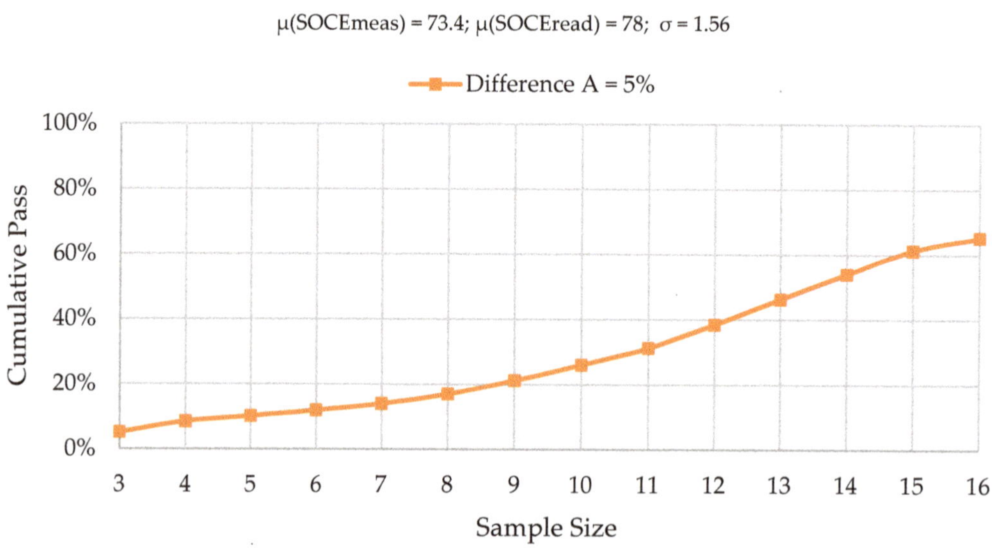

Figure 5. Cumulative pass curve as a function of sample size after iterative sampling $SOCE_{read}$ and $dSOCE_{meas}$ from generated normal distributions and applying the GTR No. 22 Part A statistics.

Table 8. Cumulative pass values.

Sample Size	3	4	5	6	7	8	9	10	11	12	13	14	15	16
Pass rate (%)	5.25	3.30	1.75	1.75	2.05	3.00	4.20	4.85	5.25	7.20	7.80	7.70	7.35	3.95
Cumulative Pass rate (%)	5.25	8.55	10.30	12.05	14.10	17.10	21.30	26.15	31.40	38.60	46.40	54.10	61.45	65.40

4. Discussion and Conclusions

The present work introduces the testing procedures for the in-vehicle battery durability assessment outlined in the new UN ECE GTR No. 22., aiming to verify its applicability on an electric vehicle and illustrating the results obtained with a testing campaign on a JRC property aged battery electric vehicle.

According to GTR No. 22, future off-vehicle charging hybrid electric vehicles and pure electric vehicles will have monitors indicating the actual state of health of the battery pack to the customers, expressed in terms of state of certified energy and state of certified range. The accuracy of these monitors will be confirmed by testing up to 16 vehicles for each vehicle family and applying a statistical procedure (Part A of the GTR 22), while the battery aging at specific years and kilometers will be controlled against minimum performance requirements by remotely collecting the available monitor values from a large number of vehicles within a family (Part B of the GTR No. 22).

To measure the remaining usable battery energy and driving range of the tested vehicle, the WLTP procedure has been applied. The measured SOCE and SOCR values have been calculated, and the results were compared with certification values. In Section 3.3, an exercise was performed to apply the GTR No. 22 Part A statistics, used to validate the accuracy of the SOCE/SOCR monitor values. Since the vehicle was registered in 2015, and the new GTR No. 22 is still not enforced, the value of the monitor has been approximated with a SOH channel retrieved from the vehicle CAN bus.

Moreover, it was possible to test only one aged vehicle.

Despite the assumptions, the test campaign proved the applicability of the testing method and statistical analysis, as described in Part A of GTR No. 22.

The study can be extended in the future by collecting experimental data points for more aging steps of the same vehicle over the course of many years or for different aged vehicles.

For what concerns the UN GTR, future developments will foresee the extension of the in-vehicle battery durability regulation to other classes of vehicles, such as heavy-duty electrified vehicles, guaranteeing that battery performance will be controlled over time and usage for this important market share of vehicles as well. Making sure each vehicle battery lasts longer would also help ease the pressure on in-demand critical raw materials needed for their production and reduce waste from used batteries [25].

Author Contributions: Conceptualization, methodology, formal analysis, data curation, E.P.; writing—original draft preparation, G.L.P.; writing—review and editing, E.P.; review and editing, G.L.P. and E.P. All authors have read and agreed to the published version of the manuscript.

Funding: This research received no external funding.

Data Availability Statement: The data presented in this study are available upon request from the corresponding author.

Acknowledgments: The authors would like to acknowledge the essential contribution of the JRC VELA laboratory staff.

Conflicts of Interest: The authors declare no conflict of interest.

Disclaimer: The opinions expressed in this manuscript are those of the authors and should in no way be considered to represent an official opinion of the European Commission.

Abbreviations

BEV	Battery electric vehicle
CAN	Controller area network
CCT	Consecutive Cycle Test
CSS	Constant speed segment
EVE	Electric Vehicles and Environment
GTR	Global Technical Regulation
HVAC	Heating, Venting, and Air Conditioning
ICE	Internal combustion engine
IWG	Informal Working Group
JRC	Joint Research Centre
MPR	Minimum performance requirement
NEDC	New European Driving Cycle
OVC-HEV	Off-vehicle charging hybrid electric vehicle
SOH	State of health
SOCE	State of certified energy
SOCR	State of certified range
STP	Shortened test procedure
UBE	Usable battery energy
UN ECE	United Nations Economic Commission for Europe
V2X	Vehicle to Everything
VELA	Vehicle emission laboratories
WLTP	Worldwide Harmonized Light-duty Test Procedure

Appendix A

The calculations and tabulated parameters of GTR No. 22 [15] used in Part A to make a decision on the accuracy of SOCE/SOCR monitor values based on confidence intervals are reported in this Appendix. Distinct statistics shall be considered for the SOCR and the SOCE monitor, here named indefinitely as *SOC*. A sufficient number of vehicles (at least 3 and not more than 16) shall be sampled from the same monitor family for testing, following a vehicle survey containing information designed to ensure that the vehicle has been correctly used and maintained according to the manufacturer's specifications [15]. For evaluating the SOCE/SOCR monitors, normalized values shall be calculated for each sample:

$$x_i = SOC_{read,i} - SOC_{measured,i}$$

where $SOC_{read,i}$ is the onboard SOCE/SOCR read, and $SOC_{measured,i}$ is the measured SOCE/SOCR of the vehicle *i*.

For the total number of *N* tests and the normalized values of the tested vehicles, x_1, x_2, ... x_N, the average X_{tests} and the standard deviation *s* shall be determined:

$$X_{tests} = \frac{(x_1 + x_2 + x_3 + \ldots + x_N)}{N}$$

$$s = \sqrt{\frac{(x_1 - X_{tests})^2 + (x_2 - X_{tests})^2 + \ldots + (x_N - X_{tests})^2}{N-1}}$$

For each *N* tests $3 \leq N \leq 16$, one of the three following decisions can be reached, where the factor A shall be set at 5 percent:

(a) Pass the family if $X_{tests} \leq A - (t_{P1,N} + t_{P2,N}) \cdot s$;
(b) Fail the family if $X_{tests} > A + (t_{F1,N} - t_{F2}) \cdot s$;
(c) Take another measurement if:

$$A - (t_{P1,N} + t_{P2,N}) \cdot s < X_{tests} \leq A + (t_{F1,N} - t_{F2}) \cdot s$$

where the parameters $t_{P1,N}$, $t_{P2,N}$, $t_{F1,N}$, and t_{F2} are taken from Table A1 [15].

Table A1. Pass/fail decision criteria for the sample size [15].

Tests (N)	PASS $t_{P1,N}$	PASS $t_{P2,N}$	FAIL $t_{F1,N}$	FAIL t_{F2}
3	1.686	0.438	1.686	0.438
4	1.125	0.425	1.177	0.438
5	0.850	0.401	0.953	0.438
6	0.673	0.370	0.823	0.438
7	0.544	0.335	0.734	0.438
8	0.443	0.299	0.670	0.438
9	0.361	0.263	0.620	0.438
10	0.292	0.226	0.580	0.438
11	0.232	0.190	0.546	0.438
12	0.178	0.153	0.518	0.438
13	0.129	0.116	0.494	0.438
14	0.083	0.078	0.473	0.438
15	0.040	0.038	0.455	0.438
16	0.000	0.000	0.438	0.438

References

1. International Energy Agency IEA. Global EV Outlook 2023. 2023. Available online: https://www.iea.org/reports/global-ev-outlook-2023 (accessed on 3 August 2023).
2. European Environment Agency. Greenhouse Gas Emissions from Transport in Europe. Available online: https://www.eea.europa.eu/ims/greenhouse-gas-emissions-from-transport (accessed on 3 August 2023).
3. European Commission. Paris Agreement COP21. 2015. Available online: https://climate.ec.europa.eu/eu-action/international-action-climate-change/climate-negotiations/paris-agreement_en (accessed on 3 August 2023).
4. European Commission. European Green Deal. Available online: https://ec.europa.eu/commission/presscorner/detail/en/ip_19_6691 (accessed on 3 August 2023).
5. European Commission. Sustainable and Smart Mobility Strategy—Putting European Transport on Track for the Future. 9 December 2020. Available online: https://eur-lex.europa.eu/legal-content/EN/TXT/?uri=CELEX:52020DC0789 (accessed on 3 August 2023).
6. European Commission. Sustainable Mobility the European Green Deal. December 2021. Available online: https://ec.europa.eu/commission/presscorner/detail/en/fs_19_6726 (accessed on 3 August 2023).
7. Bloomberg NEF. Electric Vehicle Outlook 2022. 2022. Available online: https://about.bnef.com/electric-vehicle-outlook/ (accessed on 3 August 2023).
8. Miao, Y.; Hynan, P.; von Jouanne, A.; Yokochi, A. Current Li-Ion Battery Technologies in Electric Vehicles and Opportunities for Advancements. *Energies* **2019**, *12*, 1074. [CrossRef]
9. Dechent, P.; Epp, A.; Jöst, D.; Preger, Y.; Attia, P.M.; Li, W.; Sauer, D.U. ENPOLITE: Comparing Lithium-Ion Cells across Energy, Power, Lifetime, and Temperature. *ACS Energy Lett.* **2021**, *6*, 2351–2355. [CrossRef]
10. Thomas Merichko (FEV North America Inc.). Battery Durability in Electrified Vehicle Applications: A Review of Degradation Mechanisms and Durability Testing. 2016. Available online: https://wiki.unece.org/download/attachments/29884985/EVE-18-04e.pdf?api=v2 (accessed on 3 August 2023).
11. Broussely, M.; Biensan, P.; Bonhommeb, F.; Blanchard, P.; Herreyre, S.; Nechev, K.; Staniewicz, R. Main aging mechanisms in Li ion batteries. *J. Power Sources* **2005**, *146*, 90–96. [CrossRef]
12. Keil, P.; Jossen, A.; Kowal, J. *Aging of Lithium-Ion Batteries in Electric Vehicles*; Technische Universität München—Fakultät für Elektrotechnik und Informationstechnik: München, Germany, 2017.
13. Preger, Y.; Barkholtz, H.M.; Fresquez, A.; Campbell, D.L.; Juba, B.W.; Romàn-Kustas, J.; Ferreira, S.R.; Chalamala, B. Degradation of Commercial Lithium-Ion Cells as a Function of Chemistry and Cycling Conditions. *J. Electrochem. Soc.* **2020**, *167*, 120532. [CrossRef]
14. UNECE; GRPE. Working Party on Pollution and Energy. 2023. Available online: http://www.unece.org/trans/main/wp29/meeting_docs_grpe.html (accessed on 3 August 2023).
15. UN GTR No. 22. In-Vehicle Battery Durability for Electrified Vehicles. 2023. Available online: https://unece.org/transport/documents/2022/04/standards/un-gtr-no22-vehicle-battery-durability-electrified-vehicles (accessed on 3 August 2023).
16. GRPE. *Proposal for a Final Status Report on the Development of a New UN Global Technical Regulation on In-Vehicle Battery Durability for Electrified Vehicles*; UNECE: Geneva, Switzerland, 2021.
17. UNECE. *Special Resolution No. 1 Concerning the Common Definitions of Vehicle Categories, Masses and Dimensions*; TRANS/WP.29/1045; UNECE: Geneva, Switzerland, 2005.
18. UNECE. Global Technical Regulation No. 15 Worldwide Harmonized Light Vehicles Test Procedure. 2014. Available online: https://unece.org/sites/default/files/2022-06/ECE-TRANS-180a15am6e.pdf (accessed on 3 August 2023).

19. Patrone, G.L.; Paffumi, E.; Otura, M.; Centurelli, M.; Ferrarese, C.; Jahn, S.; Brenner, A.; Thieringer, B.; Braum, D.; Hoffmann, T. Assessing the Energy Consumption and Driving Range of the QUIET Project Demonstrator Vehicle. *Energies* **2022**, *10*, 1290. [CrossRef]
20. European Commission. Vehicle Emissions Laboratories. Available online: https://ec.europa.eu/jrc/en/research-facility/vehicle-emissions-laboratory-vela (accessed on 3 August 2023).
21. De Gennaro, M.; Paffumi, E.; Martini, G.; Manfredi, U.; Vianelli, S.; Ortenzi, F.; Genovese, A. Experimental Test Campaign on a Battery Electric Vehicle: Laboratory Test Results (Part 1). *SAE Int. J. Altern. Powertrains* **2015**, *4*, 100–114. [CrossRef]
22. Paffumi, E.; De Gennaro, M.; Martini, G.; Manfredi, U.; Vianelli, S.; Ortenzi, F.; Genovese, A. Experimental Test Campaign on a Battery Electric Vehicle: On-Road Test Results (Part 2). *SAE Int. J. Altern. Powertrains* **2015**, *4*, 277–292. [CrossRef]
23. European Commission. Commission Regulation (EU) 2017/1151 of 1 June 2017 Supplementing Regulation (EC) No 715/2007 of the European Parliament and of the Council on Type-Approval of Motor Vehicles with Respect to Emissions from Light Passenger and Commercial Vehicles (Euro 5 and Euro 6) and on Access to Vehicle Repair and Maintenance Information, Amending Directive 2007/46/EC of the European Parliament and of the Council, Commission Regulation (EC) No 692/2008 and Commission Regulation (EU) No 1230/2012 and Repealing Commission Regulation (EC) No 692/2008. *Off. J. Eur. Union* **2017**, L175. Available online: https://eur-lex.europa.eu/eli/reg/2017/1151/oj (accessed on 3 August 2023).
24. UN Regulation No 154. Available online: https://eur-lex.europa.eu/legal-content/EN/TXT/?uri=CELEX%3A42021X2039 (accessed on 3 August 2023).
25. UNECE. Major Auto Markets Join Forces for Draft UN Legislation on Electric Vehicle Battery Durability. Press Release. Available online: https://unece.org/circular-economy/press/major-auto-markets-join-forces-draft-un-legislation-electric-vehicle-battery (accessed on 3 August 2023).

batteries

MDPI

Article

Intra-Layer Inhomogeneity of the Anode in Commercial Li-Ion Batteries

Tuo Fang [1], Guangsen Jiang [2], Yong Xia [1] and Pengfei Ying [1,*]

1 School of Vehicle and Mobility, Tsinghua University, Beijing 100084, China; ft22@mails.tsinghua.edu.cn (T.F.); xiayong@tsinghua.edu.cn (Y.X.)
2 Beijing Chehejia Automobile Technology Co., Ltd., Beijing 101399, China; jiangguangsen@lixiang.com
* Correspondence: ypf21@mails.tsinghua.edu.cn

Abstract: The Li intercalation reaction exhibits non-uniform behavior along the thickness direction of the electrode in a Li-ion battery. This non-uniformity, or intra-layer inhomogeneity (ILIH), becomes more serious as the charging and discharging speed increases. Substantial ILIH can lead to Li plating and the emergence of inhomogeneous inner stress, resulting in a decrease in battery service life and an increase in battery safety risks. In this study, an operando optical observation was conducted based on the color change reaction during Li intercalation in the anode. Subsequently, we introduce a novel quantitative method to assess ILIH in commercial Li-ion batteries. A specific ILIH value (K_{ILIH}) is first used in this article for ILIH characterization. An analysis of K_{ILIH} at different charging and discharging rates was conducted, alongside the exploration of K_{ILIH}-SOC trends and their underlying mechanisms. The proposed method exhibits favorable mathematical convergence and physical interpretability, as supported by the results and mechanism analysis. By enabling the assessment of ILIH evolution in response to SOC and (dis)charging rate variations, the proposed method holds significant potential for optimizing fast charging protocols in commercial batteries and contributing to the development of refined electrochemical battery models in future research.

Keywords: Li-ion battery; electric vehicle; battery safety; inhomogeneity; operando observation

1. Introduction

Lithium-ion batteries (LIBs) have already become the dominant power source for consumer electric devices [1–3]. Since the widespread use of LIBs in electric vehicles (EVs), the ever-increasing market has led to even higher demands for commercial LIBs. To address the requirements of the automobile industry and to alleviate "drive-range anxiety" among consumers, extensive efforts have been made to research and develop LIBs with higher energy density and higher charging rates [4,5]. Other efforts have also been made in the pursuit of LIBs with better cycling performance, higher thermal stability, better safety, and reduced cost [2,6,7].

In the study of LIBs, the potential for fast charging is limited by issues such as reaction inhomogeneity and Li plating. Graphite is commonly used in commercial battery anodes due to its low cost, high Li diffusivity, and low delithiation potential characteristics, but the low-voltage plateau in the charge–discharge profiles of graphite may inevitably increase the risk of Li plating during the fast charging process [8,9]. Li plating and the localized abuse caused by inhomogeneity are recognized as crucial factors leading to reduced lifetime and increased safety risks under abusive operating conditions [10–14]. For example, inhomogeneity-induced local overuse and overcharge can result in internal short circuits under mechanical load, potentially culminating in thermal runaways.

The inhomogeneity of reactions in commercial LIBs can be categorized into three types: in-plane, which refers to inhomogeneity on the surface parallel to the current collector; inter-layer, which represents the variation among electrode layers in full cells;

and intra-layer, which describes the inhomogeneity along the thickness direction of the electrode coating. Numerous studies have investigated inhomogeneity in the in-plane directions [15–18] and the inter-layer dimension [19–21]. However, limited attention has been given to inhomogeneity in the intra-layer direction [11]. Figure 1 shows typical cathode and anode configurations in commercial LIBs, in which the vertical direction of the SEM images and the schematic diagram represents the intra-layer direction mentioned above. Previous studies by Harris et al. [22], Hogrefe et al. [23], and Kuwabara et al. [24] have examined microscopic electrochemical behaviors, the dendrite growth of deposited Li, and contaminated Cu on the cross-section of LIBs, respectively. Although these studies have demonstrated the significance of intra-layer inhomogeneity (ILIH in short), more complex experimental methods, such as operando or in situ techniques, are required for the detailed and quantitative characterization of ILIH.

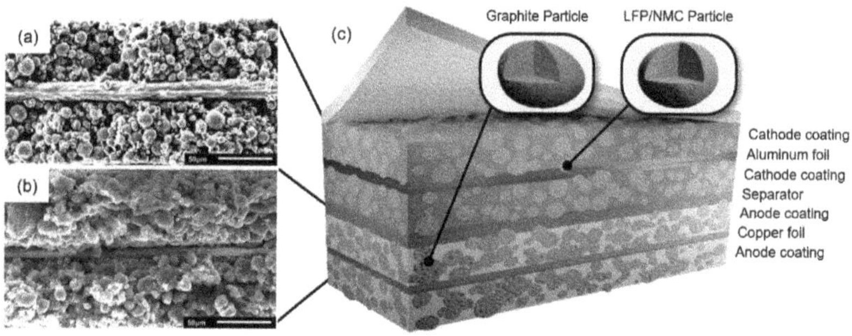

Figure 1. SEM images and a schematic illustration of the electrode structure of LIBs: (**a**,**b**) the SEM images of the cross-sections of the cathode and anode in LIBs [25]; and (**c**) schematic diagram of the electrode structure of LIBs [26].

Compared to ex situ observation and characterization techniques, in situ techniques provide a precise and reliable analysis of battery cells under different conditions (SOCs, etc.). Furthermore, operando techniques make it even more possible to observe and detect batteries during the charging and discharging processes [27]. Consequently, operando or in situ techniques have emerged as a research hotspot in LIBs since the 2010s [28].

Researchers have employed operando or in situ techniques to conduct numerous studies, including investigating Li dendrite growth [29], observing cracks on electrode particles [30], and characterizing Li concentration and current densities in electrodes or electrolytes [31–33], etc. However, there is a relative scarcity of studies focusing on the cross-section of batteries, which require operando techniques and transverse-type samples [28]. Most existing studies have only concentrated on the observation of Li plating and Li dendrite growth [22,29,31,34]. Although Li plating is crucial for studying battery failure, inhomogeneity may be more important for maintaining battery durability from an integrity perspective [10].

To investigate inhomogeneity via the operando approach, optical characterization methods have been developed. For instance, Marie et al. employed an "in situ colorimetry method", utilizing optical observation to quantify the SOC or potential of graphite electrodes [18,35]. Hogrefe et al. adopted the method to estimate the speed of lithiation by analyzing the colorimetric area of LiC_{12} and LiC_6 in the cross-section of battery cells [36]. Even though the colorimetric method has been employed in relevant studies, the limitations of this method when applied to particles in commercial batteries with poor consistency remain unsolved.

In summary, thorough studies on ILIH and its quantitative characterization are crucial for enhancing the durability and safety of LIBs. Nevertheless, to the best of the authors' knowledge, there have been very few studies that have successfully characterized ILIH during the (dis)charging process, let alone established a quantitative ILIH measuring method.

In this research, an operando optical microscopic approach was used to obtain simultaneous electrical and chromatic image data. Furthermore, a new data analysis method is proposed to characterize the Li intercalation level of anode graphite particles in commercial batteries, enabling the quantitative characterization of ILIH. The robustness of the method was verified via the results obtained from multiple battery cells subjected to different charging and discharging rates. Additionally, based on the relationship between ILIH and the SOC, the electrochemical mechanism underlying the trend of ILIH along the SOC is further discussed.

The remainder of this article is organized as follows: In Section 2, the sample preparation and experimental settings are introduced. In Section 3, the newly proposed ILIH assessment method is detailed, and its necessity is briefly explained through a preliminary analysis of the optical signal. The results of the ILIH assessment are shown in Section 4. Finally, the conclusions and outlooks of this study are provided in Section 5.

2. Experiments

In this section, the experimental details for preparing the battery samples and conducting optical observations are provided. Also, the experimental settings for the operando optical observations are illustrated.

2.1. Preparation of Battery Samples

As introduced in Section 1, the observations in this study focus on commercial batteries, which are unsuitable for optical observation. Consequently, specific battery samples were prepared beforehand. The flow chart of the sample preparation process is shown in Figure 2. For this study, the objects of investigation were pouch cell batteries with lithium-ion phosphate (LFP) as the cathode material and graphite as the anode material. As Figure 2a shows, the batteries were cut with insulated tools. Subsequently, the cathodes, anodes, and separators were peeled off layer-by-layer and cut into specific sizes. Due to the impracticality of performing the heat pressing process inside a glove box, the configuration of the electrodes and separators had to be fixed by other means. Therefore, the pouch cells were designed to be reassembled into the "G-shape" shown in Figure 2b. As Figure 2c shows, electrolyte injection, encapsulation, and the necessary inspections for leak-proofness and capacity maintenance of the pouch cells were conducted. In order to avoid atmospheric interference, the disassembly and reassembly process described above took place in the glovebox filled with argon gas. Finally, three cycles of low-rate charging and discharging were performed on each battery to ensure the electrical reliability of the battery sample for subsequent observation.

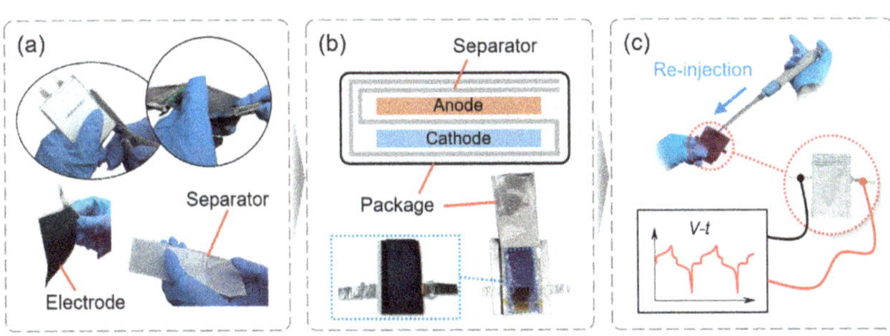

Figure 2. Flow chart of the sample preparation process based on the commercial pouch cell battery: (**a**) decomposition of the commercial pouch cell battery; (**b**) reassembly of the single-piece electrode pair pouch cell battery sample; and (**c**) re-injection of the electrolyte, encapsulation of the battery sample package, and electrical reliability testing of the battery sample.

The cathodes and anodes were cut into sizes of 25 mm × 30 mm and 30 mm × 35 mm, respectively. The samples had an approximate capacity of 20 mAh, with a cut-off voltage range from 2.5 V to 3.65 V.

The optical observation was conducted using the battery sample prepared as described in Figure 2. The operando observation was carried out in a confocal optical microscope (Lasertec Corp., Yokohama, Tokyo, ECCS). The cross-section of the prepared battery sample was continuously illuminated by an independent light source from the microscope, and the reflected light was used to generate confocal chromatic images. The schematic diagram of the sample's cross-section is shown in Figure 2b.

To obtain a high-quality cross-section, the pouch cell sample was cut using a special cutter. The cross-section of the sample was exposed in an anhydrous air environment to protect the electrolytes. Subsequently, the sample was installed in a specially designed airtight fixture. The anhydrous air environment inside the fixture was then replaced with argon gas via a vacuum pump. The exposed cross-section was utilized for the following observations. The operation mentioned above (shown in Figure 3) was vital for maintaining the chemical stability of the exposed battery throughout long-term testing.

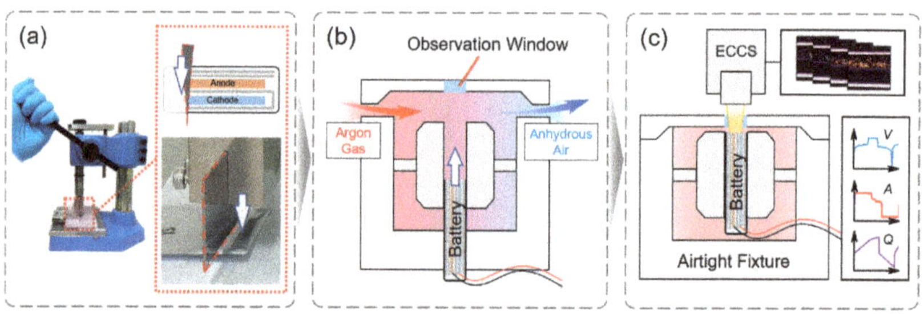

Figure 3. Flow chart of the optical observation based on confocal optical microscopy: (**a**) exposure of the cross-section of the sample; (**b**) schematic of sample installation and the gas replacement process; and (**c**) schematic of optical observation.

2.2. *Experimental Settings*

In this article, the experiments were conducted at room temperature (25 °C). The confocal microscope was utilized for the operando observation, enabling the acquisition of a series of high-resolution photochromes synchronized with the electrical data in the (dis)charging process.

The following protocol was adopted for this study: (1) constant current (CC) charging to 3.65V at 1/8C and then constant voltage (CV) charging at 3.65 V until the current is reduced to 1/20C; (2) rest for 5 min; (3) CC discharging to 2.5 V at 1/8C; (4) rest for 5 min; (5) repeat steps (1) through (4) three times, changing the CC charging and discharging rate to 1/4C, 1/2C, and 1C.

As discussed in Section 1, inhomogeneous Li intercalation occurs during high-rate charging/discharging processes, whereas the inhomogeneity in slow charging/discharging processes can be neglected. Therefore, for the sake of comparison, low-rate charging/discharging (1/8C) was set as the homogeneous baseline. Further discussions and hypotheses are introduced in the following analysis.

3. Methods

Based on the optical and electrical data obtained in the experiments, a series of data analysis processes were conducted. The corresponding methods are fully illustrated in this section. The section begins by presenting an intuitive analysis of the optical signal from the cross-sections of the batteries, highlighting the necessity of the ILIH assessment method, which is further introduced in this section.

3.1. Preliminary Analysis of the Optical Signal

In this subsection, we illustrate that the color response of the cross-sections on the anode side is highly sensitive to changes in battery SOC, and significant inhomogeneity in the intra-layer direction can be observed. However, owing to the inherent randomness of commercial batteries, the statistical results exhibit the limitation of the colorimetric method in identifying the SOCs of local regions and characterizing the ILIH values.

In this study, three battery samples were tested and observed. One sample was selected to demonstrate the characteristics of the operando optical response. Figure 4 presents the representative electrical signal of the battery in the tests, whereas Figure 5a illustrates the color reaction of the graphite lithiation process captured by optical microscopy. As the lithiation process progresses, the Lithium–Carbon composite in the anode undergoes a color change reaction, transitioning from grey to blue, orange, and gold [35,36]. This physicochemical phenomenon is a consequence of the transition between different phases within the Lithium–Carbon composite during lithiation and delithiation processes, which can be readily discerned through optical observation. At room temperature, the phases correspond to $Li_{0.04}C_6$-$Li_{0.33}C_6$-$Li_{0.5}C_6$-LiC_6, correlating with a color transition from grey to dark blue to orange (or dark red) to gold [22,37]. Moreover, distinctive differences in the graphite color in the cross-section were observed at different charging rates. The arrows in Figure 5a indicate that higher charging rates result in a larger golden area near the separator and a larger blue area further away from the separator, which can be classified as an ILIH phenomenon. On the other hand, previous studies [23] and the observations in Figure 5a indicate that the LFP cathode material exhibited insignificant optical variation throughout the lithiation and delithiation processes. Additionally, variations in the size and distribution of anode graphite particles were observed in commercial battery coatings, primarily due to the manufacturing process.

In previous studies, the colorimetric method has been employed to determine the SOC of the particles. However, due to the inherent randomness resulting from the experimental settings, sample preparation, anode material systems, etc., the colorimetric method was not suitable for evaluating the Li intercalation status of the local area of interest.

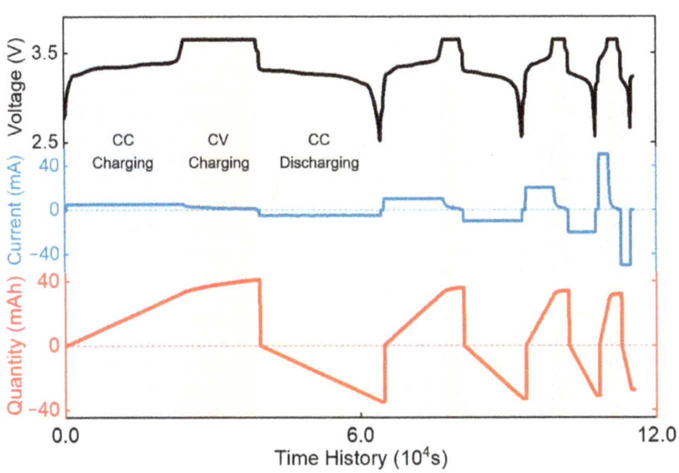

Figure 4. Original voltage–current–time data and the calculated electric charge quantity (to demonstrate the data clearly, the electric charge quantity value in the discharging process was set as negative, so the absolute value represents the battery's discharge capability).

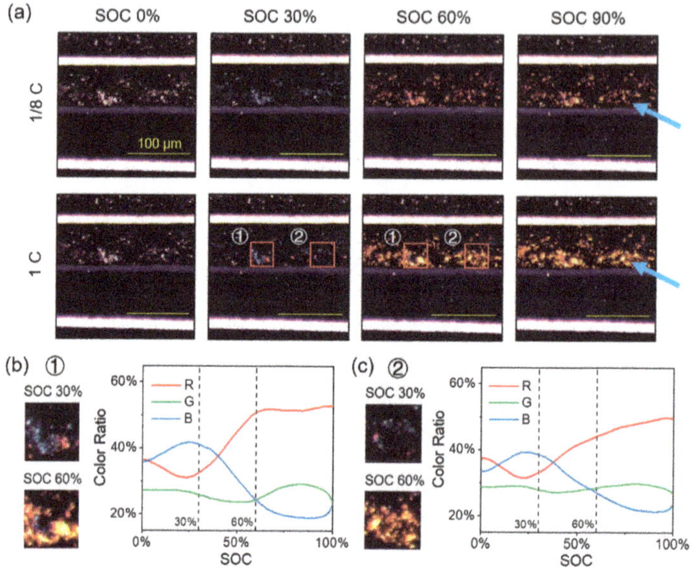

Figure 5. The image capture and preliminary analysis results: (**a**) operando in situ image captures of the cross-sections with different SOCs and charging rates (The blue arrows emphasize the noticeable inhomogeneity at higher charging rates); (**b**) primary color ratio analysis of part ①; and (**c**) primary color ratio analysis of part ②.

The primary color ratio analysis in two parts of the same battery sample is shown in Figure 5. The local image captures of part ① and ② (Figure 5a) at 30% SOC and 60% SOC are respectively zoomed in and shown in Figure 5b,c. The primary color ratios of these two parts were calculated throughout the 0.125C charging process, and the results are presented in curve graphs. In these graphs, the short-dashed lines indicate the positions of 30% and 60% SOC. The inconsistent color ratio trends of the primary colors indicate variations in the color reaction at different local areas. The differences between part ① and ② are also obvious intuitively: at the same 30% SOC condition, they exhibit a similar blueish color, whereas at the same 60% SOC condition, the particles in part ② turn orange to a similar extent. Furthermore, in part ①, some particles become bright gold, whereas others remain blue.

From the preliminary analysis, it is evident that there are variations in the color change among different parts of the cross-section. Therefore, a single absolute criterion is not suitable for quantitatively analyzing the intricate information obtained from the image. This introduces non-negligible interference in the absolute colorimetric analysis process and may bring limitations to the existing colorimetric methods. Hence, in order to evaluate the real-time ILIH value of the battery, a new method is introduced (in the following section).

3.2. ILIH Assessment Method

In this subsection, the ILIH assessment method is explained in detail. The method was based on the color change of the Lithium–Carbon composite in the anode during its Li intercalation: a grey–blue–red–gold color reaction along with the intercalation reaction of $Li_{0.04}C_6$-$Li_{0.33}C_6$-$Li_{0.5}C_6$-LiC_6 [35–37]. However, to overcome the inherent randomness mentioned above, the assessment could not rely on absolute criteria. Therefore, this paper has established a relative criterion using the optical data obtained during low-rate (dis)charging processes. Through a series of data processing operations, the quantitative assessment of ILIH was achieved.

Specifically, the data processing operations included: characterization of the Original Relative Illuminance, which managed to normalize the data in a positional sense, as introduced in Section 3.2.1; calculation of the Normalized Relative Illuminance, which normalizes the data in terms of illuminance, as introduced in Section 3.2.2; and characterization of the ILIH value, which included a linear simplification process and an angular decomposition process aimed at eliminating systematic inhomogeneity noise, as introduced in Section 3.2.3.

3.2.1. Characterization of the Original Relative Illuminance

As mentioned before, the randomness in battery electrode manufacturing leads to different sizes and non-uniform distribution of the anode graphite particles. Moreover, the thickness and position of the electrode coating also changes during charging and discharging processes, which can be seen in Supplementary Video S1. This is mainly due to the volume change of electrode particles in the (de)lithiation process and other environmental disturbances. Thus, the absolute position of the anode coating (the object of this study) cannot be guaranteed.

In order to carry out an operando comparison of the same sample under different SOCs, charging rates, battery cells, etc., it is necessary to eliminate the influence of component deformation and position shifting on image data. As a result, to obtain robust quantitative results from the image analysis of the battery samples, the following operation is required.

The Original Relative Illuminance (ORI) is calculated by analyzing the color information of the photochrome captured in optical microscopy. Figure 6a shows a typical part of the image capture when Battery-A (one of the studied battery samples) was fully charged. In this image, ①④⑥ represent the anode and ②⑤⑦ represent the cathode. Only ④①②③⑤ and the external circuit form a complete loop.

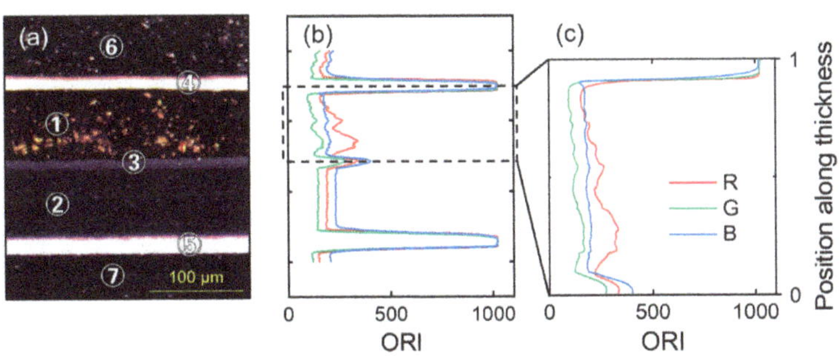

Figure 6. Data analyzing process of ORI: (**a**) Original image capture in ECCS: ① Anode Coating (in loop); ② Cathode Coating (in loop); ③ Separator; ④ Anode Current Collector (Cu); ⑤ Cathode Current Collector (Al); ⑥ Anode Coating (out of loop); and ⑦ Cathode Coating (out of loop); (**b**) ORI directly extracted from the original image capture; (**c**) the result of position normalization for the ORI.

Figure 6b presents the ORI extracted from the original image (Figure 6a) by summing up the primary color values of pixels on the same row (perpendicular to the anode coating thickness direction). The upper and lower boundaries of the anode coating were identified according to the optical characteristics of the separator (③) and current collector (④), respectively. Then, the pixels boxed in the identified area were isolated and characterized in terms of their relative position (ranging from 0 to 1) along the anode coating thickness direction.

Figure 6c shows the result of position normalization for the ORI. This process helps to eliminate interference caused by the volume change of the battery or any wobbling of the microscope.

3.2.2. Calculation of the Normalized Relative Illuminance

As mentioned earlier, non-uniformity in graphite particle size, distribution, and physicochemical characteristics can arise due to experimental settings, sample preparation, the manufacture of anode coatings, anode material systems, etc. Non-uniformity may cause inconsistent color reactions and incomparable illuminance changes, posing challenges in evaluating the Li intercalation status of the local area of interest.

In order to address the challenges posed by the inherent randomness mentioned above, this article proposed a method: normalizing the ORI of one of the primary colors to establish consistent "Color–Li intercalation" relationships across different local areas. The normalized ORI is referred to as the Normalized Relative Illuminance (NRI) throughout the rest of this article.

The formula for calculating NRI is as follows:

$$\mathrm{NRI}(x,t;Color) = \frac{\mathrm{ORI}(x,t;Color) - \min\limits_{t} \mathrm{ORI}(x,t;Color)}{\max\limits_{t} \mathrm{ORI}(x,t;Color) - \min\limits_{t} \mathrm{ORI}(x,t;Color)} \quad (1)$$

where, x is the relative position along the thickness in the anode coating (ranging from 0 to 1); t denotes the time history; $Color$ represents the RGB primary colors. Using the formula, the minimum and maximum values of ORI were calculated at each x_0 point with respect to t. Consequently, NRI serves as a rational characterization parameter for the Relative Lithium Insertion Amount (RLI) of the anode. Figure 7 illustrates the calculation process of $\mathrm{NRI}(x,t;Red)$, wherein the following equations are satisfied:

$$\mathrm{ORI}_{min}(x) = \mathrm{ORI}(x,t_{min};Red) = \min\limits_{t} \mathrm{ORI}(x,t;Red), \quad (2)$$

$$\mathrm{ORI}_{max}(x) = \mathrm{ORI}(x,t_{max};Red) = \max\limits_{t} \mathrm{ORI}(x,t;Red), \quad (3)$$

where any $x_0 \in [0,\ 1]$.

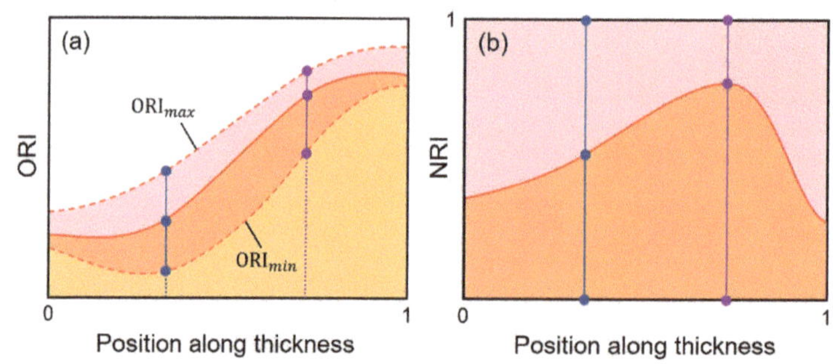

Figure 7. Schematic diagram of the NRI calculation process (the blue and purple vertical lines in the two subplots have the same proportions): (**a**) an ORI curve at time t (solid line), and the maximum and minimum curves of $\mathrm{ORI}(x,t_{max};Red)$ and $\mathrm{ORI}(x,t_{min};Red)$; and (**b**) the NRI curve derived from (**a**).

By normalizing the local illuminance of ORI, the inherent difference among the particles at different relative positions can be reduced to an acceptable extent. As a result, real-time evaluation of ILIH in the battery sample can be further achieved using NRI data.

3.2.3. Characterization of the Intra-Layer Inhomogeneity (ILIH) Value

The ILIH is specifically represented by the degree of SOC deviation at different positions along the thickness direction. In this study, the authors tried to describe ILIH with a specific "ILIH value", represented by K_{ILIH}. In the practical analysis process, the determination of this slope is based on the fitting of NRI at different positions along the thickness direction. However, it should be emphasized that the NRI value at a particular location cannot directly represent the local SOC. As mentioned before, the manufacturing of commercial batteries and the preparation of samples in the laboratory may introduce systematic deviations to the relationship between NRI and SOC. To overcome this deviation, K_{ILIH} during low-rate (dis)charging processes (1/8C in this case) is considered zero, i.e., it is considered homogeneous. Consequently, the NRI slope values in the low-rate process represent the systematic noise of the sample at specific SOCs. Therefore, the following operation is required to eliminate systematic deviations and to obtain the assessed ILIH value (K_{ILIH}).

Figure 8 presents the schematic diagrams related to NRI fitting and the ILIH characterization process. Time t_1 corresponds to medium/high-rate conditions (1/4C, 1/2C, or 1C in this case), whereas time t_0 corresponds to low-rate conditions (1/8C in this case). Notably, the battery sample is at the same SOC at both moments. As mentioned before, K_{ILIH} in low-rate conditions is considered systematic noise. So, as Figure 8a illustrates, a comparison between t_1 and t_0 can derive K_{ILIH} with a specified SOC and charging/discharging rate.

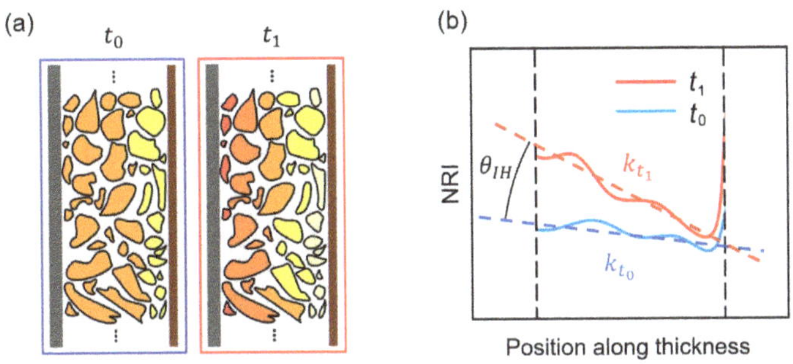

Figure 8. Schematic diagrams related to the characterization of K_{ILIH}. (**a**) Schematic diagram of the original inhomogeneity noise in the slow charging process when $SOC(t_0) = SOC(t_1)$, $K_{ILIH}(t_0) \equiv 0$; and (**b**) schematic diagram of the NRI fitting process.

Figure 8b illustrates the NRI fitting process, in which the estimation of K_{ILIH} is made. Systematic noise needs to be excluded from the image analysis results at t_1, so an angular decomposition formula was derived as follows:

$$\tan\left(\theta_{k_{t_0}}\right) = k_{t_0} \tag{4}$$

$$\tan\left(\theta_{k_{t_1}}\right) = k_{t_1} \tag{5}$$

$$SOC(t_0) = SOC(t_1) \tag{6}$$

$$K_{ILIH}(t_1) := \tan\theta_{IH} = \tan\left(\theta_{k_{t_1}} - \theta_{k_{t_0}}\right) = \frac{k_{t_1} - k_{t_0}}{1 + k_{t_1}k_{t_0}} \tag{7}$$

in which case, as t_0 is in the low-rate charging process, $K_{ILIH}(t_0)$ is considered to be zero.

To summarize, the ILIH value assessment process is outlined in a flow chart, as shown in Figure 9. The nominalization of position and illuminance ensures reliability at both the position and color levels, respectively. This process converts the qualitative and unstable image results into quantitative and robust K_{ILIH} results.

Significantly, because the images, ORI, NRI, and K_{ILIH} are time series data, a data-navigating video can be generated during processing. Furthermore, the data-navigating video can assist in error checking during the data analysis process. An example of a data-navigating video is provided in Supplementary Video S2, and a brief introduction and analysis of the video can be found in the Supplementary Material. A detailed analysis of the K_{ILIH} results will be discussed in the next section.

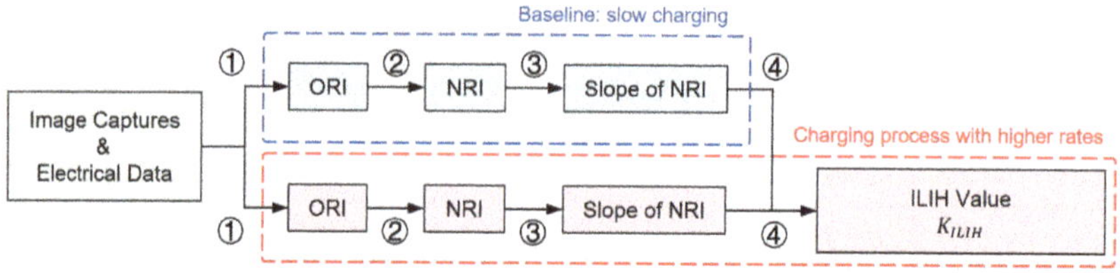

Figure 9. Flow chart of the ILIH assessment process: ① nominalization of position; ② nominalization of illuminance; ③ linear simplification; and ④ elimination of inhomogeneity noise.

4. Results and Discussions

By applying the ILIH assessment method demonstrated above, a series of K_{ILIH} results were obtained, all of which are illustrated and discussed in this section. In Section 4.1, the K_{ILIH} results are presented and discussed, and the robustness of the introduced method is verified for different battery samples with different charging rates. In Section 4.2, the K_{ILIH} trend and its underlying physicochemical mechanism are illustrated and discussed in detail, providing support to the interpretability of the ILIH assessment method.

4.1. Results of ILIH Value Assessment

As shown in Figure 5 and Supplementary Video S2, the Red Channel exhibits the most significant changes during the (dis)charging process. Therefore, processing the Red Channel data would be highly beneficial in characterizing ILIH. Consequently, a series of K_{ILIH}-SOC data (via the Red Channel) from different batteries and charge/discharge rates are sorted out.

Figure 10 illustrates the K_{ILIH}-SOC plot during the charging process of the three batteries, whereas Figure 11 shows the plot during the discharging process. Due to slight electrochemical instability in Battery-C, the charging/discharging rates did not precisely match the pre-set values in the tests, and the 1C rate was not properly executed. However, because the three cycles were still comparable (0.172C is close to 0.125C, and 0.345C and 0.69C are two times and four times the 0.172C baseline), Battery-C was still worth analyzing. For the sake of clarity, the three different charging rates are uniformly expressed as "0.125C", "0.25C", and "0.5C" throughout this study. Additionally, the cross-section of the Battery-B sample exhibited lower reflectance in the ECCS, resulting in dimmer original photochromatic captures that were more easily affected by systematic noise, leading to greater fluctuations in its K_{ILIH}-SOC plot.

In the charging process, K_{ILIH} has negative values because the zero point of position along the thickness is defined on the separator side. It is important to note that the positive and negative values of K_{ILIH} represent the direction of inhomogeneity, whereas the absolute value refers to the extent of inhomogeneity. Figure 10 shows a similar K_{ILIH} trend, including a near-zero plateau at low SOCs, a rapid change at medium–high SOCs, and restoration at high SOCs.

During the discharging process, shown in Figure 11, K_{ILIH} returns to zero as the SOC decreases. Battery-A and Battery-B exhibit similar trends, whereas Battery-C behaves differently, with barely any change in K_{ILIH} during the discharging process. The divergent trend is attributed to the different end points of K_{ILIH}-SOC in the charging process, where Battery-C has large positive end points instead of negative ones. In the experiments, Battery-C undergoes an extremely short CV charging process compared to Battery-A and Battery-B, which may result in abnormal charging end points. The authors believe that the abnormal behavior of Battery-C in this case may also result from the sample's slight electrochemical instability.

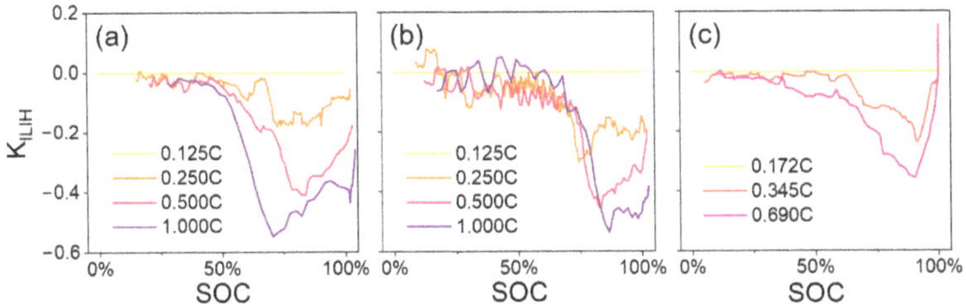

Figure 10. K_{ILIH}-SOC plot of the charging process: (**a**) Battery-A; (**b**) Battery-B; and (**c**) Battery-C.

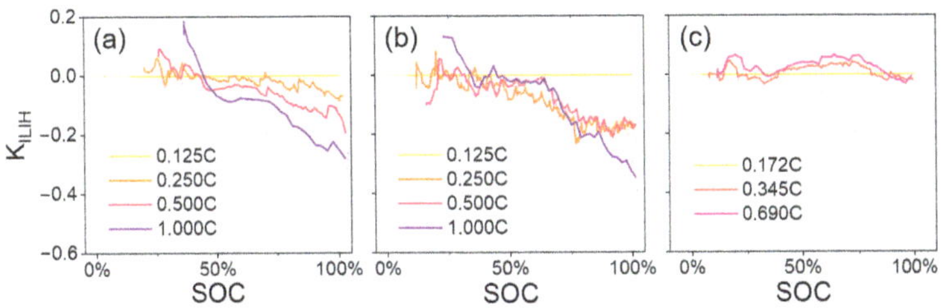

Figure 11. K_{ILIH}-SOC plot of the discharging process: (**a**) Battery-A; (**b**) Battery-B; and (**c**) Battery-C.

By combining the K_{ILIH}-SOC curves obtained at the same charging rates, as shown in Figure 12a–c, a comparison between different battery cells can be made. The peak K_{ILIH} (in the negative domain) was -0.24 ± 0.06 at 0.25C and -0.41 ± 0.05 at 0.5C, whereas in the 1C condition (excluding Battery-C), the peak value was approximately -0.55. The peak K_{ILIH} results are summarized in Figure 12d. The relationship exhibits an increasing trend with increasing charging rate. In conclusion, the quantitative ILIH characterization method demonstrates robustness across different battery cells and charging rates.

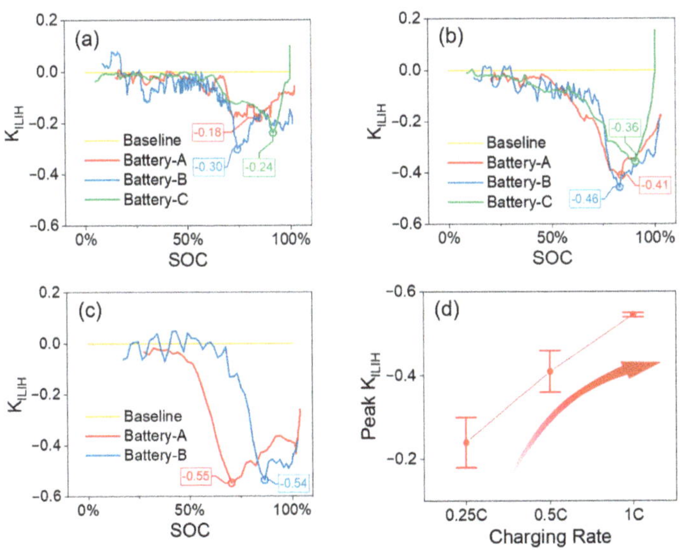

Figure 12. Results of peak K_{ILIH} in the charging process: (**a**) K_{ILIH} at 1/4C charging rate; (**b**) K_{ILIH} at 1/2C charging rate; (**c**) K_{ILIH} at 1C charging rate; and (**d**) the relationship between peak K_{ILIH} and the charging rates.

4.2. ILIH Value Trend Analysis

To provide a detailed analysis of the charging K_{ILIH}-SOC curve trend, a specific curve is singled out in Figure 13. By focusing on the 0.5C single CC–CV–CC cycle of Battery-A as an example, the K_{ILIH}-SOC curve can be divided into six parts. For each part, a Relative Lithium Insertion Amount vs. Position graph is shown in Figure 13 to illustrate the underlying mechanism. The Relative Lithium Insertion Amount (RLI) represents the relative insertion amount at each position along the thickness based on the assessment approach proposed in this study. It is important to note that the rectangular dashed frame in Figure 13 represents the fully charged status in the real-world low-rate charging process, which is not equivalent to the theoretical fully inserted state of Li.

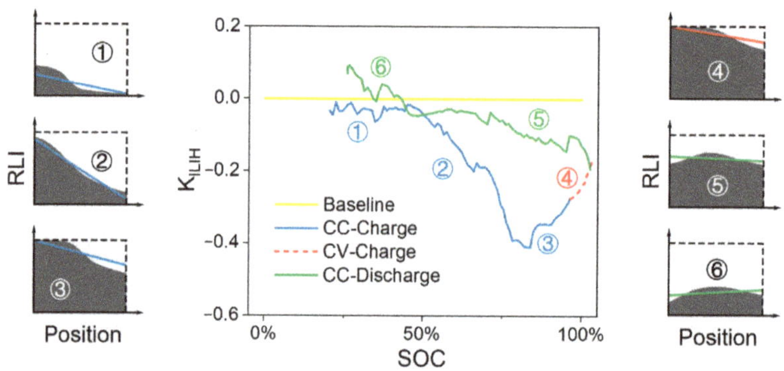

Figure 13. K_{ILIH}-SOC plot from the experiment of Battery-A, with a 0.5C charging process, and Relative Lithium Insertion Amount (RLI) vs. Position schematic diagrams are shown from the concept.

The six parts and the underlying mechanism are as follows:

①: Only anode particles near the separator start Li insertion, but because the fitting of NRI covers the entire anode, K_{ILIH} (or the fitted NRI slope) changes minimally.

②: Li inserts into particles evenly, resulting in a rapid change in K_{ILIH} towards a significant negative value within a narrow SOC range.

③: Anode particles near the separator reach the maximum insertion amount, whereas the insertion of particles near the current collector continues, causing K_{ILIH} to decrease.

④: In the CV charging process, K_{ILIH} continues to decrease.

⑤ and ⑥: In the CC discharging process, Li begins to take off from the separator side particle. So, K_{ILIH} decreases in the negative domain until it reaches zero, and then it continues to increase.

Based on the underlying mechanism, it becomes evident that enhancing anode materials and refining the manufacturing process through parameter optimization, such as microstructure and material system adjustments, may elevate the diffusion rate within the anode coating layer. These improvements hold the promise of mitigating ILIH during the reaction process and, consequently, improving the stability of fast-charging battery systems.

Although no literature strictly supports the mechanism of the demonstrated K_{ILIH}-SOC trend, the mechanism proposed in this paper was inspired by Guo et al. [38] and Krachkovskiy et al. [32]. Nevertheless, the K_{ILIH} trend and its underlying mechanism effectively support the interpretability of the ILIH assessment method.

5. Conclusions and Outlooks

5.1. Conclusions

This study successfully conducted an operando observation and characterization of the cross-section of commercial battery anodes. The newly introduced analysis method enables the quantification of intra-layer inhomogeneity in operating cells. A detailed analysis of the relationship between the ILIH value (K_{ILIH}) and SOC was performed. The peak inhomogeneity values between different cells and charging rates exhibit a great degree of convergence. Thus, the new method has been demonstrated to be robust in different conditions and overcomes the limitations of the colorimetric method.

5.2. Outlooks

The proposed method holds great potential for future studies, with the following possibilities:

(1) Adoption of the new method for graphite anodes (or various cathodes) in various battery material systems.

(2) Conducting large-scale factorial experiments to establish functional relationships between the trend of ILIH values, SOCs, charging rates, and the current ILIH value. Then, the relationship functions can be used as guidance for setting fast charging protocols. The newly guided protocols may result in a longer lifespan and better health conditions of commercial batteries compared to protocols solely scaled from Li plating limits.

(3) By combining the new method for characterizing ILIH with Digital Image Correlation (DIC) techniques for electrode strain measurement, the relationships between strain and Li intercalation inhomogeneity can be studied extensively. Furthermore, by using nanoindentation to measure the mechanical properties of particles and other materials in the cross-section, an inner stress study can be conducted to further elucidate the relationships between particle cracking, stress (or strain) distribution, and the inhomogeneity of Li intercalation.

Supplementary Materials: The following supporting information can be downloaded at: https://www.mdpi.com/article/10.3390/batteries9090463/s1.

Author Contributions: Conceptualization, P.Y.; methodology, T.F. and P.Y.; software, T.F. and P.Y.; validation, T.F. and P.Y.; formal analysis, T.F. and P.Y.; investigation, T.F. and G.J.; resources, G.J.; data curation, G.J.; writing—original draft preparation, T.F.; writing—review and editing, Y.X. and P.Y.; visualization, T.F.; supervision, Y.X. and P.Y.; project administration, P.Y.; funding acquisition, Y.X. All authors have read and agreed to the published version of the manuscript.

Funding: This study was sponsored by the National Natural Science Foundation of China (Grant No. 52172405). This study was also partially supported by the University Research Program of Ford (Grant No. 001068-2019-7170R-H).

Data Availability Statement: The data presented in this study are available on request from the corresponding author. The data are not publicly available due to confidentiality reasons.

Conflicts of Interest: The authors declare no conflict of interest.

References

1. Whittingham, M.S. Lithium Batteries and Cathode Materials. *Chem. Rev.* **2004**, *104*, 4271–4302. [CrossRef] [PubMed]
2. Duan, J.; Tang, X.; Dai, H.; Yang, Y.; Wu, W.; Wei, X.; Huang, Y. Building Safe Lithium-Ion Batteries for Electric Vehicles: A Review. *Electrochem. Energ. Rev.* **2020**, *3*, 1–42. [CrossRef]
3. Kim, T.; Song, W.; Son, D.-Y.; Ono, L.K.; Qi, Y. Lithium-Ion Batteries: Outlook on Present, Future, and Hybridized Technologies. *J. Mater. Chem. A* **2019**, *7*, 2942–2964. [CrossRef]
4. Li, W.; Erickson, E.M.; Manthiram, A. High-Nickel Layered Oxide Cathodes for Lithium-Based Automotive Batteries. *Nat. Energy* **2020**, *5*, 26–34. [CrossRef]
5. Zeng, X.; Li, M.; Abd El-Hady, D.; Alshitari, W.; Al-Bogami, A.S.; Lu, J.; Amine, K. Commercialization of Lithium Battery Technologies for Electric Vehicles. *Adv. Energy Mater.* **2019**, *9*, 1900161. [CrossRef]
6. Pang, P.; Wang, Z.; Deng, Y.; Nan, J.; Xing, Z.; Li, H. Delayed Phase Transition and Improved Cycling/Thermal Stability by Spinel LiNi0.5Mn1.5O4 Modification for $LiCoO_2$ Cathode at High Voltages. *ACS Appl. Mater. Interfaces* **2020**, *12*, 27339–27349. [CrossRef]
7. Pang, P.; Wang, Z.; Tan, X.; Deng, Y.; Nan, J.; Xing, Z.; Li, H. $LiCoO_2$@$LiNi_{0.45}Al_{0.05}Mn_{0.5}O_2$ as High-Voltage Lithium-Ion Battery Cathode Materials with Improved Cycling Performance and Thermal Stability. *Electrochim. Acta* **2019**, *327*, 135018. [CrossRef]
8. Nitta, N.; Wu, F.; Lee, J.T.; Yushin, G. Li-Ion Battery Materials: Present and Future. *Mater. Today* **2015**, *18*, 252–264. [CrossRef]
9. Waldmann, T.; Hogg, B.-I.; Wohlfahrt-Mehrens, M. Li Plating as Unwanted Side Reaction in Commercial Li-Ion Cells—A Review. *J. Power Sources* **2018**, *384*, 107–124. [CrossRef]

10. Harris, S.J.; Lu, P. Effects of Inhomogeneities—Nanoscale to Mesoscale—On the Durability of Li-Ion Batteries. *J. Phys. Chem. C* **2013**, *117*, 6481–6492. [CrossRef]
11. Sasaki, T.; Villevieille, C.; Takeuchi, Y.; Novák, P. Understanding Inhomogeneous Reactions in Li-Ion Batteries: Operando Synchrotron X-Ray Diffraction on Two-Layer Electrodes. *Adv. Sci.* **2015**, *2*, 1500083. [CrossRef] [PubMed]
12. Nakamura, T.; Watanabe, T.; Kimura, Y.; Amezawa, K.; Nitta, K.; Tanida, H.; Ohara, K.; Uchimoto, Y.; Ogumi, Z. Visualization of Inhomogeneous Reaction Distribution in the Model $LiCoO_2$ Composite Electrode of Lithium Ion Batteries. *J. Phys. Chem. C* **2017**, *121*, 2118–2124. [CrossRef]
13. He, J.; Meng, J.; Huang, Y. Challenges and Recent Progress in Fast-Charging Lithium-Ion Battery Materials. *J. Power Sources* **2023**, *570*, 232965. [CrossRef]
14. Huang, Q.; Ni, S.; Jiao, M.; Zhong, X.; Zhou, G.; Cheng, H.-M. Aligned Carbon-Based Electrodes for Fast-Charging Batteries: A Review. *Small* **2021**, *17*, 2007676. [CrossRef]
15. Yari, S.; Van Bael, M.K.; Hardy, A.; Safari, M. Non-Uniform Distribution of Current in Plane of Large-Area Lithium Electrodes. *Batter. Supercaps* **2022**, *5*, e202200217. [CrossRef]
16. Zhu, X.; Revilla, R.I.; Jaguemont, J.; Van Mierlo, J.; Hubin, A. Insights into Cycling Aging of LiNi0.80Co0.15Al0.05O2 Cathode Induced by Surface Inhomogeneity: A Post-Mortem Analysis. *J. Phys. Chem. C* **2019**, *123*, 30046–30058. [CrossRef]
17. Mühlbauer, M.J.; Schökel, A.; Etter, M.; Baran, V.; Senyshyn, A. Probing Chemical Heterogeneity of Li-Ion Batteries by in Operando High Energy X-Ray Diffraction Radiography. *J. Power Sources* **2018**, *403*, 49–55. [CrossRef]
18. Maire, P.; Kaiser, H.; Scheifele, W.; Novák, P. Colorimetric Determination of Lithium-Ion Mobility in Graphite Composite Electrodes. *J. Electroanal. Chem.* **2010**, *644*, 127–131. [CrossRef]
19. Vamvakeros, A.; Matras, D.; Ashton, T.E.; Coelho, A.A.; Dong, H.; Bauer, D.; Odarchenko, Y.; Price, S.W.T.; Butler, K.T.; Gutowski, O.; et al. Cycling Rate-Induced Spatially-Resolved Heterogeneities in Commercial Cylindrical Li-Ion Batteries. *Small Methods* **2021**, *5*, 2100512. [CrossRef]
20. Petz, D.; Mühlbauer, M.J.; Baran, V.; Schökel, A.; Kochetov, V.; Hofmann, M.; Dyadkin, V.; Staron, P.; Vaughan, G.; Lienert, U.; et al. Lithium Distribution and Transfer in High-Power 18650-Type Li-Ion Cells at Multiple Length Scales. *Energy Storage Mater.* **2021**, *41*, 546–553. [CrossRef]
21. Senyshyn, A.; Mühlbauer, M.J.; Dolotko, O.; Hofmann, M.; Ehrenberg, H. Homogeneity of Lithium Distribution in Cylinder-Type Li-Ion Batteries. *Sci. Rep.* **2015**, *5*, 18380. [CrossRef] [PubMed]
22. Harris, S.J.; Timmons, A.; Baker, D.R.; Monroe, C. Direct In Situ Measurements of Li Transport in Li-Ion Battery Negative Electrodes. *Chem. Phys. Lett.* **2010**, *485*, 265–274. [CrossRef]
23. Hogrefe, C.; Waldmann, T.; Hölzle, M.; Wohlfahrt-Mehrens, M. Direct Observation of Internal Short Circuits by Lithium Dendrites in Cross-Sectional Lithium-Ion In Situ Full Cells. *J. Power Sources* **2023**, *556*, 232391. [CrossRef]
24. Kuwabara, J.; Sato, K. In Situ Observation of the Electrochemical Dissolution and Deposition of Copper Contaminations in Li-Ion Batteries. *ECS Trans.* **2017**, *75*, 47. [CrossRef]
25. Zhu, J.; Zhang, X.; Luo, H.; Sahraei, E. Investigation of the Deformation Mechanisms of Lithium-Ion Battery Components Using in-Situ Micro Tests. *Appl. Energy* **2018**, *224*, 251–266. [CrossRef]
26. Tian, X.; Ying, P.; Xia, Y. In-Situ Mechanical Characterization of Compression Response of Anode Coating Materials through Inverse Approach. In Proceedings of the SAE 2022 International Automotive Safety, Security and Testing Congress, Shanghai, China, 7–8 July 2022.
27. Paul, P.P.; McShane, E.J.; Colclasure, A.M.; Balsara, N.; Brown, D.E.; Cao, C.; Chen, B.-R.; Chinnam, P.R.; Cui, Y.; Dufek, E.J.; et al. A Review of Existing and Emerging Methods for Lithium Detection and Characterization in Li-Ion and Li-Metal Batteries. *Adv. Energy Mater.* **2021**, *11*, 2100372. [CrossRef]
28. Meyer, L.; Saqib, N.; Porter, J. Review—Operando Optical Spectroscopy Studies of Batteries. *J. Electrochem. Soc.* **2021**, *168*, 090561. [CrossRef]
29. Steiger, J.; Kramer, D.; Mönig, R. Mechanisms of Dendritic Growth Investigated by In Situ Light Microscopy during Electrodeposition and Dissolution of Lithium. *J. Power Sources* **2014**, *261*, 112–119. [CrossRef]
30. Yui, Y.; Hayashi, M.; Nakamura, J. In Situ Microscopic Observation of Sodium Deposition/Dissolution on Sodium Electrode. *Sci. Rep.* **2016**, *6*, 22406. [CrossRef]
31. Schweikert, N.; Hofmann, A.; Schulz, M.; Scheuermann, M.; Boles, S.T.; Hanemann, T.; Hahn, H.; Indris, S. Suppressed Lithium Dendrite Growth in Lithium Batteries Using Ionic Liquid Electrolytes: Investigation by Electrochemical Impedance Spectroscopy, Scanning Electron Microscopy, and In Situ 7Li Nuclear Magnetic Resonance Spectroscopy. *J. Power Sources* **2013**, *228*, 237–243. [CrossRef]
32. Krachkovskiy, S.A.; Foster, J.M.; Bazak, J.D.; Balcom, B.J.; Goward, G.R. Operando Mapping of Li Concentration Profiles and Phase Transformations in Graphite Electrodes by Magnetic Resonance Imaging and Nuclear Magnetic Resonance Spectroscopy. *J. Phys. Chem. C* **2018**, *122*, 21784–21791. [CrossRef]
33. Agrawal, S.; Bai, P. Operando Electrochemical Kinetics in Particulate Porous Electrodes by Quantifying the Mesoscale Spatiotemporal Heterogeneities. *Adv. Energy Mater.* **2021**, *11*, 2003344. [CrossRef]
34. Guo, Z.; Zhu, J.; Feng, J.; Du, S. Direct In Situ Observation and Explanation of Lithium Dendrite of Commercial Graphite Electrodes. *RSC Adv.* **2015**, *5*, 69514–69521. [CrossRef]

35. Maire, P.; Evans, A.; Kaiser, H.; Scheifele, W.; Novák, P. Colorimetric Determination of Lithium Content in Electrodes of Lithium-Ion Batteries. *J. Electrochem. Soc.* **2008**, *155*, A862. [CrossRef]
36. Hogrefe, C.; Waldmann, T.; Molinero, M.B.; Wildner, L.; Axmann, P.; Wohlfahrt-Mehrens, M. Cross-Sectional In Situ Optical Microscopy with Simultaneous Electrochemical Measurements for Lithium-Ion Full Cells. *J. Electrochem. Soc.* **2022**, *169*, 050519. [CrossRef]
37. Dahn, J.R. Phase Diagram of Li x C 6. *Phys. Rev. B* **1991**, *44*, 9170–9177. [CrossRef] [PubMed]
38. Guo, Y.; Smith, R.B.; Yu, Z.; Efetov, D.K.; Wang, J.; Kim, P.; Bazant, M.Z.; Brus, L.E. Li Intercalation into Graphite: Direct Optical Imaging and Cahn–Hilliard Reaction Dynamics. *J. Phys. Chem. Lett.* **2016**, *7*, 2151–2156. [CrossRef]

batteries

MDPI

Article

A Coupled Nonlinear Viscoelastic–Viscoplastic Thermomechanical Model for Polymeric Lithium-Ion Battery Separators

Royal Chibuzor Ihuaenyi [1], Jie Deng [2], Chulheung Bae [2] and Xinran Xiao [1,*]

[1] Department of Mechanical Engineering, Michigan State University, Lansing, MI 48910, USA
[2] Department of Electrification Subsystems and Power Supply, Ford Motor Company, Dearborn, MI 48124, USA
* Correspondence: xinran@egr.msu.edu

Abstract: One of the major concerns in ensuring lithium-ion battery (LIB) safety in abuse scenarios is the structural integrity of the battery separator. This paper presents a coupled viscoelastic–viscoplastic model for predicting the thermomechanical response of polymeric battery separators in abuse scenarios under combined mechanical and thermal loadings. The viscoplastic model is developed based on a rheological framework that considers the mechanisms involved in the initial yielding, change in viscosity, strain softening and strain hardening of polymeric separators. The viscoplastic model is then coupled with a previously developed orthotropic nonlinear thermoviscoelastic model to predict the thermomechanical response of polymeric separators before the onset of failure. The model parameters are determined for Celgard®2400, a polypropylene (PP) separator, and the model is implemented in the LS-DYNA® finite element (FE) package as a user-defined subroutine. Punch test simulations are employed to verify the model predictions under biaxial stress states. Simulations of uniaxial tensile stress–strain responses at different strain rates and temperatures are compared with experimental data to validate the model predictions. The model predictions of the material anisotropy, rate and temperature dependence agree well with experimental observations.

Keywords: lithium-ion batteries; battery separator; viscoelasticity; viscoplasticity; constitutive modeling; simulation

1. Introduction

The advancement and commercialization of lithium-ion batteries (LIBs) have enabled a new generation of electrical vehicles and high-density energy storage systems. However, ensuring the safe operation of these high-energy-density batteries has been a key challenge that hinders the large-scale application and deployment of LIBs for future e-mobility and battery energy storage systems (BESSs). Therefore, understanding battery failure mechanisms and reducing safety risks is critical. One of the major mechanisms that lead to thermal runaway in LIBs is an internal short circuit (ISC) [1–4]. In LIBs, the failure of the battery separator creates a pathway for the electrodes to come in contact with each other, and this triggers an ISC. The separator failure can be caused by abusive loadings such as crash or impact, natural expansion during intercalation, or piercing by deposited lithium dendrites [1–4]. To improve the safety of LIBs and prevent thermal runaway scenarios due to ISC, it is necessary to develop accurate models to predict the response of battery separators in their entire range of deformation and under combined mechanical and thermal loadings.

Although the mechanical and thermal responses of commonly used microporous polymeric membranous battery separators have been explored in recent years [5–15], there are still some challenges associated with developing accurate models for predicting the response of such separators. These are due to the complexities involved in the thermomechanical behaviors of these separators. For example, commonly used polymer separators are dry-processed. This process results in strong anisotropy in materials. These separators have two in-plane principal directions known as the machine direction (MD) and the transverse direction (TD) [5–11]. As polymers, they are rate-dependent [5–7] and temperature-dependent [6,8]. However, unlike typical polymers, dry-processed separators have a unique thermal expansion/shrinkage behavior [12,13]. Additionally, in the working condition of LIBs, separators are immersed in electrolyte solutions that are known to act as plasticizers and weaken their mechanical response [7,14–16].

In previous works, we have developed orthotropic viscoelastic models for predicting the response of separators within the linear range [17], with the effects of temperature and electrolyte [18], and in the nonlinear viscoelastic range [19]. In the literature, a few works have attempted to develop models for simulating the separator responses under large deformations. Zhao et al. presented a constitutive model based on the Rich–Hill elastoplastic theory for large deformations of a polypropylene (PP) separator [20]. The load–displacement response of a polyethylene (PE) separator was modeled by Zhang et al. using the anisotropic crushable foam model in LS-DYNA [21]. Furthermore, a representative volumetric element modeling method was proposed by Xie et al. [22] to predict the tensile properties of PP separators. Recently, an elastic–hyperelastic–viscoplastic model was developed by Makki et al. [23] for predicting the mechanical response of battery separators under large deformations. The model considered the rate dependence in the plastic range through a damage rate sensitivity parameter. However, no rate-dependent results were presented. In summary, although the above models can represent the large deformation behavior of the separators and some with rate dependency, they did not address the temperature dependence and the thermal expansion/shrinkage of the materials, which are critical for simulations of the separator response under combined mechanical and thermal loadings.

The stress–strain response of polymeric materials beyond the deformation recoverable limit (yield point) can be described by the viscoplasticity (or rate-dependent plasticity) model. In the literature, various viscoplastic models have been developed for polymeric materials. An elasto-viscoelastic–viscoplastic model was developed by Findlay et al. [24] to model the stress–strain response of polymeric materials. A nonlinear elasto-viscoelastic–viscoplastic model was developed from the thermodynamic principles by Schapery [25,26] to model the response of fiber-reinforced and unreinforced plastics, as well as semicrystalline polymers. Lai and Bakker [27] proposed an elasto-viscoelastic–viscoplastic integral model for polymeric materials. The developed model was validated using creep and recovery tests for high-density PP. Coupled elasto-viscoelastic–viscoplastic models based on rheological frameworks have been developed and implemented for predicting the stress–strain response of various polymers in their entire deformation range [28–33]. Also, a group of consistency viscoplasticity models that incorporate the time-dependent response in the large strain deformation region in a rate-dependent yield surface was introduced by Wang et al. [34] and has been studied extensively by others [35–38]. Another group of viscoplastic models is the overstress model based on the ideas presented by Perzyna [39]. The main postulate of these models is that the current stress state can be outside the yield surface. The Perzyna model [39] and the Duvaut–Lions model [40] are the most commonly used formulations. Both models have been widely implemented in small strain [41–49] as well as finite strain problems [50–53].

Finally, phenomenological constitutive models have been developed to predict the flow stress of polymeric materials. G'sell and Jonas [54] developed a constitutive relationship to predict the flow stress behavior of polymeric materials at different constant strain rates. This model has been implemented in its original and modified forms in different works to predict the stress–strain response of polymeric materials at different strain rates and temperatures [55–58]. The Johnson–Cook model [59] and its variations have also been applied in the literature to describe the tensile and compressive response of polymeric materials [60,61]. Furthermore, a flow stress model was proposed by Nasraoui et al. [62] for the strain rate and the temperature-dependent response of polymeric materials. The model was validated against uniaxial compression tests for PMMA (polymethylmethacrylate) under quasi-static and dynamic loadings. Another flow stress model known as the DSGZ model was developed by Duan et al. [63] to predict the response of glassy and semicrystalline polymers. Recently, a phenomenological model for predicting the flow stress of thermoplastics was developed by Zhu et al. [64]. The model was validated against uniaxial tensile and compression tests for PEEK and PC.

The above constitutive models form a foundation for modeling the thermomechanical behavior of polymeric separators under large deformation.

This work presents a phenomenological flow stress-based viscoplastic model. The viscoplastic model is coupled with the orthotropic nonlinear thermoviscoelastic model developed in our previous work [19] to accurately predict the stress–strain response of polymeric battery separators from small strain to the onset of failure. The model parameters were calibrated for Celgard®2400 PP separators. The model's predictions of the material anisotropy, rate and temperature dependences are compared with experimental data.

2. Theoretical Background

2.1. Viscoelasticity

The stress–strain response of viscoelastic materials under small deformations is formulated through the hereditary representation of the Boltzmann superposition principle, whose stiffness-based form is mathematically expressed as [65–69]:

$$\sigma(t) = \int_0^t G(t-\tau)\frac{d\varepsilon(\tau)}{d\tau}d\tau \tag{1}$$

where σ is the stress, ε is the strain and $G(t)$ is the stress relaxation stiffness function.

Furthermore, when the applied loading exceeds the linear viscoelastic limit, material nonlinearities arise in the mechanical response of viscoelastic materials. Beyond the linear viscoelastic limit, the creep or stress relaxation responses become functions of both time and the applied constant stress or strain. The stress–strain response of viscoelastic materials beyond the limit of linear viscoelasticity is commonly represented by the Schapery hereditary integral formulation [70]. For uniaxial loading, the Schapery model in its stiffness form is represented as:

$$\sigma(t) = h_\infty(\varepsilon)\cdot G_\infty\cdot\varepsilon(t) + h_1(\varepsilon)\cdot\int_0^t \Delta G[\rho(t)-\rho(\tau)]\frac{dh_2(\varepsilon)\varepsilon(\tau)}{d\tau}d\tau \tag{2}$$

where G_∞ and $\Delta G(t)$ are the equilibrium and linear viscoelastic transient relaxation modulus. h_∞, h_1, h_2, are strain-dependent nonlinearity parameters, and ρ is the reduced time defined as:

$$\rho(t) = \int_0^t \frac{dt'}{a_\varepsilon[\varepsilon(t')]} \tag{3}$$

$$\rho(\tau) = \int_0^\tau \frac{dt'}{a_\varepsilon[\varepsilon(t')]} \tag{4}$$

In Equations (3) and (4), a_ε is the strain-dependent time–strain shifting factor. It is vital to note that when the applied deformation is within the linear viscoelastic range, the values of the nonlinearity parameters h_∞, h_1, h_2, and a_ε become equal to 1, and Equation (2) reduces to the hereditary integral representation for linear viscoelastic materials (Equation (1)).

In previous works [17–19], we have developed orthotropic linear and nonlinear thermoviscoelastic models to predict the thermomechanical response of polymeric battery separators accounting for their constitutive behaviors using Equations (1) and (2) as the basis of the constitutive models. In these works, the stress–strain response of the separator was expressed in a rheological framework using the generalized Maxwell model (Figure 1).

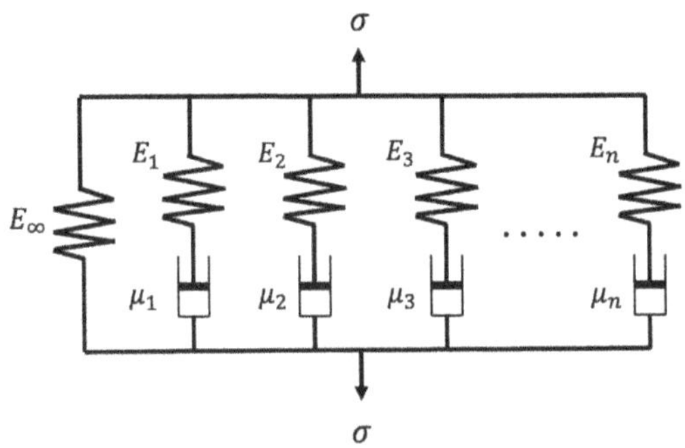

Figure 1. Generalized Maxwell Model.

The stress relaxation response of the generalized Maxwell model is generally expressed as a Prony series represented as:

$$G(t) = G_\infty + \sum_{i=1}^{n} G_i \cdot exp\left[-\frac{t}{\tau_i}\right] \quad (5)$$

where G_∞ is the equilibrium or fully relaxed modulus, G_i is the stress relaxation constant and τ_i is the relaxation time. However, for the nonlinear model, it was assumed that the components of the rheological network are nonlinear springs and dashpots whose nonlinearities are introduced through the strain-dependent nonlinearity parameters of the Schapery model.

To introduce material orthotropy in the models, the stress relaxation function was expressed as a viscoelastic analog of the elastic stiffness matrix [71]. This was achievable by applying the elastic–viscoelastic correspondence principle [68,69]. A symmetric stiffness matrix (Equation (6)) was assumed in the linear model [17,18]. The non-symmetric stiffness matrix (Equation (7)) was introduced in the nonlinear viscoelastic model [19]. From a previous study [5], it was observed that at strain values well beyond the linear viscoelastic limit, the condition of symmetry in the viscoelastic response breaks down.

$$G = \frac{1}{[1 - v_{12}v_{21}]} \cdot \begin{bmatrix} G_{11}(t) & v_{12}G_{22}(t) & 0 \\ v_{12}G_{22}(t) & G_{22}(t) & 0 \\ 0 & 0 & [1 - v_{12}v_{21}]G_{66}(t) \end{bmatrix} \quad (6)$$

$$G = \frac{1}{[1 - v_{12}v_{21}]} \cdot \begin{bmatrix} G_{11}(t) & v_{12}G_{22}(t) & 0 \\ v_{21}G_{11}(t) & G_{22}(t) & 0 \\ 0 & 0 & [1 - v_{12}v_{21}]G_{66}(t) \end{bmatrix} \quad (7)$$

In Equations (6) and (7), G_{11}, G_{22} and G_{66} are the relaxation moduli in the MD, TD and in-plane shear, respectively. v_{12} and v_{21} are the major and minor Poison's ratios.

The time–temperature superposition principle (TTSP) [65,72] was employed to account for the temperature effect within the range of 20–110 °C in the linear model and the range of 20–60 °C in the nonlinear model. Furthermore, a time–temperature–solvent superposition method was developed as an extension of the TTSP to account for the plasticization effect of electrolyte solutions in the linear thermoviscoelastic model [18]. Consequently, to implement Equations (1) and (2) in finite element modeling for loadings with arbitrary stress, strain and temperature histories, the hereditary integrals were evaluated using a developed discretization algorithm that uses the Prony series representation of the generalized Maxwell model as a kernel function [17,19].

The model parameters for the orthotropic linear and nonlinear viscoelastic models were calibrated for the Celgard®2400 PP separator, and the models were implemented as user-defined subroutines in the LS-DYNA® FE package, verified and validated against experimental data.

2.2. Viscoplasticity

The formulation of the viscoplastic flow stress model is based on the rheological framework as shown in Figure 2. Beyond the initial yield threshold, the governing relationship is simplified using a rheological system, which consists of a sliding frictional element, a dashpot and two nonlinear springs. The sliding frictional element represents the initial yield threshold, the dashpot accounts for the changes in the viscosity of the material due to the accumulation of plastic strains and changes in the strain rate, and the two springs account for the strain-softening and strain-hardening mechanisms.

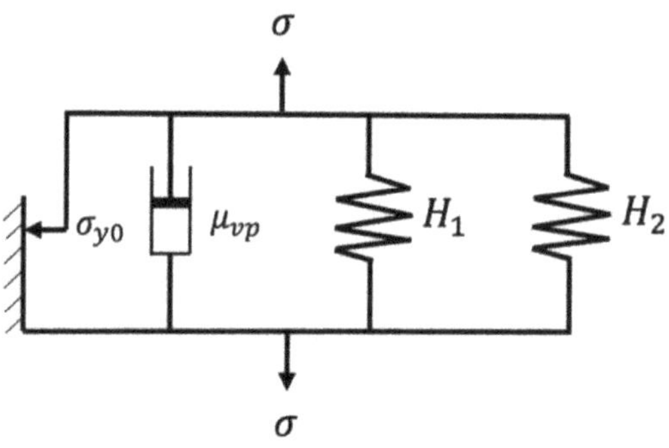

Figure 2. Rheological framework for viscoplastic model.

The elements in the rheological model are arranged in such a way that the total stress can be determined by a summation of the stresses in the individual elements. Hence, after the initial yield threshold is exceeded, the stress as a function of the accumulated plastic strain and the constant strain rate at a constant temperature is expressed as:

$$\sigma(\varepsilon_p, \dot{\varepsilon}) = \sigma_{y0}(\dot{\varepsilon}) + \mu_{vp}(\dot{\varepsilon})\cdot\{1 - exp(-k(\dot{\varepsilon})\cdot\varepsilon_p)\} + H_1(\dot{\varepsilon})\cdot\varepsilon_p + H_2(\dot{\varepsilon})\cdot\varepsilon_p^m \quad (8)$$

where σ_{y0} is the initial yield stress. With an increase in the strain rate and a decrease in temperature, σ_{y0} increases in value and vice versa. μ_{vp} and k are the first and second viscosity parameters, respectively. μ_{vp} accounts for the effect of the rate of change in the viscosity of the material under permanent deformations on the flow stress due to changes in the strain rate and temperature, while k regulates the rate at which softening and or hardening occurs after the initial yield stress has been exceeded. H_1 is the strain-rate- and temperature-dependent softening coefficient that accounts for the decrease in flow stress with an increase in strain. H_2 and m are the hardening coefficients that combine to depict the increase in flow stress under a further increase in strain.

The viscoplastic model parameters were obtained through experimental calibration, which is presented next.

3. Experimental Procedure

Uniaxial tensile tests were carried out with Celgard®2400 samples at strain rates of 0.1/s, 0.01/s, 0.001/s and 0.0001/s. The tests were also carried out at constant temperatures ranging from 20 °C to 50 °C. The in-plan dimensions of each specimen were 45 × 5 mm [17]. The tests were carried out using an RSA G2 rheometer as shown in Figure 3. For tests carried out below ambient conditions (20 °C), a liquified nitrogen (LN2) tank (Figure 3b) was connected to the rheometer to cool each tested sample to the desired temperature. The experimentally determined stress–strain curves for the samples cut along the MD and TD are presented in Figures 4 and 5, respectively. Furthermore, the typical post-test geometry of the MD and TD samples are presented in Figure 6 to show the deformation mechanisms the samples undergo under uniaxial tension.

(a) (b)

Figure 3. Experimental setup for uniaxial tensile tests: (**a**) sample fixture; (**b**) liquified nitrogen tank connected to the rheometer.

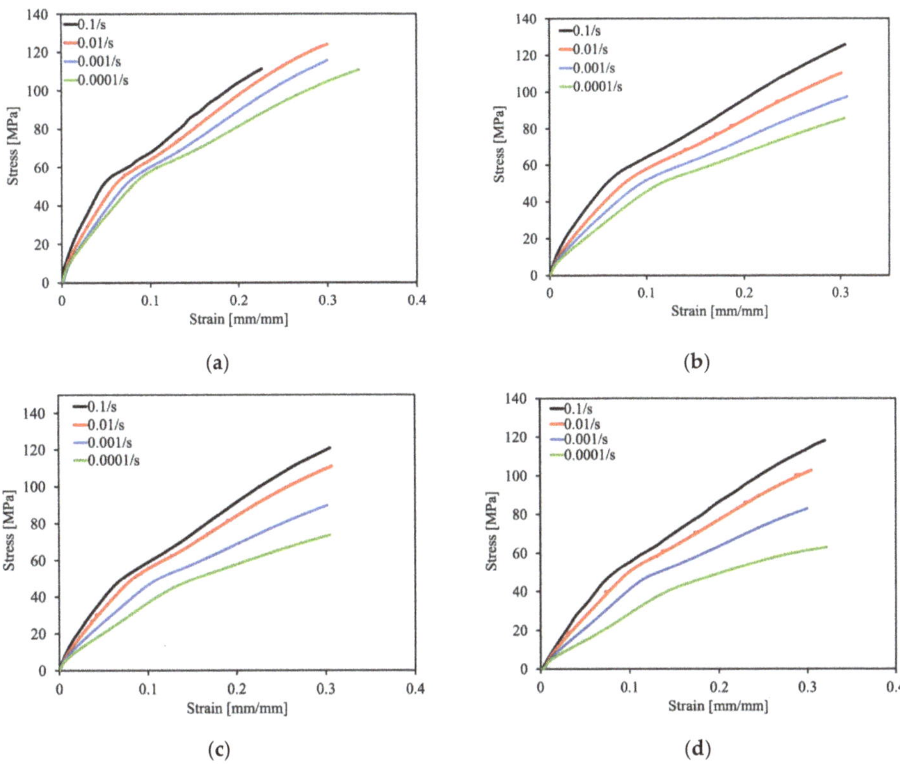

Figure 4. Uniaxial tensile stress–strain curves for samples cut along the MD carried out at different strain rates and temperatures of (**a**) 20 °C, (**b**) 30 °C, (**c**) 40 °C and (**d**) 50 °C.

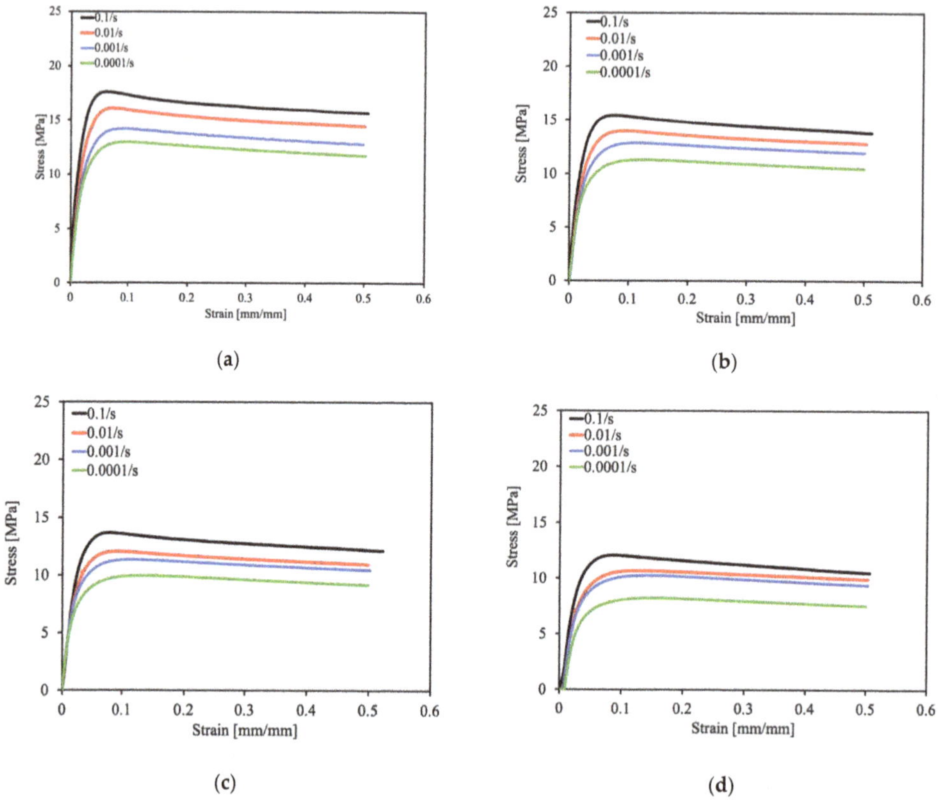

Figure 5. Uniaxial tensile stress–strain curves for samples cut along the TD carried out at different strain rates and temperatures of (**a**) 20 °C, (**b**) 30 °C, (**c**) 40 °C and (**d**) 50 °C.

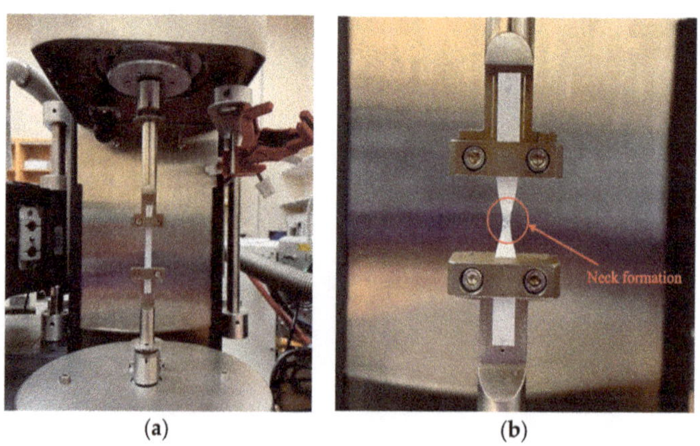

Figure 6. Typical post-test geometry of samples under tension in the (**a**) MD and (**b**) TD.

From the results, the effects of the strain rate and temperature are manifested in the stress–strain response of the separator in the MD and TD. With the increasing strain rate, there is an observed increase in the stiffness of the material and vice versa. Also, with increasing temperature, the material softens, and there is an observed reduction in its stiffness. The results also show that the stress–strain curves for the samples cut along the TD are that of a typical semicrystalline polymer with five major phases in its deformation process. The first phase consists of fully recoverable linear viscoelastic deformation. In the second phase, the material transitions into a nonlinear viscoelastic deformation region. Phase three involves the evolution of the yield mechanism, after which deformations become irrecoverable. Phases four and five constitute the strain-softening and strain-hardening mechanisms, respectively. For semicrystalline polymers, strain softening is attributed to broken crystallization, and strain hardening is related to recrystallization [64]. Also, a neck formation is observed during the TD sample deformation process. The necking region is characterized by a rapid decrease in the cross-sectional area at a particular point along the gage length of the specimen. This necking phenomenon is expressed in the post-test sample geometry (Figure 6b). However, the material response in the MD is different from that of a typical semicrystalline material in the sense that there is no clearly identifiable yield point following phases one and two of its deformation history. Furthermore, in the MD, there is an observed continuous hardening in the material response under very large strains. These distinct differences in the uniaxial stress–strain response of the material in the MD and TD introduce complexities that have to be accounted for in the constitutive modeling of polymeric separators under large strains. The uniaxial tensile test results give us an understanding of the total stress–strain behavior of the separator material in its different orientations and present a good case for model calibration and validation.

4. Model Parameter Identification

4.1. Strain-Rate-Dependent Model Parameters at the Reference Temperature

The strain-rate-dependent viscoplastic model parameters are calibrated using the uniaxial tensile test results carried out at different strain rates at the reference temperature (20 °C). This temperature value is selected intentionally as it is the reference temperature for the viscoelastic model that will be coupled with the viscoplastic model. The parameters to be determined are the initial yield stress (σ_{y0}), first and second viscosity parameters (μ_{vp} and k), strain-hardening and strain-softening coefficients (H_1, H_2 and m). The initial yield stress was determined based on the need to identify it within the boundaries of the range of prediction of the nonlinear viscoelastic model (0–5% strain) [19]. Also, identifying the initial yield point of polymeric materials is not as straightforward as it is with metallic materials. Hence, the offset method was employed to identify the initial yield stresses in the MD and TD of the polymeric separator at different strain rates. From the test results for samples cut along the MD, the yield stresses were identified by taking a 1% offset, while the yield stresses for samples cut along the TD were determined by taking a 0.3% offset. The rationale behind taking a larger offset in the MD than in the TD is supported by the superiority in the mechanical response of the material in the MD compared to that in the TD. Hence, it is expected that the material will yield lower stress and strain values in the TD than in the MD.

To determine the other model parameters (μ_{vp}, k, H_1, H_2 and m), a scripted optimization fitting method that minimizes the error between the fitted results and experimental data to generate the model parameters for each constant strain rate value was used. The obtained model parameters for the MD and TD are summarized in Tables 1 and 2, respectively. Comparisons between the experimental and the fitted curves are presented in Figure 7.

Table 1. Viscoplastic model parameters for MD response.

Strain Rate (s^{-1})	σ_{y0} (MPa)	μ_{vp} (MPa)	k	H_1 (MPa)	H_2 (MPa)	m
0.0001	19.33	22.19	25.21	0	228.77	1
0.001	21.20	18.11	34.44	0	281.24	1
0.01	25.61	16.84	41.29	0	307.71	1
0.1	34.45	9.70	115.36	0	343.93	1

Table 2. Viscoplastic model parameters for TD response.

Strain Rate (s^{-1})	σ_{y0} (MPa)	μ_{vp} (MPa)	k	H_1 (MPa)	H_2 (MPa)	m
0.0001	7.39	6.06	51.25	−4.79	2.76	2
0.001	8.61	6.13	52.74	−5.98	4.24	2
0.01	9.83	6.74	63.94	−8.06	7.85	2
0.1	10.33	7.35	77.86	−8.74	8.49	2

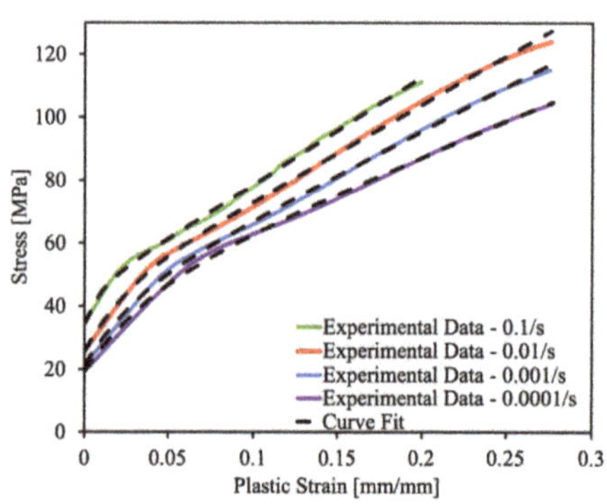

(**a**)

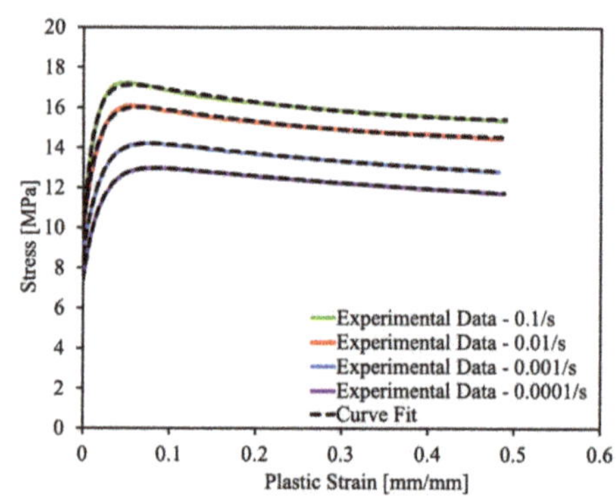

(**b**)

Figure 7. Curve fitting for identifying viscoplastic model parameters at 20 °C in the (**a**) MD and (**b**) TD.

Figure 7 shows that the fitted curves overlap with the experimental curves nicely. This verifies the accuracy in the determination of the model parameters given in Tables 1 and 2. However, from Tables 1 and 2, we have unique sets of values for each viscoplastic model parameter at different strain rate values. To implement the viscoplastic model in predictive modeling, given arbitrary strain rate values within the considered range, simple equations relating the viscoplastic parameters as functions of strain rate have to be defined. From the current data generated, we have to define five sets of equations for the TD model parameters and four sets of equations for the MD model parameters. To reduce the number of equations needed and the number of input parameters for the model implementation overall, a parameter reduction method is applied. A strain-rate-scaling factor ($a_{\dot{\varepsilon}}(\dot{\varepsilon})$) is introduced to modify Equation (8) such that the stress evolution after the yield threshold is exceeded at the reference temperature is expressed as:

$$\sigma(\varepsilon_p, \dot{\varepsilon}) = \sigma_{y0}(\dot{\varepsilon}) + \left\{\mu_{vp}^0 \cdot \{1 - exp(-k(\dot{\varepsilon}) \cdot \varepsilon_p)\} + H_1^0 \cdot \varepsilon_p + H_2^0 \cdot \varepsilon_p^m\right\} \cdot a_{\dot{\varepsilon}}(\dot{\varepsilon}) \quad (9)$$

where μ_{vp}^0, H_1^0 and H_2^0 are the first viscosity parameter and strain-softening and strain-hardening coefficients at the reference strain rate value set as 0.0001/s, respectively. The reduced viscoplastic model parameters for the MD and TD responses are summarized in Tables 3 and 4, respectively.

Table 3. Reduced viscoplastic model parameters for MD response.

Strain Rate (s^{-1})	σ_{y0} (MPa)	μ_{vp} (MPa)	k	H_1 (MPa)	H_2 (MPa)	m	$a_{\dot{\varepsilon}}$
0.0001	19.33	22.19	25.21	0	228.77	1	1
0.001	21.20	22.19	25.21	0	228.77	1	1.09
0.01	25.61	22.19	25.21	0	228.77	1	1.16
0.1	34.45	22.19	25.21	0	228.77	1	1.12

Table 4. Reduced viscoplastic model parameters for TD response.

Strain Rate (s^{-1})	σ_{y0} (MPa)	μ_{vp} (MPa)	k	H_1 (MPa)	H_2 (MPa)	m	$a_{\dot{\varepsilon}}$
0.0001	7.39	6.06	51.25	−4.79	2.76	2	1
0.001	8.61	6.06	52.74	−4.79	2.76	2	0.98
0.01	9.83	6.06	63.94	−4.79	2.76	2	1.06
0.1	10.33	6.06	77.86	−4.79	2.76	2	1.16

Figure 8 compares the fitted curves using the reduced model parameters with the experimental data. The model provided good representations for most of the experimental curves with the exception of MD at 0.1/s and 0.01/s conditions. As shown in Figure 8a, the model deviated from the experimental curves over the plastic strain range of 0.05 to 0.15 for these two conditions. This local deviation is due to a reduced number of parameters. Overall, the model is adequate for representing the experimental curves.

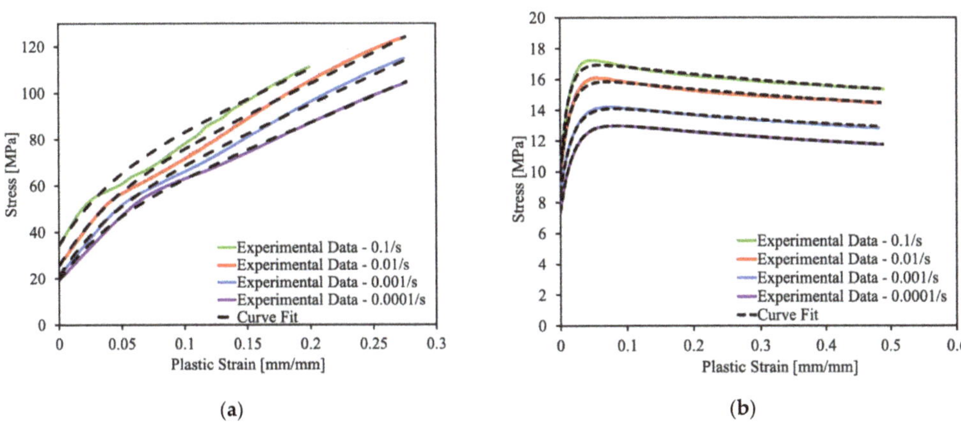

Figure 8. Curve fitting for reduced viscoplastic model parameters at 20 °C in the (**a**) MD and (**b**) TD.

Comparing Tables 3 and 4, it is important to note that the second viscosity parameter (k) has a constant value in the MD but a variable one in the TD. This is due to the difference in the stress–strain responses in the MD and TD. The TD stress–strain curves (Figure 5) show that the rate of initiation of the softening and hardening mechanisms after the initial yield stress has been exceeded is not constant for the material response at different strain rates. Hence, keeping this value constant will lead to the divergence of the fitted curves from the experimental data.

The strain-rate-dependent parameters are plotted vs. the logarithm of the strain rate and fitted using the least-squares regression method to generate simple linear and polynomial relationships that can be introduced into the constitutive model to span out the values of the parameters within the considered range of strain rates. These fittings as well as their mathematical expressions are presented in Figures 9 and 10 for the MD and TD parameters, respectively.

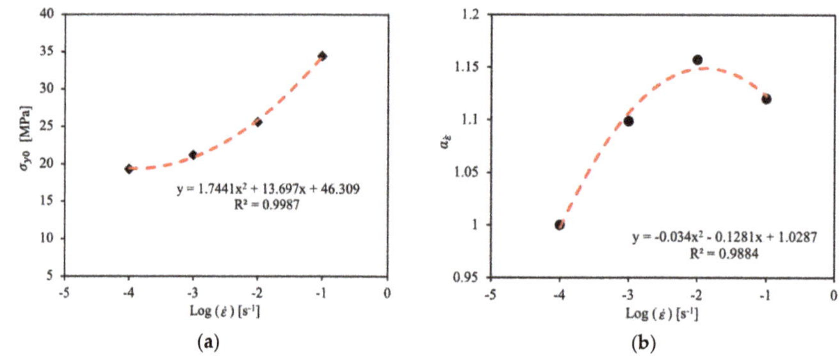

Figure 9. Least-squares regression fittings for viscoplastic parameters: (**a**) initial yield stress and (**b**) strain-rate-scaling factor for MD response.

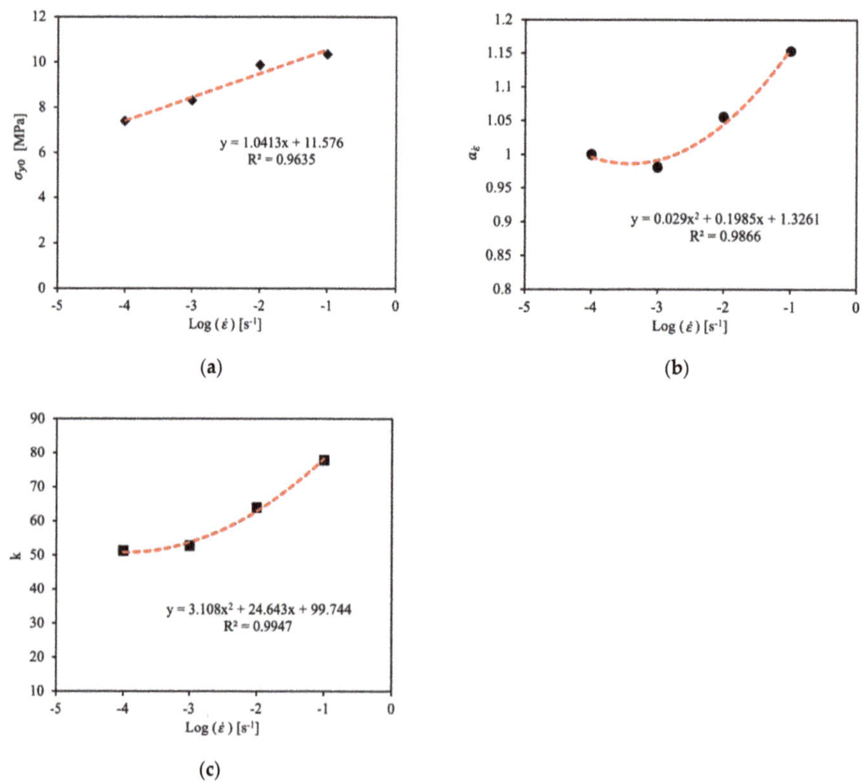

Figure 10. Least-squares regression fittings for viscoplastic parameters: (**a**) initial yield stress, (**b**) strain-rate-scaling factor and (**c**) viscosity coefficient for TD response.

4.2. Temperature Dependence and Parameter Identification

To account for temperature dependence in the developed viscoplastic model, a decaying term $h(T)$, modified from the G'sell–Jonas model [54], is introduced into Equation (9) such that:

$$\sigma(\varepsilon_p, \dot{\varepsilon}, T) = \left\{\sigma_{y0}(\dot{\varepsilon}) + \left\{\mu_{vp}^0 \cdot \{1 - exp(-k(\dot{\varepsilon})\cdot\varepsilon_p)\} + H_1^0\cdot\varepsilon_p + H_2^0\cdot\varepsilon_p^m\right\} \cdot a_{\dot{\varepsilon}}(\dot{\varepsilon}) \right\} \cdot h(T) \quad (10)$$

$$h(T) = exp\left[a\cdot\left(\frac{1}{T} - \frac{1}{T_0}\right)\right] \quad (11)$$

where a is a material parameter that accounts for the effect of temperature on the flow stress of the material, and T and T_0 are the current and reference temperatures, respectively. The modified term in Equation (10) is different from the original term used in the G'sell–Jonas model [54] in the sense that it introduces the reference temperature to give a clearer meaning to $h(T)$. At the reference temperature, $h(T)$ becomes equal to 1 and Equation (10) reduces to Equation (9), which only accounts for the post-yield response at the reference temperature.

To implement the fully developed model, the values of the parameter a have to be determined for the material response in the MD and TD. At the same strain value (ε), the value of the flow stress (σ) is related to the strain rate and current temperature (T). To determine parameter a at a constant large strain value and strain rate, some points on the stress–strain curves are selected as $(\varepsilon, \sigma_1), (\varepsilon, \sigma_2),$ *and* (ε, σ_i) at different temperatures T_1, T_2 and T_i, respectively, where i is the maximum number of constant temperature values considered. At the reference temperature, the reference flow stress value is obtained by using Equation (9). Therefore, the stress at the current temperature can be expressed as:

$$\sigma(\varepsilon_p, \dot{\varepsilon}, T) = \sigma(\varepsilon_p, \dot{\varepsilon}, T_0)\cdot h(T) \quad (12)$$

In this case, the reference temperature was selected as 20 °C (293 K), and the reference stress is already known. Hence, parameter a can be determined by fitting the relationship between the flow stress and temperature with Equation (12). A similar methodology for determining the parameter a was applied by Zhu et al. [64] in the constitutive modeling of thermoplastics. The analysis for the determination of parameter a was carried out at different large strain values. For MD samples, the analysis was carried out at 20% constant strain, and the determined parameter a values were different at different strain rates. However, for the TD samples, the analysis was carried out at 20%, 30% and 40% constant strain values, and the values of parameter a determined were similar for the different strain values and the different strain rates. Hence, the average value of 1100.64 was selected as parameter a to calibrate the material response in the TD. Figure 11 presents an example of the fitting procedure for the determination of parameter a for the MD response. The determined values of parameter a for the MD response showed a linear relationship with the strain rate as shown in Figure 12. The determined values of parameter a for the material response in the MD and TD are summarized in Table 5.

With parameters presented in Tables 3–5, the viscoplatic model for Celgard®2400 has been established. It should be noted that the parameter values determined here are for Celgard®2400 only. They are not valid for other types of separators. Nevertheless, these values can be treated as reference values to determine the likely range of these parameters.

At this stage, the viscoplastic model parameters are only determined for the MD and TD directions. The parameters for the shear direction can be determined with the off-axis tensile test following the method presented above.

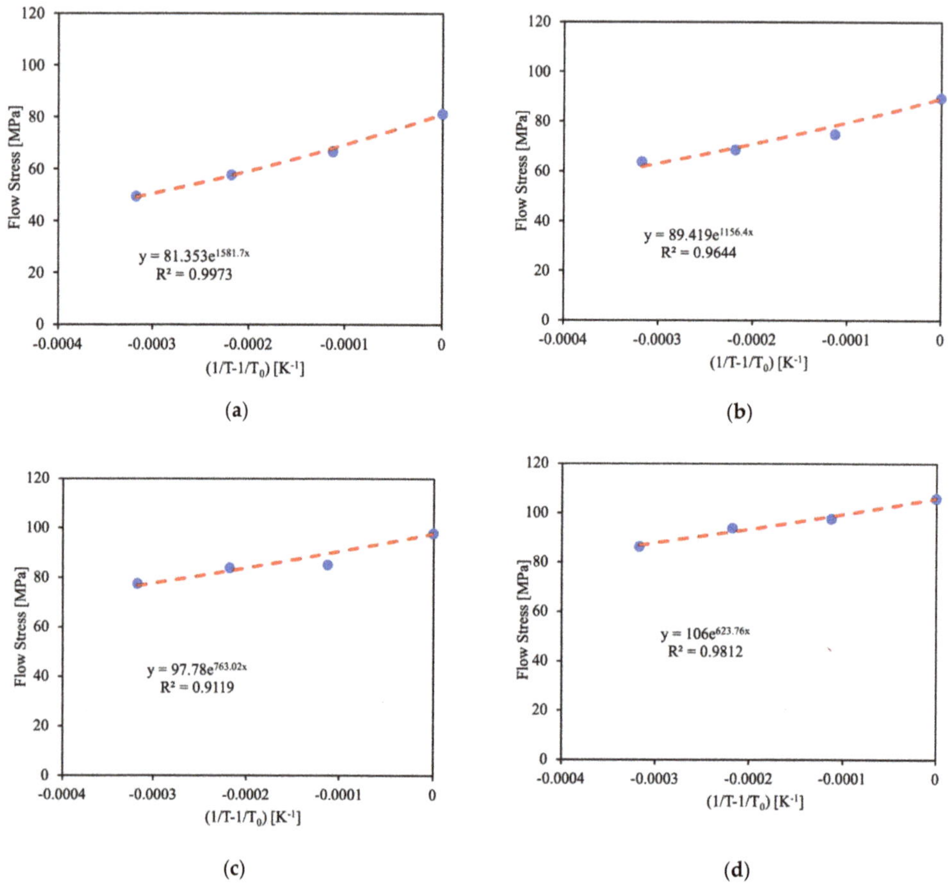

Figure 11. Least-squares regression analysis at 20% strain for the determination of parameter *a* for MD response at (**a**) 0.0001/s, (**b**) 0.001/s, (**c**) 0.01/s and (**d**) 0.1/s strain rates.

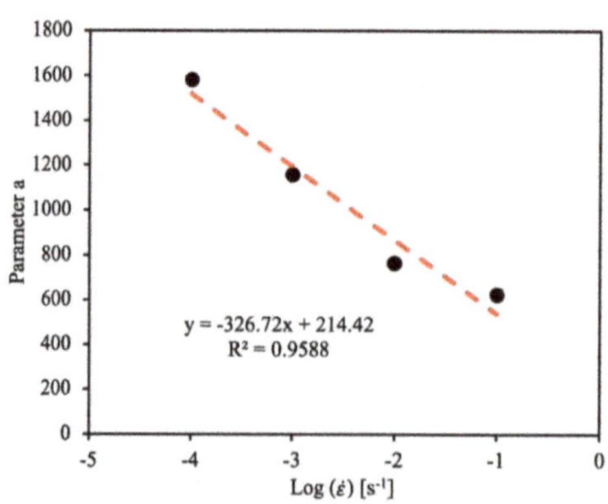

Figure 12. Linear least-squares regression fitting of parameter *a* for MD response.

Table 5. Determined parameter *a* values.

Strain Rate (s^{-1})	MD 20%	TD 20%	TD 30%	TD 40%	TD Average
0.0001	1581.70	1298.60	1293.60	1305.6	1299.27
0.001	1156.40	918.87	910.48	908.12	912.49
0.01	763.02	1144.5	1144.50	1154.30	1149.47
0.1	632.76	965.24	1063.4	1095.4	1041.35

5. Results and Discussions

5.1. Model Implementation

The fully developed model was implemented as a user-defined material model for shell element simulations in LS-DYNA®. To couple the orthotropic nonlinear viscoelastic model with the viscoplastic model, a uniaxial von Mises yield criterion was introduced to indicate the onset of the yielding mechanism. The uniaxial yield function is expressed as:

$$\Phi = \sigma^{Trial} - \sigma_{y0} \tag{13}$$

where σ^{Trial} is the viscoelastic trial stress. The onset of yielding is marked when the value of the yield function is equal to zero. Once the yield threshold is exceeded, the flow stress evolution becomes governed by the viscoplastic mechanism.

5.2. Biaxial Punch Simulations

Among various forms of biaxial loadings, the punch test has been identified as the most appropriate form of representing the out-of-plane deformation of a battery in crash scenarios [73]. In our previous work [17], punch simulations have been used to validate the orthotropic linear viscoelastic model against experimental results up to a 0.5% strain level. Unfortunately, the experimental results are unavailable at larger strains. In the current work, we compare the punch simulation results using the nonlinear viscoelastic model with the results generated using the linear model.

The FE model for the punch test is presented in Figure 13. In experiments, the separator was contained with a pair of annulus flanges with an inner opening of a 33.8 mm diameter (R = 16.9 mm). The separator was modeled with an outer radius of 40mm, the nodes at a radial distance in the interval 16.9 mm < R < 40 mm were constrained in the z-direction only, and the nodes at R = 40 mm were fully constrained. The punch was modeled as a rigid semi-sphere with a radius of 25.4 mm and given a prescribed velocity of 0.2 mm/s in the z-direction. The simulations were scaled so that 1ms corresponds to 1s in real time. The static coefficient of friction was chosen as 0.4, and the dynamic coefficient of friction was set as 0.2. The simulations were carried out at a constant temperature of 20 °C.

Figure 14 compares the force–displacement curves predicted by the two models. The two curves coincide at small displacements (d < 3 mm) and start to separate at d > 3 mm. Due to the instability issue, the simulation with the nonlinear model stopped at d = 4.67 mm. Therefore, the comparison for the punch simulation between the two models is limited to d = 4.67 mm.

To observe the differences in the material response from small to large deformations, the punch test results are compared at punch displacement d = 1.5 mm and d = 4.67 mm. These displacement values correspond to 7.5 s and 23.3 s of simulation time, respectively. The predicted stress and average strain contours at the above-mentioned punch displacements are presented in Figures 15–18.

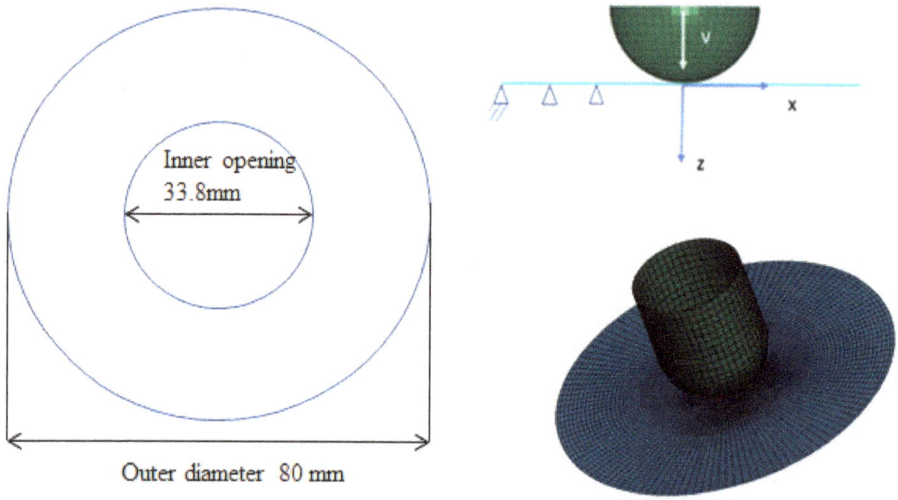

Figure 13. Punch test FE model.

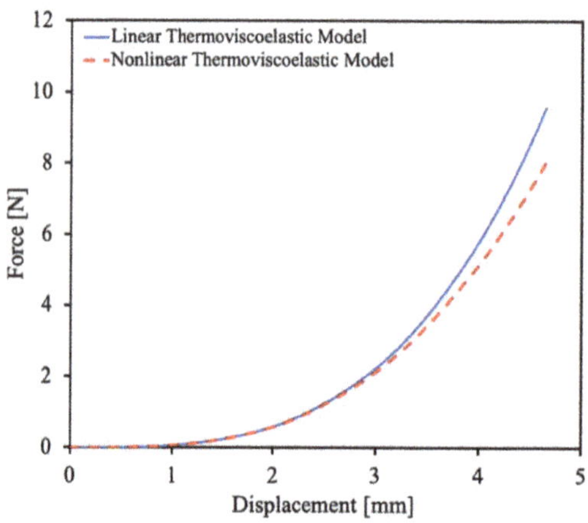

Figure 14. Force–displacement curves from punch test simulations.

Figures 15 and 16 compare the stress and strain contours from the punch test simulations by the orthotropic linear and nonlinear viscoelastic models at d = 1.5 mm. The different viscoelastic responses in the MD and TD directions observed in the experiment are captured by the two models presented here. At this stage, the stress and strain patterns predicted by the two models are indistinguishable. As expected, the nonlinear model reduces to the linear model at small deformations.

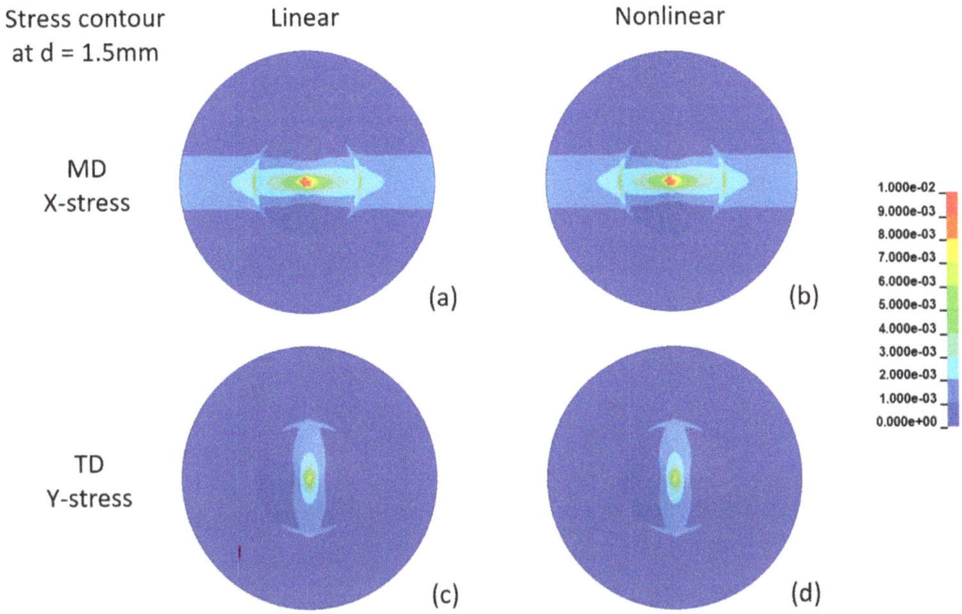

Figure 15. The MD and TD stress contours in the separator during the punch test at d = 1.5 mm, predicted with the orthotropic linear and nonlinear viscoelastic models. (**a**) MD—linear model, (**b**) MD—nonlinear model, (**c**) TD—linear model and (**d**) TD—nonlinear model.

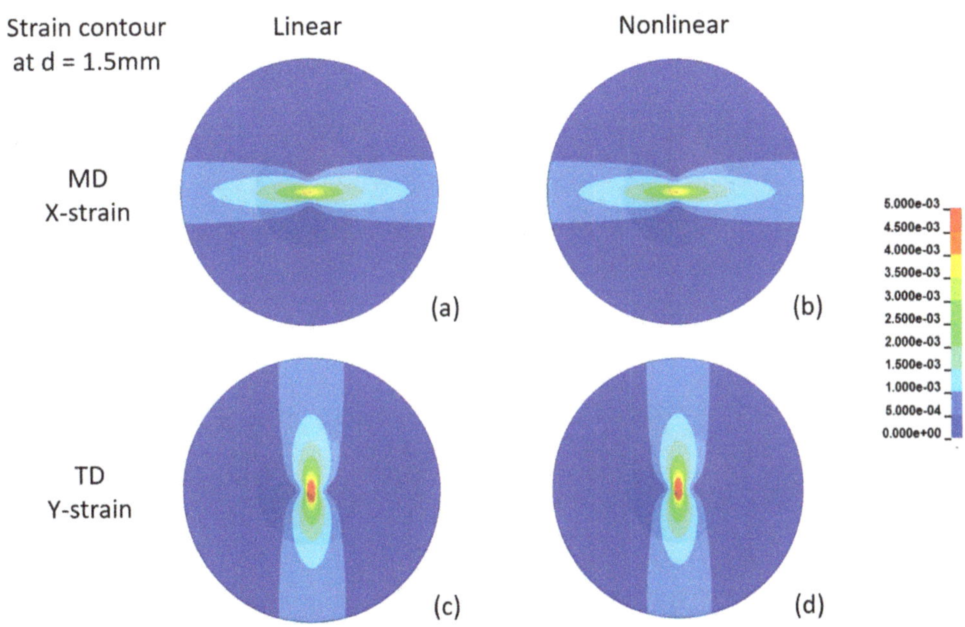

Figure 16. The MD and TD strain contours in the separator during the punch test at d = 1.5 mm, predicted with the orthotropic linear and nonlinear viscoelastic models. (**a**) MD—linear model, (**b**) MD—nonlinear model, (**c**) TD—linear model and (**d**) TD—nonlinear model.

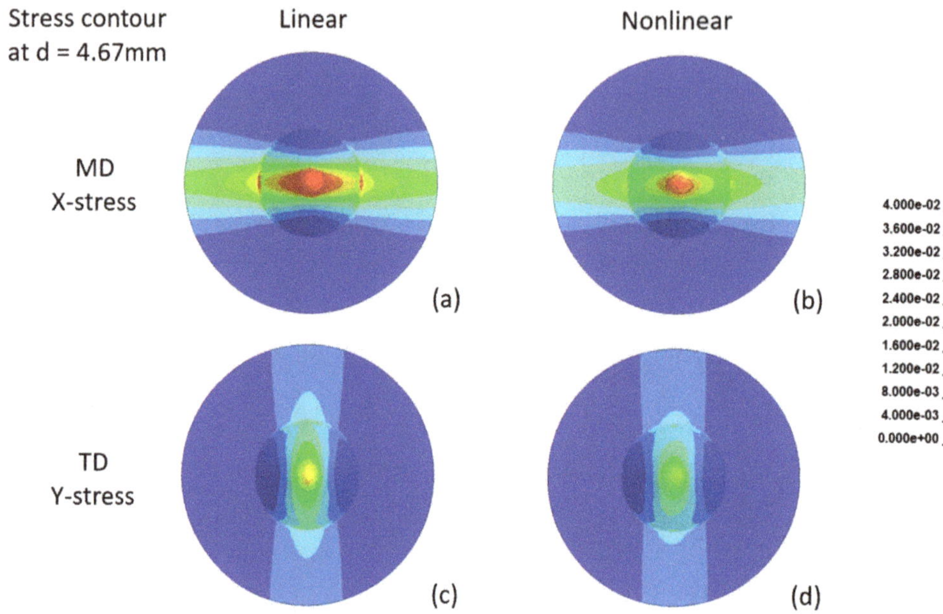

Figure 17. The MD and TD stress contours in the separator during the punch test at d = 4.67 mm, predicted with the orthotropic linear and nonlinear viscoelastic models. (**a**) MD—linear model, (**b**) MD—nonlinear model, (**c**) TD—linear model and (**d**) TD—nonlinear model.

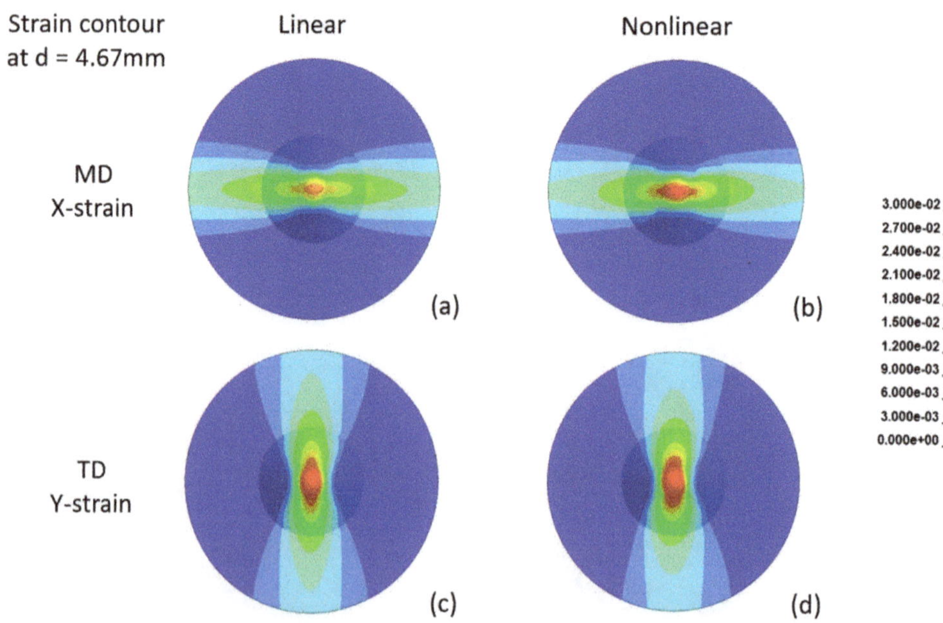

Figure 18. The MD and TD strain contours in the separator during the punch test at d = 4.67 mm, predicted with the orthotropic linear and nonlinear viscoelastic models. (**a**) MD—linear model, (**b**) MD—nonlinear model, (**c**) TD—linear model and (**d**) TD—nonlinear model.

Figure 15a,b compare the x-stress (MD) patterns. The x-stress contours display a circular shape in the area in contact with the punch and then become elliptical shapes elongated more and more along the x-axis with increasing distance from the center. Figure 15c,d compare the y-stress (TD) at d = 1.5 mm. The stress contours are ecliptics elongated along the y-axis. The value of the maximum stress in the TD is about 62% of that of the MD. Figure 16a,b compare the x-strain of the two model predictions at d = 1.5 mm. The x-strain contours display a kidney shape. The maximum x-strain from the nonlinear model is slightly higher than the value of the linear model. The y-strain contours are compared in Figure 16c,d. These contours also display a kidney shape but orthogonal to that of x-strain. The maximum y-strain is 1.62 times the x-strain. The maximum stress and strain values and the ratios of TD/MD are summarized in Table 6.

Table 6. The maximum stress and strain values in punch test simulations predicted by orthotropic linear and nonlinear viscoelastic models.

	d = 1.5 mm			d = 4.67 mm		
Maximum Stress (MPa)	**Linear Model**	**Nonlinear Model**	**Nonlinear/ Linear Ratio**	**Linear Model**	**Nonlinear Model**	**Nonlinear/ Linear Ratio**
MD	13.32	12.74	0.96	51.95	38.99	0.75
TD	8.35	7.96	0.95	31.79	24.14	0.76
TD/MD Ratio	0.63	0.62		0.61	0.62	
Maximum Strain (%)						
MD	0.43	0.44	1.02	2.63	2.98	1.14
TD	0.70	0.71	1.02	3.44	3.73	1.08
TD/MD Ratio	1.62	1.62		1.31	1.25	

Figures 17 and 18 compare the stress and strain contours predicted by the two models at d = 4.67 mm. With increasing punch displacement, the stress and strain patterns of the two models remain similar, but the difference in the maximum stress and strain values increases. Table 6 shows that at d = 4.67 mm, the TD/MD ratio for the stress remains the same, whereas the ratio for strains decreases. Table 6 also presents the ratio of nonlinear and linear model predictions. At d = 1.5 mm, the ratios of the maximum stress and strain values predicted by the two models are close to 1.0. At d = 4.67 mm, the maximum stresses predicted by the nonlinear model are about 75% of those of the linear model, whereas the maximum MD and TD strains are 114% and 108% of the linear model, respectively.

In summary, the nonlinear model reduces to the linear model at small strains and predicts a more compliant response at larger strains. This trend is as expected. The stress patterns predicted by the two models are similar, indicating that the ratio between the MD and TD stresses remains the same under biaxial loading in the tested range.

5.3. Uniaxial Tensile Tests

In this section, the model predictions are validated by comparing simulation results with uniaxial tensile test results carried out at different strain rates and temperatures for samples cut along the MD and TD. All simulations were carried out using a one-element model, with boundary conditions as shown in Figure 19, and the temperature history was applied at the nodal points of the element.

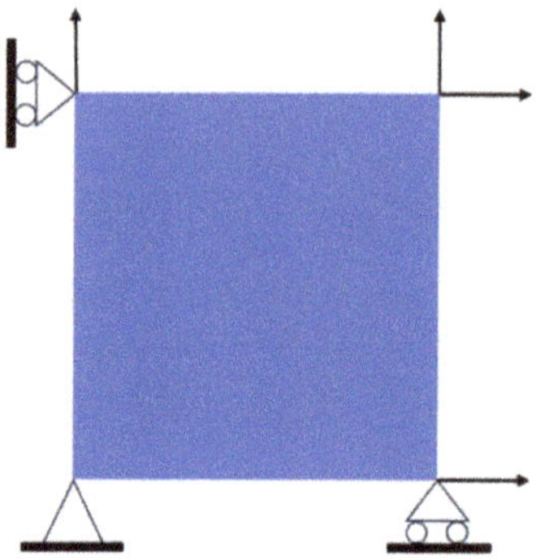

Figure 19. Shell element with boundary conditions.

Figures 20 and 21 present the comparison between the experimental and the simulation results for the uniaxial stress–strain response in the MD and TD, respectively. The results show that the stress–strain curves predicted by the coupled nonlinear viscoelastic–viscoplastic model are in good agreement with the experimental results and capture the trend in the anisotropy, rate and temperature dependence of the stress–strain response of the polymeric separator.

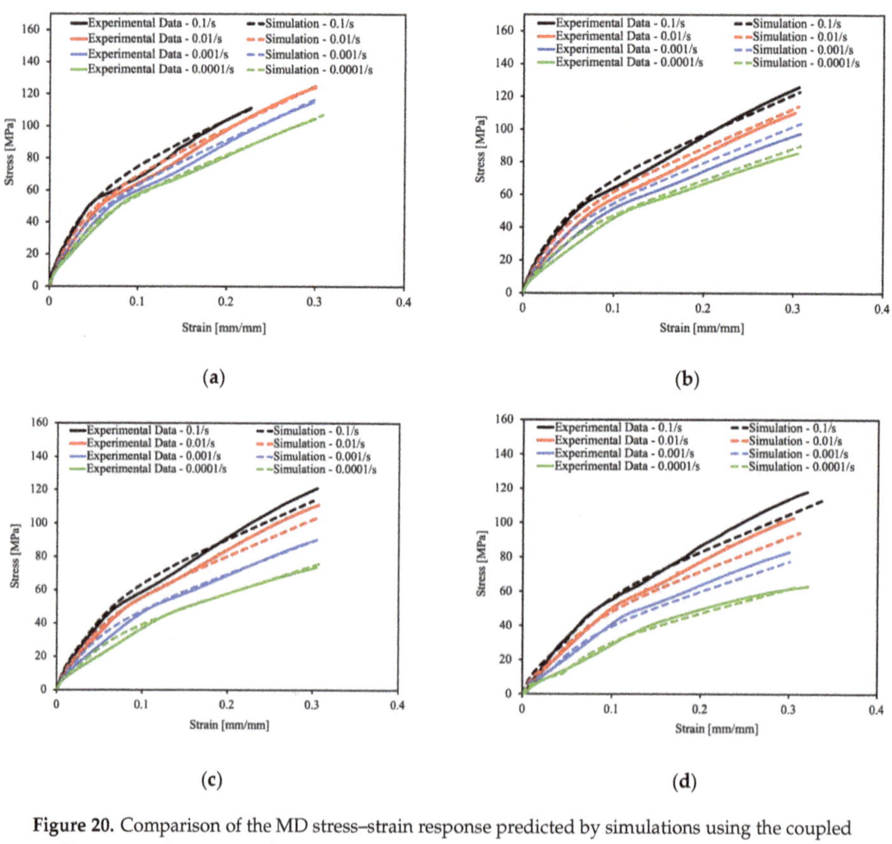

Figure 20. Comparison of the MD stress–strain response predicted by simulations using the coupled nonlinear viscoelastic–viscoplastic material model with experimental data from uniaxial tensile tests carried out at different strain rates and temperatures of (**a**) 20 °C, (**b**) 30 °C, (**c**) 40 °C and (**d**) 50 °C.

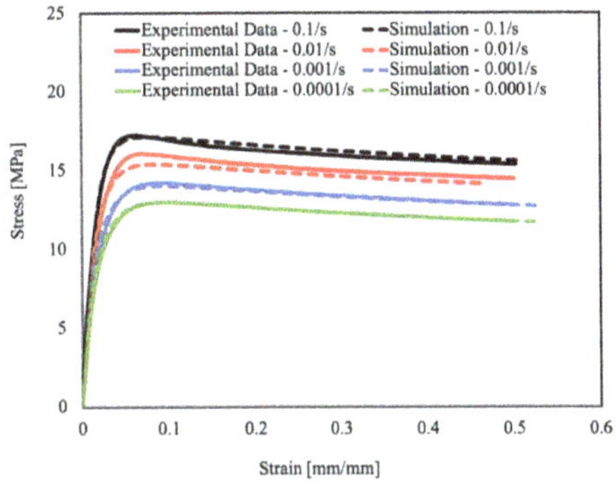

(a)

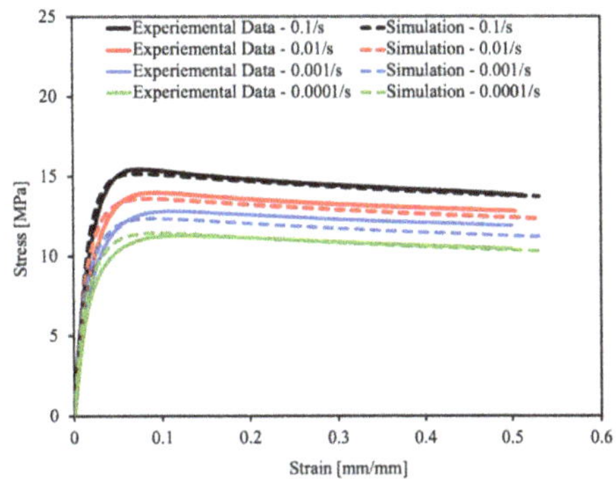

(b)

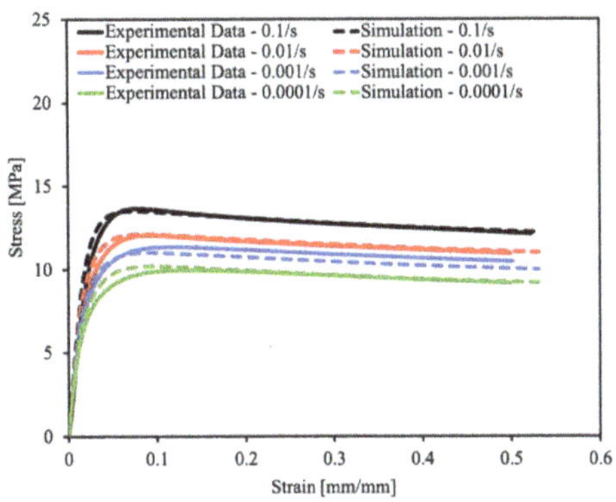

(c)

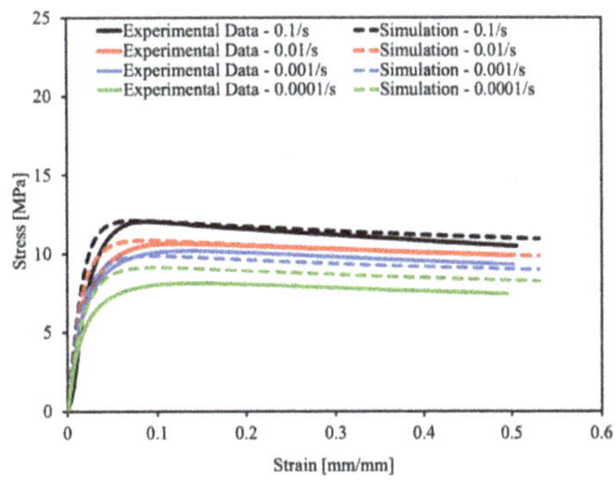

(d)

Figure 21. Comparison of the TD stress–strain response predicted by simulations using the coupled nonlinear viscoelastic–viscoplastic material model with experimental data from uniaxial tensile tests carried out at different strain rates and temperatures of (**a**) 20 °C, (**b**) 30 °C, (**c**) 40 °C and (**d**) 50 °C.

6. Conclusions

The development of a coupled viscoelastic–viscoplastic model for predicting the thermomechanical response of polymeric battery separators has been presented in this work. The viscoplastic model is based on a rheological system consisting of a sliding frictional element, representing the initial yielding mechanism, a viscoplastic dashpot to account for the changes in the viscosity of the material with the accumulation of large deformations and two springs representing the strain-softening and -hardening mechanisms. The formulation for the stress–strain response of the polymeric material after the yielding takes the form of a modified G'sell–Jonas model. The viscoplastic model was coupled with a developed orthotropic nonlinear viscoelastic model to predict the stress–strain response of the material in its deformation range before the onset of failure.

This model was implemented in LS-DYNA® as a user material model. The model parameters were determined for a PP separator. The developed model was validated against uniaxial tensile tests carried out at constant strain rates and temperatures. The results show that the model predictions of the material anisotropy, rate dependence and temperature dependence of the separator in its range of deformation before the onset failure agree well with the experimental data. Punch test simulations were also carried out to verify the nonlinear viscoelastic model predictions under biaxial loadings.

Author Contributions: Data curation, Formal analysis, Investigation, Writing—original draf, R.C.I.; Project administration, J.D. and C.B.; Funding acquisition, C.B. and X.X.; Writing—review & editing, J.D. and X.X. All authors have read and agreed to the published version of the manuscript.

Funding: This research was funded by Ford-MSU Innovation Alliance, grant number ENG26MSU.

Data Availability Statement: Data are available upon approval by Ford Motor Campany.

Conflicts of Interest: The authors declare no conflict interest.

References

1. Liu, B.; Jia, Y.; Yuan, C.; Wang, L.; Gao, X.; Yin, S.; Xu, J. Safety issues and mechanisms of lithium-ion battery cell upon mechanical abusive loading: A review. *Energy Storage Mater.* **2020**, *24*, 85–112. [CrossRef]
2. Deng, J.; Bae, C.; Denlinger, A.; Miller, T. Progress in battery safety modeling. *Prog. Energy* **2022**, *4*, 043001. [CrossRef]
3. Liu, B.; Duan, X.; Yuan, C.; Wang, L.; Li, J.; Finegan, D.P.; Feng, B.; Xu, J. Quantifying and modeling of stress-driven short-circuits in lithium-ion batteries in electrified vehicles. *J. Mater. Chem. A* **2021**, *9*, 7102–7113. [CrossRef]
4. Feng, X.; Ouyang, M.; Liu, X.; Lu, L.; Xia, Y.; He, X. Thermal runaway mechanism of lithium-ion battery for electric vehicles: A review. *Energy Storage Mater.* **2018**, *10*, 246–267. [CrossRef]
5. Yan, S.; Deng, J.; Bae, C.; He, Y.; Asta, A.; Xiao, X. In-plane orthotropic property characterization of a polymeric battery separator. *Polym. Test.* **2018**, *72*, 46–54. [CrossRef]
6. Avdeev, I.; Martinsen, M.; Francis, A. Rate-and temperature-dependent material behavior of a multilayer polymer battery separator. *J. Mater. Eng. Perform.* **2014**, *23*, 315–325. [CrossRef]
7. Xu, J.; Wang, L.; Guan, J.; Yin, S. Coupled effect of strain rate and solvent on dynamic mechanical behaviors of separators in lithium-ion batteries. *Mater. Des.* **2016**, *95*, 319–328. [CrossRef]
8. Zhu, J.; Wierzbicki, T.; Li, W. A review of safety-focused mechanical modeling of commercial lithium-ion batteries. *J. Power Sources* **2018**, *378*, 153–168. [CrossRef]
9. Zhang, X.; Sahraei, E.; Wang, K. Li-ion battery separators, mechanical integrity and failure mechanisms leading to soft and hard internal shorts. *Sci. Rep.* **2016**, *6*, 32578. [CrossRef]
10. Cannarella, J.; Liu, X.; Leng, C.Z.; Sinko, P.D.; Gor, G.Y.; Arnold, C.B. Mechanical properties of a battery separator under compression and tension. *J. Electrochem. Soc.* **2014**, *161*, F3117. [CrossRef]
11. Sheidaei, A.; Xiao, X.; Huang, X.; Hitt, J. Mechanical behavior of a battery separator in electrolyte solutions. *J. Power Sources* **2011**, *196*, 8728–8734. [CrossRef]
12. Yan, S.; Deng, J.; Bae, C.; Xiao, X. Thermal expansion/shrinkage measurement of battery separators using a dynamic mechanical analyzer. *Polym. Test.* **2018**, *71*, 65–71. [CrossRef]
13. Love, C.T. Thermomechanical analysis and durability of commercial micro-porous polymer Li-ion battery separators. *J. Power Sources* **2011**, *196*, 2905–2912. [CrossRef]
14. Chen, J.; Hu, H.; Li, S.; He, Y. Evolution of mechanical properties of polypropylene separator in liquid electrolytes for lithium-ion batteries. *J. Appl. Polym. Sci.* **2018**, *135*, 46441. [CrossRef]
15. Gor, Y.; Cannarella, J.; Leng, C.Z.; Vishnyakov, A.; Arnold, C.B. Swelling and softening of lithium-ion battery separators in electrolyte solvents. *J. Power Sources* **2015**, *294*, 167–172. [CrossRef]
16. Yan, S.; Xiao, X.; Huang, X.; Li, X.; Qi, Y. Unveiling the environment-dependent mechanical properties of porous polypropylene separators. *Polymer* **2014**, *55*, 6282–6292. [CrossRef]
17. Yan, S.; Deng, J.; Bae, C.; Kalnaus, S.; Xiao, X. Orthotropic viscoelastic modeling of polymeric battery separator. *J. Electrochem. Soc.* **2020**, *167*, 090530. [CrossRef]
18. Ihuaenyi, R.C.; Yan, S.; Deng, J.; Bae, C.; Sakib, I.; Xiao, X. Orthotropic thermo-viscoelastic model for polymeric battery separators with electrolyte effect. *J. Electrochem. Soc.* **2021**, *168*, 090536. [CrossRef]
19. Ihuaenyi, R.C.; Deng, J.; Bae, C.; Xiao, X. An orthotropic nonlinear thermoviscoelastic model for polymeric battery separators. *J. Electrochem. Soc.* **2023**, *170*, 010520. [CrossRef]
20. Zhao, G.; Xi, H.; Yang, J. Transversely Isotropic Constitutive Model of the Polypropylene Separator Based on Rich–Hill Elastoplastic Constitutive Theory. *J. Electrochem. Energy Convers. Storage* **2021**, *18*, 020911. [CrossRef]
21. Zhang, X.; Sahraei, E.; Wang, K. Deformation and failure characteristics of four types of lithium-ion battery separators. *J. Power Sources* **2016**, *327*, 693–701. [CrossRef]
22. Xie, W.; Wu, L.; Liu, W.; Dang, Y.; Tang, A.; Luo, Y. Modelling electrolyte-immersed tensile property of polypropylene separator for lithium-ion battery. *Mech. Mater.* **2021**, *152*, 103667. [CrossRef]
23. Makki, M.; Ayoub, G.; Lee, C.W. Modeling the anisotropic behavior of highly orthotropic lithium-ion batteries polymer separators. *Int. J. Solids Struct.* **2023**, *264*, 112102. [CrossRef]
24. Findley, W.N.; Lai, J.S.; Onaran, K. *Creep and Relaxation of Nonlinear Viscoelastic Materials*; Dover Publications, Inc.: New York, NY, USA, 1976.
25. Schapery, R.A. Nonlinear viscoelastic and viscoplastic constitutive equations based on thermodynamics. *Mech. Time-Depend. Mater.* **1997**, *1*, 209–240. [CrossRef]

26. Schapery, R.A. Nonlinear viscoelastic and viscoplastic constitutive equations with growing damage. *Int. J. Fract.* **1999**, *97*, 33–66. [CrossRef]
27. Lai, J.; Bakker, A. An integral constitutive equation for nonlinear plasto-viscoelastic behavior of high-density polyethylene. *Polym. Eng. Sci.* **1995**, *35*, 1339–1347. [CrossRef]
28. Brusselle-Dupend, N.; Lai, D.; Feaugas, X.; Guigon, M.; Clavel, M. Mechanical behavior of a semicrystalline polymer before necking. Part I: Characterization of uniaxial behavior. *Polym. Eng. Sci.* **2001**, *41*, 66–76. [CrossRef]
29. Brusselle-Dupend, N.; Lai, D.; Feaugas, X.; Guigon, M.; Clavel, M. Mechanical behavior of a semicrystalline polymer before necking. Part II: Modeling of uniaxial behavior. *Polym. Eng. Sci.* **2003**, *43*, 501–518. [CrossRef]
30. Kletschkowski, T.; Schomburg, U.; Bertram, A. An endochronic viscoplastic approach for materials with different behavior in tension and compression. *Mech. Time-Depend. Mater.* **2004**, *8*, 119–135. [CrossRef]
31. Zhang, C.; Moore, I.D. Nonlinear mechanical response of high-density polyethylene. Part I: Experimental investigation and model evaluation. *Polym. Eng. Sci.* **1997**, *37*, 404–413. [CrossRef]
32. Zhang, C.; Moore, I.D. Nonlinear mechanical response of high-density polyethylene. Part II: Uniaxial constitutive modeling. *Polym. Eng. Sci.* **1997**, *37*, 414–420. [CrossRef]
33. Peric, D.; Dettmer, W. A computational model for generalized inelastic materials at finite strains combining elastic, viscoelastic and plastic material behavior. *Eng. Comput.* **2003**, *20*, 768–787. [CrossRef]
34. Wang, W.M.; Sluys, L.J.; Borst, R.R. Viscoplasticity for instabilities due to strain softening and strain-rate softening. *Int. J. Numer. Methods Eng.* **1997**, *40*, 3839–3864. [CrossRef]
35. Ristinmaa, M.; Ottosen, N.S. Consequences of dynamic yield surface in viscoplasticity. *Int. J. Solids Struct.* **2000**, *37*, 4601–4622. [CrossRef]
36. Carosio, A.; Willam, K.; Etse, G. On the consistency of viscoplastic formulations. *Int. J. Solids Struct.* **2000**, *37*, 7349–7369. [CrossRef]
37. Heeres, O.M.; Suiker, A.S.J.; De Borst, R. A comparison between the Perzyna viscoplastic model and the consistency viscoplastic model. *Eur. J. Mech. A/Solids* **2002**, *21*, 1–12. [CrossRef]
38. Zaera, R.; Fernandez-Saez, J. An implicit consistent algorithm for the integration of thermoviscoplastic constitutive equations in adiabatic conditions and finite deformations. *Int. J. Solids Struct.* **2006**, *43*, 1594–1612. [CrossRef]
39. Perzyna, P. Fundamental Problems in Viscoplasticity. *Adv. Appl. Mech.* **1966**, *9*, 243–377.
40. Duvaut, G.; Lions, J.L. *Les Inequations en Mecanique et en Physique*; Dunod: Paris, France, 1972.
41. Zienkiewicz, O.C.; Cormeau, I.C. Visco-Plasticity-Plasticity and Creep in Elastic Solids—A unified numerical solution approach. *Int. J. Numer. Methods Eng.* **1974**, *8*, 821–845. [CrossRef]
42. Hughes, T.J.; Taylor, R.L. Unconditionally stable algorithms for quasi-static elasto/visco-plastic finite element analysis. *Comput. Struct.* **1978**, *8*, 169–173. [CrossRef]
43. Cormeau, I. Numerical stability in quasistatic elasto/viscoplasticity. *Int. J. Numer. Methods Eng.* **1975**, *9*, 109–127. [CrossRef]
44. Simo, J.C.; Kennedy, J.G.; Govindjee, S. Non-smooth multisurface plasticity and viscoplasticity. Loading/unloading conditions and numerical algorithms. *Int. J. Numer. Methods Eng.* **1988**, *26*, 2161–2185. [CrossRef]
45. Chaboche, J.L. Constitutive equations for cyclic plasticity and cyclic viscoplasticity. *Int. J. Plast.* **1989**, *5*, 247–302. [CrossRef]
46. Peric, D. On a class of constitutive equations in viscoplasticity: Formulation and computational issues. *Int. J. Numer. Methods Eng.* **1993**, *36*, 1365–1393. [CrossRef]
47. Ristinmaa, M.; Ottosen, N.S. Viscoplasticity based on an additive split of the conjugated forces. *Eur. J. Mech. A/Solids* **1998**, *17*, 207–235. [CrossRef]
48. Runesson, K.; Ristinmaa, M.; Mahler, L. Comparison of viscoplasticity formats and algorithms. *Mech. Cohesive-Frict. Mater.* **1999**, *4*, 75–98. [CrossRef]
49. Caggiano, A.; Martinelli, E.; Schicchi, D.S.; Etse, G. A modified Duvaut-Lions zero-thickness interface model for simulating the rate-dependent bond behavior of FRP-concrete joints. *Compos. Part B Eng.* **2018**, *149*, 260–267. [CrossRef]
50. Ibrahimbegovic, A.; Chorfi, L. Viscoplasticity model at finite deformations with combined isotropic and kinematic hardening. *Comput. Struct.* **2000**, *77*, 509–525. [CrossRef]
51. Nedjar, B. Frameworks for finite strain viscoelastic-plasticity based on multiplicative decompositions. Part I: Continuum formulations. *Comput. Methods Appl. Mech. Eng.* **2002**, *191*, 1541–1562. [CrossRef]
52. Shutov, V.; Kreißig, R. Finite strain viscoplasticity with nonlinear kinematic hardening: Phenomenological modeling and time integration. *Comput. Methods Appl. Mech. Eng.* **2008**, *197*, 2015–2029. [CrossRef]
53. Kowalczyk-Gajewska, K.; Pieczyska, E.A.; Golasinski, K.; Maj, M.; Kuramoto, S.; Furutab, T. A finite strain elastic-viscoplastic model of Gum Metal. *Int. J. Plast.* **2019**, *119*, 85–101. [CrossRef]
54. G'sell, C.; Jonas, J.J. Determination of the plastic behaviour of solid polymers at constant true strain rate. *J. Mater. Sci.* **1979**, *14*, 583–591. [CrossRef]
55. Morin, D.; Haugou, G.; Lauro, F.; Bennani, B.; Bourel, B. Elasto-viscoplasticity behaviour of a structural adhesive under compression loadings at low, moderate and high strain rates. *J. Dyn. Behav. Mater.* **2015**, *1*, 124–135. [CrossRef]
56. Múgica, J.I.; Aretxabaleta, L.; Ulacia, I.; Aurrekoetxea, J. Rate-dependent phenomenological model for self-reinforced polymers. *Compos. Part A Appl. Sci. Manuf.* **2016**, *84*, 96–102. [CrossRef]

57. Schoßig, M.; Bierögel, C.; Grellmann, W.; Mecklenburg, T. Mechanical behavior of glass-fiber reinforced thermoplastic materials under high strain rates. *Polym. Test.* **2008**, *27*, 893–900. [CrossRef]
58. Torres, J.P.; Frontini, P.M.; Aretxabaleta, L. Experimental characterization and computational simulations of the low-velocity impact behaviour of polypropylene. *Polym. Int.* **2013**, *62*, 1553–1559. [CrossRef]
59. Johnson, R.; Cook, W.H. A constitutive model and data for materials subjected to large strains, high strain rates, high temperatures. In Proceedings of the 7th International Symposium on Ballistics, Hague, The Netherlands, 19–21 April 1983; pp. 541–547.
60. Chen, F.; Ou, H.; Lu, B.; Long, H. A constitutive model of polyether-ether-ketone (PEEK). *J. Mech. Behav. Biomed. Mater.* **2016**, *53*, 427–433. [CrossRef]
61. Garcia-Gonzalez, D.; Rusinek, A.; Jankowiak, T.; Arias, A. Mechanical impact behavior of polyether–ether–ketone (PEEK). *Compos. Struct.* **2015**, *124*, 88–99. [CrossRef]
62. Nasraoui, M.; Forquin, P.; Siad, L.; Rusinek, A. Influence of strain rate, temperature and adiabatic heating on the mechanical behaviour of poly-methyl-methacrylate: Experimental and modelling analyses. *Mater. Des.* **2012**, *37*, 500–509. [CrossRef]
63. Duan, Y.; Saigal, A.; Greif, R.; Zimmerman, M.A. A uniform phenomenological constitutive model for glassy and semicrystalline polymers. *Polym. Eng. Sci.* **2001**, *41*, 1322–1328. [CrossRef]
64. Zhu, H.; Ou, H.; Popov, A. A new phenomenological constitutive model for thermoplastics. *Mech. Mater.* **2021**, *157*, 103817. [CrossRef]
65. Ferry, J.D. *Viscoelastic Properties of Polymers*; John Wiley & Sons: Hoboken, NJ, USA, 1980.
66. Ward, I.M.; Sweeney, J. *An Introduction to the Mechanical Properties of Solid Polymers*; John Wiley & Sons: Hoboken, NJ, USA, 2004.
67. Brinson, H.F.; Brinson, L.C. *Polymer Engineering Science and Viscoelasticity. An Introduction*; Springer: New York, NY, USA, 2008; pp. 99–157.
68. Wineman, A.S.; Rajagopal, K.R. *Mechanical Response of Polymers: An Introduction*; Cambridge University Press: Cambridge, UK, 2000.
69. Christensen, R.M. *Theory of Viscoelasticity: An Introduction*, 2nd ed.; Academic Press: New York, NY, USA, 1982.
70. Schapery, R.A. On the characterization of nonlinear viscoelastic materials. *Polym. Eng. Sci.* **1969**, *9*, 295–310. [CrossRef]
71. Daniel, I.M.; Ishai, O. *Engineering Mechanics of Composite Materials*; Oxford University Press: Oxford, UK, 2006.
72. Williams, M.L.; Landel, R.F.; Ferry, J.D. The temperature dependence of relaxation mechanisms in amorphous polymers and other glass-forming liquids. *J. Am. Chem. Soc.* **1955**, *77*, 3701–3707. [CrossRef]
73. Kalnaus, S.; Kumar, A.; Wang, Y.; Li, J.; Simunovic, S.; Turner, J.A.; Gorney, P. Strain distribution and failure mode of polymer separators for Li-ion batteries under biaxial loading. *J. Power Sources* **2018**, *378*, 139–145. [CrossRef]

batteries

MDPI

Article

Stress Distribution Inside a Lithium-Ion Battery Cell during Fast Charging and Its Effect on Degradation of Separator

Mustapha Makki, Cheol W. Lee * and Georges Ayoub

Department of Industrial and Manufacturing Systems Engineering, University of Michigan-Dearborn, Dearborn, MI 48128, USA
* Correspondence: cheol@umich.edu

Abstract: The automotive industry is rapidly transitioning to electric vehicles (EVs) in response to the global efforts to reduce greenhouse gas emissions. Lithium-ion battery (LIB) has emerged as the main tool for energy storage in electric vehicles. A widespread adoption of EVs, however, requires a fast-charging technology that can significantly reduce charging time while avoiding any unsafe conditions including short circuits due to failure of the separator in an LIB cell. Therefore, it is necessary to understand the mechanical stresses during fast charging and their long-term effect on the integrity of the separator. This paper presents a novel hybrid model for the prediction of the stress distribution in the separator of a pouch cell under various charging speeds, ambient temperatures, and pack assembly conditions, such as compressive pressures. The proposed hybrid model consists of three sub-models, namely, an electrochemical cell model, a lumped-parameter model, and a solid mechanics model. A robust parameter identification scheme is implemented to determine the model parameters using the experimental data. The separator within the test setup will experience maximum von Mises stress of 74 MPa during 4C charging, i.e., when the charge current in A is four times as high as the capacity of the battery cell in Ah. To assess the evolution of the damage in the separator under the estimated stress during fast charging, creep and fatigue tests are conducted on the separator. Their results indicate a progressive accumulation of damage in the separator, further emphasizing the importance of understanding and mitigating mechanical degradation in separator materials.

Keywords: hybrid model; creep and fatigue; parameter identification

1. Introduction

Due to the volatility of oil markets and strict mandates to limit greenhouse gas emissions, the automotive industry has been swiftly switching towards producing electrical vehicles as well as hybrid electric vehicles (HEVs). As the market share of EVs increases, there will be a greater demand for high efficiency energy storage systems such as lithium-ion batteries (LIBs). This is partly due to the high specific energy and energy density of LIBs, which can be two times those of other cell technologies such as nickel-metal hydride [1].

An LIB cell is composed of four main components, i.e., anode, cathode, separator, and electrolyte [2]. In an LIB cell, a critical component is the separator, which is a porous membrane positioned between the electrodes. The essential function of the separator is to prevent direct contact between the electrodes or an internal short while allowing ionic flows through the electrolyte [3]. The separator, therefore being considered the most crucial component for safety, should meet several requirements in regard to its thermal and mechanical properties. First, the separator should be chemically stable under the reductive and oxidative conditions encountered during battery operations [4]. Second, the separator should exhibit low thermal shrinkage, e.g., in an amount of less than 5% after being soaked at 90 °C for 60 min [5]. This requirement exists to avoid any exposure of electrode edges due to shrinkage of the separator at high temperatures, thereby preventing internal shorts and thermal runaways. Moreover, the separator should melt above an elevated temperature to avoid a thermal runaway when the cell is exposed to high temperatures. A shutdown temperature of 130 °C or lower is usually required. Lastly, the separator should be mechanically robust to withstand the mechanical stresses during cell and pack assembly, operations, and any other abuse conditions [6]. Both tensile and puncture strengths of the separator should be high enough to avoid internal shorts under such conditions [7]. A gradual and up to 50% decrease in ductility of the separator was observed after the battery cell underwent 400, 800, and 1600 charge–discharge cycles involving 4C-rate charging, i.e., charging at a rate with current that is four times as high as the battery capacity [8]. A scanning electron micrograph (SEM) analysis of the cycled separators revealed accumulated damage in the separator in the form of pore closure along the thickness direction and fiber fracture and cracks along the machine direction. The degradation was attributed to the damage caused by mechanical creep which can accumulate over the cycles. Given its integral role in battery safety and functionality, an accurate characterization and modeling of the stress distribution in the separator is necessary for addressing the safety concerns under fast-charging conditions.

During battery operation, LIBs are subject to electrochemical processes, thermal energy, and mechanical stresses that may lead to safety concerns [9]. Both the electrochemical process and heat generation induce mechanical loads in the battery components. The former process induces a volumetric change of the active material due to the intercalation processes of lithium-ions, while the latter causes thermal expansion of cell components [10]. Both mechanical loads can squeeze the separator [11], which can lead to shrinkage of its pores, reduced ion transfer, and increase in internal resistance [12]. However, no practical method exists for in situ measurement of the stress inside the battery. Multiphysics modeling based on a finite element method is often adopted instead to predict the stress distribution. Xiao et al. [13] proposed a multiphysics and multiscale model for prediction of the stress in a polymeric separator by coupling a 1-dimensional (1-D) macroscopic battery model and a meso-scale battery sub-model. Shi et al. [14] built a multiphysics cell model for a pouch cell to predict the stress in the separator. A 1-D thermo-electrochemical cell model calculates lithium concentration and generates heat while a macroscale 2-D model of a pouch cell determines its stress distributions. However, such cell models by themselves lack the capability to predict the stress inside a cell according to its external conditions imposed by the battery pack. A battery cell will be subject to varying compressive pressure and ambient thermal conditions according to both design and use parameters of the battery pack. Swelling of a battery cell during charging and discharging and varying compressive pressure when the cell is constrained inside a battery pack are often modeled by a lumped-parameter modeling approach. Oh et al. [15] developed a thermal swelling model to predict the swelling in the battery due to temperature variation. In their other works, Oh et al. [16] used a phenomenological force and swelling model to predict the dynamic swelling in the battery. Xu et al. [17] developed a five Kelvin-Voigt elements model to characterize the viscoelastic behavior of the separator. This model enables the prediction of creep for two specific commercial separators. In their additional research, Xu et al. [18] formulated an anisotropic homogeneous model for predicting the mechanical behavior of an LIB under bending, indentation, and compression loads. Moreover, the numerical model took into account the state of charge dependence on the mechanical behavior of the lithium-ion battery (LIB). Liu et al. [19] created a multiphysics model that captures the mechanical behavior of cylindrical cells. The model also addresses short-circuit and thermal runaway behaviors. Duan et al. [20] developed a 2-D coupled multiphysics model to identify various internal short circuit modes. The model is capable of describing the entire evolution process of the battery starting from the initial deformation, triggering of internal short circuit, evolution of internal short circuit, up to the eventual thermal runaway. Yuan et al. [21] developed a generalized short-circuit criterion for two common failure modes, using volumetric strain and equivalent strain of the separator as key parameters. In this paper, a computationally efficient hybrid model is developed by linking the cell model in [14] with a lumped-parameter model to account for the pack parameters in addition, thereby allowing optimal design of both cell and pack parameters for safety.

The proposed hybrid model consists of three main components: (1) a 1-D electrochemical model, (2) a lumped-parameter model (LPM), and (3) a 2-D solid mechanics model. The 1-D electrochemical model predicts the heat generated in the battery cell and lithium-ion concentration which are input to the LPM. The LPM represents the mechanical and thermal behaviors of a battery cell in a battery pack array. The predicted lithium-ion concentration and temperature are input to the solid mechanics model for prediction of the stress in the separator. The LPM parameters are determined via non-linear regression based on the temperature, mechanical load, and displacement data collected from a fast charging and discharging test on an LIB cell. The hybrid model, thus developed, allows calculating the stress in the separator during fast charging. New insights into damage accumulation in the separator under fast cyclic charging are obtained by performing creep and fatigue tests on the separator under the predicted stress.

This paper is organized as follows: Section 2 describes the hybrid model proposed in this study. Then, Section 3 provides detailed explanation on the material properties of the tested LIB cell and the testing setup. Section 4 presents the experimental results along with the parameter identification process. In Section 5, a simulation study is conducted using the developed model to check the effect of charging rate, pack assembly condition, and ambient temperature on the stress induced in the separator. Section 6 focuses on presenting the results of creep and fatigue tests conducted under the predicted stress during fast charging. Finally, the conclusion is presented in Section 7.

2. Hybrid Model

During a charge and discharge cycle, an LIB cell is subject to electrochemical, thermal, and mechanical phenomena occurring simultaneously. Each phenomenon is described by a set of equations and coupled with others. The framework of the proposed hybrid model is divided into three components. The first component is a 1-D model for predicting transport of lithium-ions, electrochemical reactions, and generated heat given charge and discharge rates. The 1-D model is composed of three domains of anode, separator, and cathode. The second model component is a lumped-parameter model (LPM) that will predict the load, displacement, and temperature in the battery cell and its fixture. This model is composed of three parts: a thermal model for heat transfer, a spring model for the mechanical forces, and a linking equation which links the thermal model to the spring model. The LPM takes both predicted heat and state of charge (SOC) from the 1-D electrochemical model as inputs. Additional input variables of the LPM include ambient temperature and pre-torque on the fixture. The third component is a 2-D solid mechanics model which will calculate the stress distribution in the separator given (1) cell core temperature and boundary pressure predictions from the LPM and (2) lithium concentration in the electrodes as predicted by the 1-D battery model. Figure 1 shows all three components with the input/output variables. In what follows, a detailed description of each model component is provided.

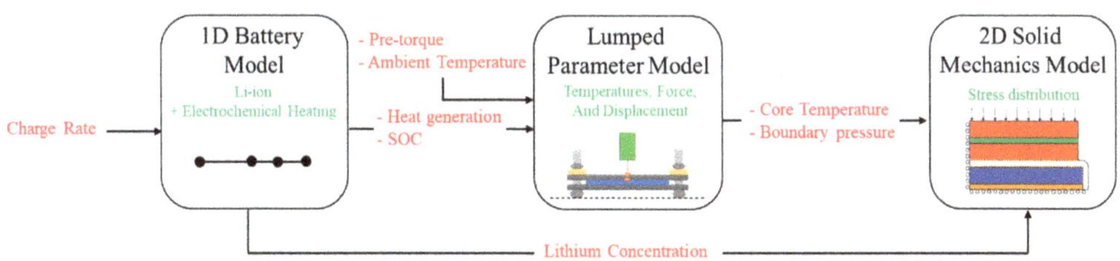

Figure 1. Flowchart of the hybrid model.

2.1. One-Dimensional (1-D) Battery Model

The 1-D battery model, shown in Figure 2, accounts for the electrochemical reaction and heat generation in a battery cell. An LIB cell Li-ion is composed of three components placed in the electrolyte, namely, cathode, anode, and separator.

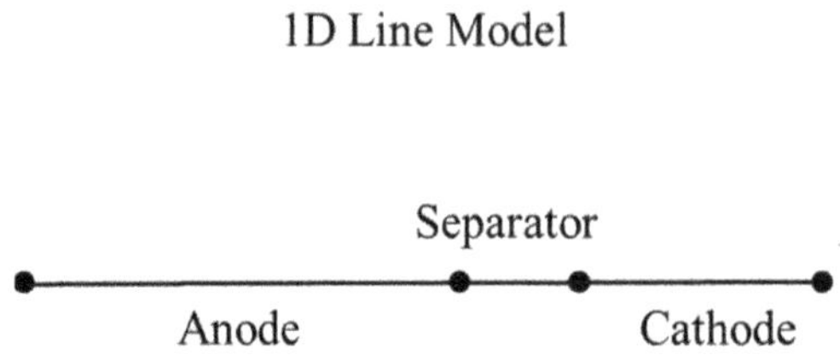

Figure 2. One-dimensional (1-D) electrochemical model.

The volume fraction of the electrodes is defined by:

$$\varepsilon_l^i = 1 - \varepsilon_s^i - \varepsilon_{filler}^i \tag{1}$$

where ε_s^i is the active material volume fraction, ε_l^i is the electrolyte volume fraction (porosity), and ε_{filler}^i denotes the binder and filler volume fraction. The superscript i denotes the electrode under consideration (p for positive electrode and n for negative electrode). In what follows, the governing equations for the diffusion, kinetics, and charge conservation [22,23] are summarized for the solid and electrolyte phases.

2.1.1. Solid Phase

The following equations apply to the solid phase of the anode and cathode. The active material in the electrodes is assumed to be composed of spherical particles. The Fick's second law describes the transport in the solid electrode particles and predicts the change of concentration field with time in spherical coordinates:

$$\frac{\partial c_s^i}{\partial t} = \frac{D_s^i}{r^2}\frac{\partial}{\partial r}\left(r^2\frac{\partial c_s^i}{\partial r}\right) \tag{2}$$

where r is the particle radius coordinate, c_s^i is the lithium concentration in the solid particle phase, and D_s^i is the diffusion coefficient of lithium-ion in solid electrode particles. The equation assumes a constant diffusion coefficient of lithium-ions in the solid phase. According to the symmetry condition, the first boundary condition yields $-D_s^i\frac{\partial c_s^i}{\partial r}\Big|_{r=0} = 0$ since there is no flux at the center of the particle. The second boundary condition relates the pore wall flux across the interface and the rate of diffusion of lithium-ions into the solid/liquid interface, yielding $-D_s^i\frac{\partial c_s^i}{\partial r}\Big|_{r=r_p} = J^i$, where r_p is the particle radius and J^i is the pore wall flux of lithium-ions defined through the Butler–Volmer equation (Equation (6)).

The conservation of charge in the solid phase is governed by Ohm's law:

$$\frac{\partial}{\partial x}\left(\tau_{eff}^i\frac{\partial \phi_s^i}{\partial x}\right) = a_v F_r J^i \tag{3}$$

where τ_{eff}^i is the effective conductivity determined by $\tau_{eff}^i = {\varepsilon_s^i}^{1.5}\tau_s^i$, ϕ_s^i is the potential in the solid phase, a_v is the particle specific surface area, and F_r denotes the Faraday's constant. The first boundary condition applies to the interface between the current collector and the positive electrode, where the charge flux is equal to the current density I_{app}, i.e., $-\tau_{eff}^i\frac{\partial \phi_s^i}{\partial x}\Big|_{x=0} = I_{app}$. The second boundary condition exists at the interface between the electrodes and separators which corresponds to no charge flux:

$-\tau^{i}_{eff}\frac{\partial\phi^{i}_{s}}{\partial x}\Big|_{x=L_N} = -\tau^{i}_{eff}\frac{\partial\phi^{i}_{s}}{\partial x}\Big|_{x=L_N+L_{sep}} = 0$, where L_N and L_{sep} are the negative electrode and separator thicknesses, respectively.

2.1.2. Liquid Phase

The following equations apply to the electrolyte in the electrodes and separator. To account for lithium species transport in the liquid, the concentrated solution theory is adopted [22,24,25]. The material balance for a salt in liquid phase is given by:

$$\varepsilon^{v}_{l}\frac{\partial c^{v}_{l}}{\partial t} = \frac{\partial}{\partial x}\left(D^{i}_{eff,l}\frac{\partial c^{v}_{l}}{\partial x}\right) + (1-t_{+})a_{v}J^{i} \tag{4}$$

where ε^{v}_{l} denotes the electrolyte volume fraction in the component v, c^{v}_{l} is the lithium concentration in the electrolyte, $D^{i}_{eff,l}$ is the effective diffusion coefficient in the electrolyte given by $D^{i}_{eff,l} = \varepsilon^{i\,1.5}_{l}D^{i}_{l}$, and t_{+} is the transport number. The superscript v denotes the component under consideration (p for cathode, n for anode, and sep for separator). For the separator, the pore wall flux is set to zero. No mass flux occurs at the two ends of the cell, hence $\frac{\partial c^{v}_{l}}{\partial x}\Big|_{x=0} = 0$ and $\frac{\partial c^{v}_{l}}{\partial x}\Big|_{x=L} = 0$.

The charge balance in the liquid phase is based on Ohm's law:

$$-\frac{\partial}{\partial x}\left(\kappa_{eff,i}\frac{\partial\phi^{i}_{s}}{\partial x}\right) + \frac{2RT(1-t_{+})}{F}\frac{\partial}{\partial x}\left(\kappa_{eff,i}\frac{\partial(\ln c^{v}_{l})}{\partial x}\right) = a_{v}F_{r}J^{i} \tag{5}$$

where $\kappa_{eff,i}$ is the effective conductivity, φ_{l} is the potential in the liquid phase, R is the ideal gas constant, and T is the temperature. At the two ends of the cell, there is no current flow in the liquid phase, i.e., $\frac{\partial\phi^{i}_{l}}{\partial x}\Big|_{x=0} = \frac{\partial\phi^{i}_{l}}{\partial x}\Big|_{x=L} = 0$.

2.1.3. Charge Transfer

The pore wall flux at the interface between the electrode and electrolyte is determined by the Butler–Volmer equation which describes the relationship between the electrical current, electrode potential, and local concentration at the interface. Specifically, the rate of transfer of lithium-ions from the electrolyte to an electrode is controlled by deviation of the surface potential from its equilibrium value as follows:

$$J^{i} = a_{v}i_{ex}\left(\exp\left(\frac{\alpha_{a}F_{r}\eta^{i}}{RT}\right) - \exp\left(\frac{-\alpha_{c}F_{r}\eta^{i}}{RT}\right)\right) \tag{6}$$

where i_{ex} is the exchange current density, η_{i} is the overpotential of the intercalation reaction, and α_{a} and α_{c} are the charge transfer coefficients for oxidation and reduction, respectively, which are assumed to be 0.5. Both the forward rate of the anodic process and the backward rate of the cathodic process are given in Equation (6). The exchange current density is given by:

$$i_{ex} = F_{r}k_{i}\left(\frac{c^{i}_{l}}{c_{l,ref}}\right)^{\alpha_{c}}\left(c^{i}_{s,max} - c^{i}_{s,surf}\right)^{\alpha_{c}}\left(c^{i}_{s,surf}\right)^{\alpha_{a}} \tag{7}$$

where k_{i} is the reaction rate constant, c^{i}_{l} is the concentration of lithium-ions in the electrolyte, $c^{i}_{s,surf}$ is the concentration of lithium-ions on the surface of the solid electrode, and $c^{i}_{s,max} - c^{i}_{s,surf}$ is the available space for lithium intercalation.

The overpotential is given by:

$$\eta^{i} = \phi^{i}_{s} - \phi^{i}_{l} - U^{i} \tag{8}$$

where ϕ_s^i is the potential of the solid electrode, ϕ_l^i is the potential of the electrolyte, and U^i is the open circuit potential.

2.1.4. Heat Generation

Two main mechanisms of heat generation are considered in this study, namely, reversible entropy heat and irreversible joule heat. The corresponding equation for heat generation rate, Q_{gen}, is as follows:

$$Q_{gen} = \left(\phi_s^i - \phi_l^i - U^i + T\frac{\partial U^i}{\partial T}\right)J \tag{9}$$

where $(\phi_s - \phi_l - U_i)J$ represents the joule heat while $T\frac{\partial U_i}{\partial T}J$ represents the entropy heat.

2.2. Lumped-Parameter Model

In a lumped-parameter model, a battery cell is represented based on equivalent material properties such as thermal conductivity and specific heat [15,26,27]. In addition, a uniform heat generation is assumed throughout the battery cell [28–30].

2.2.1. Thermal Model

A schematic diagram of the pouch cell sandwiched between two fixture plates is shown in Figure 3. The lumped-parameter model assumes a 1-D thermal behavior of the pouch cell with a conduction boundary condition via contacts with the plates. Moreover, a convective boundary condition exists on both the top and bottom of the plates.

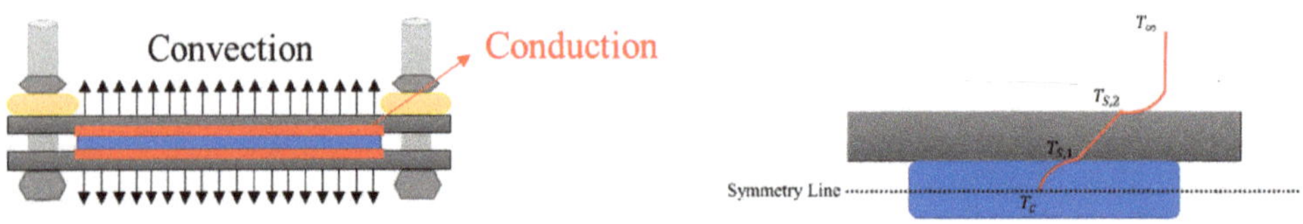

Figure 3. Schematic representation of the pouch cell inside the fixture.

The transient one-dimensional conductive heat transfer equation is as follows:

$$C_{cell}\frac{dT_C(t)}{dt} = Q_{gen}(t) - 2 \times \frac{K_{cell}A_{cell}}{\frac{L_{cell}}{2}}(T_C(t) - T_{S,1}(t)) \tag{10}$$

where C_{cell} is the cell heat capacity, $Q_{gen}(t)$ is the heat generated in the battery, T_C is the core temperature of the battery, and $T_{S,1}(t)$ is the surface temperature of the battery while K, A, and L denote the thermal conductivity, area, and length, respectively. The convective heat transfer at the top boundary of the plate can be expressed as:

$$Q_{conv} = h_{air}A_{plate}(T_{S,2}(t) - T_\infty) \tag{11}$$

Moreover, a two-state representation of the 1-D thermal model is given by:

$$\frac{K_{cell}A_{cell}}{\frac{L_{cell}}{2}}(T_C(t) - T_{S,1}(t)) = \frac{K_{plate}A_{plate}}{\frac{L_{plate}}{2}}\left(T_{S,1}(t) - \frac{T_{S,1}(t) + T_{S,2}(t)}{2}\right) \tag{12}$$

$$\frac{K_{cell}A_{cell}}{\frac{L_{cell}}{2}}(T_C(t) - T_{S,1}(t)) = h_{air}A_{plate}(T_{S,2}(t) - T_\infty) \tag{13}$$

where $T_{S,2}(t)$ is the surface temperature of the plate, T_∞ is the ambient temperature, and h_{air} is the air convection coefficient.

2.2.2. Spring Model

A spring model was developed to represent the mechanical behavior of a battery cell in the fixture. Figure 4 illustrates the behavior of the cell at constrained conditions. Two equivalent stiffness elements are adopted, i.e., the equivalent stiffness s_{Total}, representing the cell and the plate, and the equivalent stiffness s_{Rod}, representing the rod. These equivalent stiffness values refer to the relationship between the force applied on the fixture and the displacement of the fixture along the vertical axis. A linear model is adopted to predict the dependence of the force on the displacement (Branch A):

$$F_{Branch\ A} = s_{Total}\left(\delta_{Total}^{c,\%}\right) \tag{14}$$

where s_{Total} is the linear equivalent stiffnesses, and $\delta_{Total}^{c,\%}$ represents the constrained displacement of the battery and plates. The superscript c represents the constrained displacement conditions. There are four rods connected in parallel at the four edges of the plate. The total force in Branch B can be described by:

$$F_{Branch\ B} = 4 \times s_{Rod}\left(\Delta_{bolt} + \delta_b^{f,\%} - \delta_{Total}^{c,\%}\right) \tag{15}$$

where Δ_{bolt} is the displacement in the rod due to the tightening of the bolt and $\delta_b^{f,\%}$ is the free unconstrained displacement of the battery. The total measured displacement is given by:

$$\delta^{EXP,\%} = \delta_b^{f,\%} - \delta_{Total}^{c,\%} + \delta_0^c \tag{16}$$

where δ_0^c is the initial displacement.

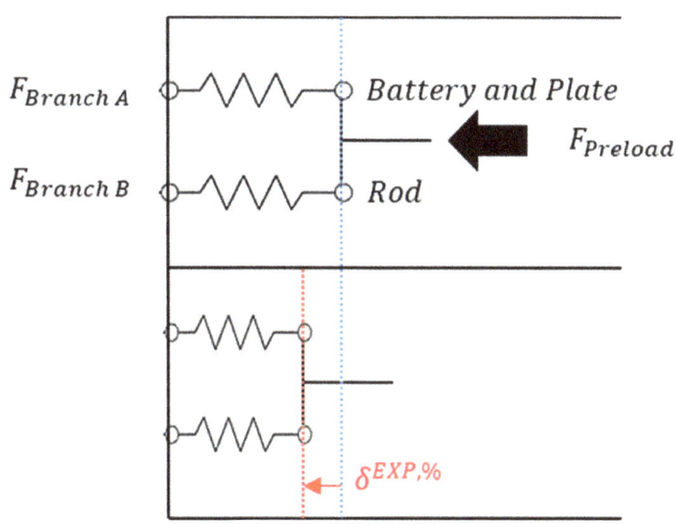

Figure 4. Spring model.

At equilibrium, the plate and the battery cell will be compressed while the four rods will be stretched. Therefore, the plates and the battery cell will be under compressive force and the rods will be subject to tensile force. At mechanical equilibrium, these two forces will be equal:

$$F_{Branch\ A} = F_{Branch\ B} \tag{17}$$

2.2.3. Coupling between Thermal and Mechanical Models

In the proposed LPM, the total swelling of the battery cell, ΔL_{Total}, consists of (1) swelling due to thermal expansion, $\Delta L_{Thermal}$, and (2) swelling due to Li-ion intercalation process, $\Delta L_{Intercalation}$. Assuming uniform heat generation across the cell, the temperature distribution in the battery will follow a parabolic shape (See Figure 3). The thermal swelling normal to the battery cell is given by:

$$\Delta L_{Thermal} = \alpha \int_0^{L_{Cell}} (T(x) - T_S)dx \tag{18}$$

where ΔL_T is the thermal swelling (due to the cell heating/cooling), L_{Cell} is the thickness of the cell, α is the lumped coefficient of thermal expansion, and T is the temperature. The subscript S denotes the surface initial state. Integrating Equation (18) based on the parabolic temperature distribution across the cell gives:

$$\Delta L_{Thermal}(t) = \alpha \times L_{Cell} \times \left(\frac{2}{3}(T_C(t) - T_S(t)) + \left(T_S(t) - T_{Ref}\right)\right) \tag{19}$$

where $T_C(t)$, $T_S(t)$, and T_{Ref} denote cell-core temperature, cell-surface temperature, and reference temperature, respectively. Swelling due to the Li-ion intercalation during charging and discharging is represented by a linear equation:

$$\Delta L_{Intercalation} = \xi.\mathrm{SOC} \tag{20}$$

where ξ is the coefficient of expansion according to the state-of-charge (SOC) of the cell. An average SOC over thickness was used in this model.

The thermal model in Section 2.2.1 and the spring model in Section 2.2.2 are linked via the free swelling term as follows:

$$\begin{gathered}\delta_b^{f,\%} = \Delta L_{Thermal} + \Delta L_{Intercalation} = \\ \alpha_{Battery} L_{Battery}\left(\frac{2}{3}(T_c - T_{s,1}) + \left(T_{s,1} - T_{Ref}\right)\right) + \xi.\mathrm{SOC}\end{gathered} \tag{21}$$

2.3. *Solid Mechanics Model*

A 2-D solid mechanics model is adopted to predict the stress in a battery cell. The model was developed in COMSOL Multiphysics 5.6. Schematic representation of the mechanics model for a pouch cell is shown in Figure 5. The part modeled is highlighted using a dotted box. The negative current collector, anode electrode, separator, cathode electrode, and positive current collector are shown in different colors. Two types of boundary conditions exist in the model, namely, internal and external boundary conditions. During battery production, the separator is tightly wrapped around the electrodes after electrodes and current collectors are bonded together. Therefore, a set of bonded electrode and current collector is considered as a continuous medium. In contrast, a contact-type boundary condition, allowing a slip at the interface, was assigned between cathode and separator and between anode and separator, respectively. External boundary conditions include a symmetric boundary condition on the left and bottom side of the model and a free boundary on the right side. Three main inputs were used in the solid mechanic model; a distributed normal load is applied at the top according to the predicted force on the battery and fixture plates by the LPM, i.e., $F_{Branch\ A}$ (see Figure 4). Additionally, the core temperature from the LPM is assumed to be uniform throughout the solid mechanics model since the cell is very thin [14], and an average lithium concentration was used for the electrodes.

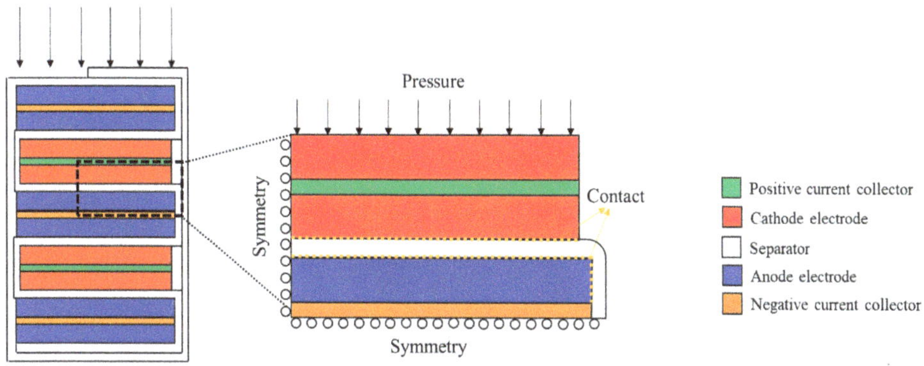

Figure 5. Schematic representation of a 2-D mechanics model.

Remark 1. *The 2D mechanics model includes an extra cathode on the top for counter bending of the existing cathode. Without an extra cathode on top of the positive current collector, the cathode below will form an asymmetric beam which tends to deflect upward away from the separator due to intercalation-induced stresses.*

The constitutive equations of the model include those based on mechanical strains ε_{ij}^{me}, thermal strain ε_{ij}^{T} [14], and intercalation-induced strain ε_{ij}^{D} [13,31] to predict the stress in the battery components as follows:

$$\begin{aligned}\varepsilon_{ij} &= \varepsilon_{ij}^{me} + \varepsilon_{ij}^{T} + \varepsilon_{ij}^{D} \\ &= \frac{1}{E^{v}}\left((1+\nu^{v})\sigma_{ij} + \nu\sigma_{kk}\delta_{ij}\right) + \alpha_{ij}\Delta T\delta_{ij} + \frac{1}{3}\Delta c\Omega\delta_{ij}\end{aligned} \tag{22}$$

where E^{v} is the Young's modulus, ν^{v} is the Poisson's ratio, σ_{ij} is the stress tensor, and δ_{ij} is the Kronecker delta, α_{ij}^{v} is the thermal expansion coefficient, ΔT is the change in temperature from a reference value, Δc is the change in the diffusion species concentration from a reference (stress-free) value, and Ω^{i} is the partial molar volume [31,32]. The partial molar volume can be calculated using [31]:

$$\Omega^{i} = \frac{3\varepsilon}{\Delta y\, C_{s,Max}} \tag{23}$$

Remark 2. *Equation (22) only considers elastic deformation and does not consider any plastic deformation. This is a safe assumption since the amount of deformation in this study is very small and does not exceed the elastic limit [8,14,33–35]. It should be also noted that the strain induced by intercalation, denoted by ε_{ij}^{D} is exclusively applicable to the positive and negative electrodes. Additionally, Equation (22) considers a constant Young's modulus in the mechanical strains for the electrodes. The Young's modulus was considered independent of the lithium concentration. In [36], the authors showed that elastic stiffening increases the elastic modulus, which increases the stress. However, to maintain simplicity in this work, no concentration-dependent material properties were considered.*

3. Experimental Setup

3.1. Battery Cell

A pouch LIB cell with a size of 110 mm by 74 mm is tested in this study. The battery cell with a capacity of 3.5 Ah is built on mesophase graphite (SMG) anode and $LiNi_{0.6}Co_{0.2}Mn_{0.2}O_2$ (NMC 622) cathode, 1 M $LiPF_6$ in ethylene carbonate (EC): ethyl methyl carbonate (EMC) (3:7 by volume) for electrolyte, and wet-processed polyethylene (PE) single-layer separator.

More details about the PE separator are as follows. The PE separator (12 μm thick) is fabricated through a co-extrusion blowing process. The separator porosity is 43% with a crystallinity ranging between 60% and 80%. The tensile strength in the machine direction (MD) is 235 MPa while that in the transversal direction (TD) is 225 MPa. The PE separator properties are summarized in Table 1.

Table 1. PE separator properties.

Thickness (μm)	Crystallinity (%)	Porosity (%)	Heat Shrinkage MD (% @105C, 8 h)	Heat Shrinkage TD (% @105C, 8 h)	Tensile Strength MD (MPa)	Tensile Strength TD (MPa)
12	60–80	43	4.0	4.5	235	225

In a battery cell, the pores in the electrodes and separator are filled with electrolyte. Therefore, the effective thermal expansion coefficient and Young's modulus of the electrodes and separator were calculated considering their porosities. The effective thermal expansions of the electrodes and separator were estimated by the rule of mixture as follows:

$$\alpha_{eff} = \sum_i \alpha_i V_i \tag{24}$$

where α_i and V_i are the thermal expansion coefficient and the volume fraction of the ith phase, respectively. The effective Young's modulus can be calculated in a similar manner.

The material properties for the 1-D battery cell model and the 2-D solid mechanics model are listed in Table 2. The use of '*' next to a value indicates that it is an effective material property.

Table 2. Material properties for 1-D battery cell and 2-D solid mechanics models.

Model Parameters	Units	Cu	Anode	Separator	Cathode	Al	Electrolyte
			Dimensional Parameters				
Thickness	μm	10	71	12	48.5	20	
Length	mm	71	71		69	69	
			Electrochemical Parameters				
Density	kg/m^3		2230	900	4750		1210
Active volume fraction			0.5686		0.6698		
Porosity			0.37454	0.43	0.26322		
Stochiometric maximum concentration	mol/m^3		49,000		31,507		
Diffusion coefficient	m^2s^{-1}		1.4523×10^{-13}		5×10^{-13}		
Particle radius	μm		4		5		
Partial molar volume	m^3/mol		4.4196×10^{-5}		2.8418×10^{-5}		
			Thermal Parameters				
Thermal expansion coefficient	1/K	17×10^{-6}	6.03×10^{-6} *	82.5×10^{-6} *	9.62×10^{-6} *	23×10^{-6}	
			Mechanical Parameters				
Young's modulus	GPa	110	6.82 *	0.645 *	6.69 *	70	
Poisson's ratio		0.35	0.3	0.45	0.3	0.33	0.25

3.2. Cell Fixture and Instrumentation

A fixture, shown in Figure 6, was used to emulate a battery module assembly. The fixture was built by holding two steel plates (each with a size of 125 mm × 99 mm × 12.7 mm) with four threaded rods (9.5 mm in diameter). A thermocouple was attached on the top of the battery cell to measure the cell surface temperature while another thermocouple was installed on the top of the upper plate. An additional thermocouple was used to monitor the ambient temperature. Four compression load cells (Omega LC9150-500-1K) were placed between the upper plate and hex nuts. A high precision displacement sensor (Keyence LK-G10) was placed 10 mm above the upper plate to measure the expansion of the battery cell. The whole setup was placed inside a thermal chamber whose temperature was set to 20 °C. After the battery cell was charged at a constant current—constant voltage (CC–CV) to 3.6 volts at 0.5 C, with a 5% current cutoff (0.0875 A), it was allowed to rest for 1 h for thermal equilibrium. Then, a pre-torque in the amount of 1 Nm was applied on each hex nut to compress the battery cell.

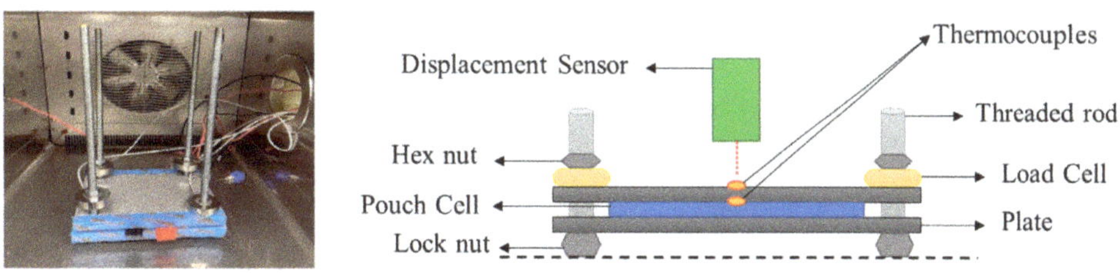

Figure 6. Experimental setup.

4. Identification of LPM Parameters

This section describes how the parameters of the lumped-parameter model (LPM) are identified.

4.1. Experiments for Measurement of Stiffness

Both force and displacement were measured under several torques to find the equivalent spring constant of battery cell and plates. The fixture shown in Figure 6 was used for the experiment. The battery was initially charged to 3.6 volts and placed between the plates. No thermocouples were used in this experiment. Four compression load cells (Omega LC9150-500-1K) were placed between the upper plate and hex nuts. The torque on each hex nut was increased by an increment of 0.2 N·m from 0 to 1 N·m and by an increment of 0.1 N·m from 1 to 1.2 N·m. Force and displacement measurements were recorded at each increment. The equivalent stiffness was fitted on the force and the displacement experimental data are shown in Figure 7a. The spring constant for the battery was found to be $s_{Total} = 9.0892 \times 10^6$ N·m^{-1}.

Another experiment was carried out to find the stiffness coefficient of the threaded rod. A rod was placed vertically with a lock-nut at the bottom. An Omega load cell was placed between the lock-nut and a hex nut. As the applied torque on the hex nut increased from 0 to 1.6 N·m, the force and elongation of the rod were measured. The rod spring constant was fitted on the force and the elongation data are shown in Figure 7b. The turn angle was used to calculate the elongation in the rod using the pitch. Note this spring constant corresponds to the portion of the rode between the two nuts. The Young's modulus can then be found using:

$$S_{Rod} = \frac{EA}{L} \tag{25}$$

Using the Young's Modulus, we can then calculate the spring constant of the rod for its length in the fixture. The spring constant for the rod was found to be $S_{Rod} = 1.0311 \times 10^7$ N·m^{-1}.

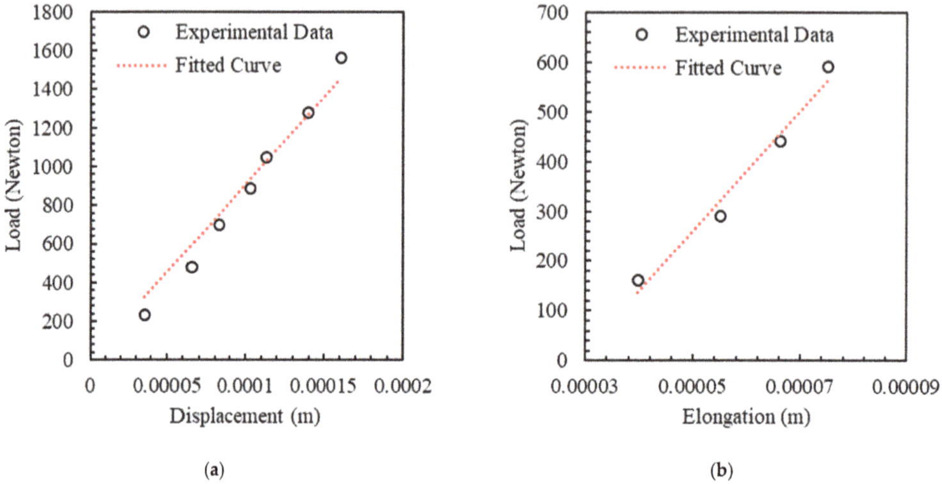

Figure 7. (**a**) Force versus displacement for the equivalent battery and plate spring constant and (**b**) rod spring constant.

4.2. Charging and Discharging Test

The force, displacement, and temperatures during charging and discharging were measured after the battery cell was installed in the fixture described in Section 3.2. The collected data were used for tuning the LPM parameters as described in Section 4.3. The test was performed under ambient temperature of 20 °C.

The battery was initially charged to 3.6 volts with a constant current of 1.75 A (0.5C), and then the voltage was held at 3.6 volts until the current dropped to 5% of its initial value. Figure 8a shows the voltage and current at the start of discharge from 3.6 volts. The data consist of three parts based on constant current–constant voltage (CC–CV) protocols. The first part of the curve shows CC–CV discharging at 0.5C from 3.6 to 3 volts (0–2178 s). The second part is a period of CC–CV fast charging at 4C from 3 to 4.2 volts (2179–3407 s), and the third part corresponds to CC–CV discharging at 0.5C from 4.2 to 3 volts (3408–9470 s). Figure 8 also shows variations of load and displacement overlayed on the voltage curve. Starting with the displacement figure (see Figure 8b), the discharge between 3.6 volts and 3 volts decreases the displacement (height) of the top surface of the fixture due to shrinkage of the battery cell. During the period of fast charging at 4C, a sharp increase in the displacement due to rapid expansion of the battery cell can be observed. During the voltage-hold period after fast charging, a small increase in the displacement is followed by a decrease. This is due to an initial increase in the temperature during charging followed by a decrease, causing cell shrinkage, during the hold period. The maximum displacement reached during fast charging was 45.8 μm. During discharging, the displacement decreases sharply between 4.2 volts and 3.7 volts, followed by a small decrease between 3.7 and 3.6 volts and another sharp decrease between 3.6 and 3 volts. The load follows the same trend of the displacement curve. After tightening the bolts with a 1 N·m torque, the initial load applied on the battery was 1314 N at the beginning of the test. The load decreases to 983 N during the first discharging, then increases significantly during charging to reach 2333 N at maximum followed by a small decrease to 2213 N during the hold period. During the second discharging period, the load decreases to 1070 N.

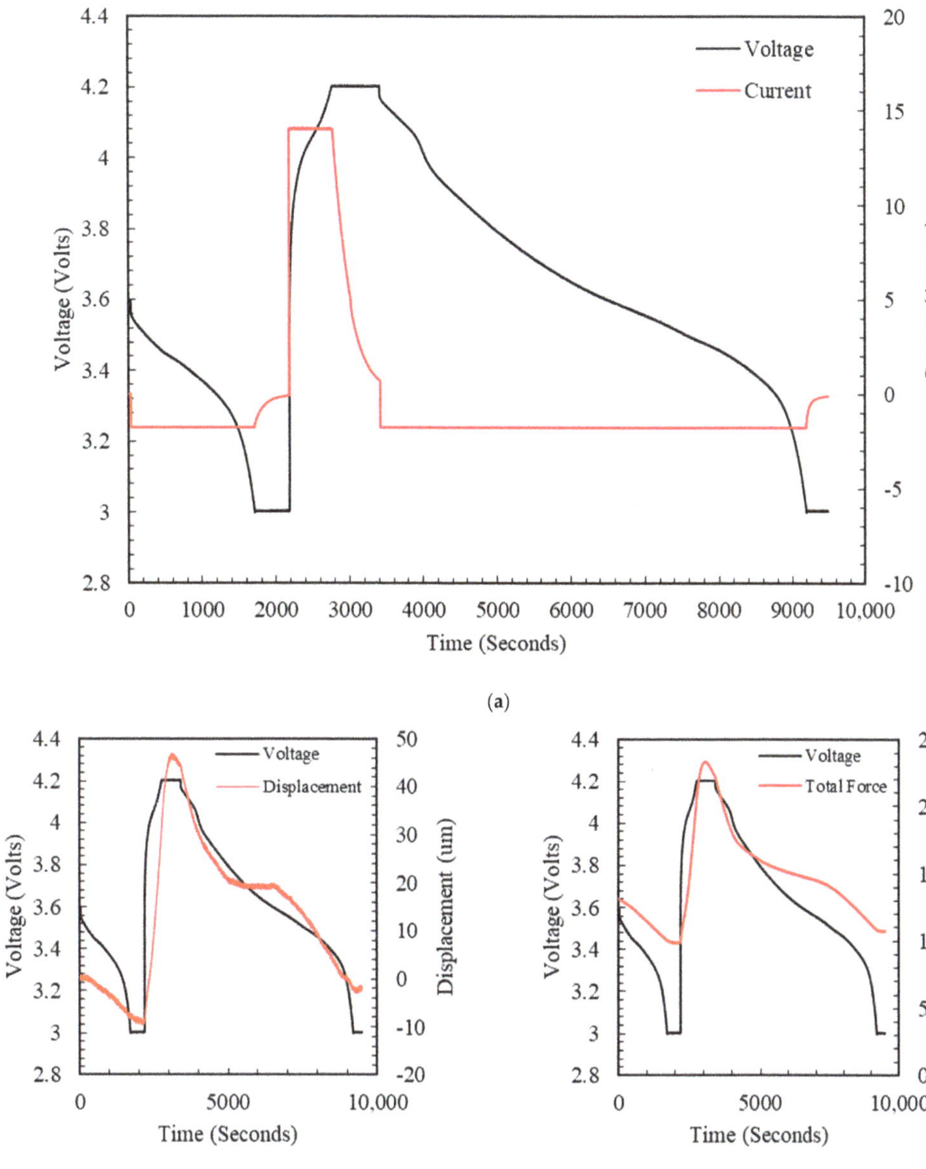

Figure 8. Variations of volage, current, displacement, and force over time. (**a**) Voltage and current. (**b**) Voltage and displacement. (**c**) Voltage and load.

Figure 9 presents the battery-surface and plate-surface temperatures measured during the fast-charging period. During fast charging at 4C, an increase can be seen in both temperatures. At an ambient temperature of 20 °C, the battery-surface and plate-surface temperatures reached 21 and 20.8 °C, respectively. During the following hold period, both temperatures decreased approaching the ambient temperature.

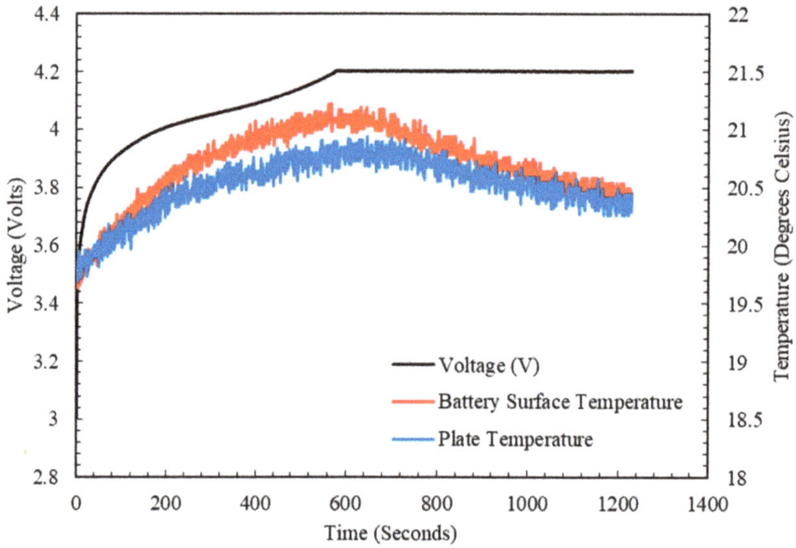

Figure 9. Battery-surface and plate-surface temperatures during fast charging.

4.3. LPM Parameters

Some parameters of the LPM were obtained from the literature while others were identified by fitting the model to the experimental data. The thermal conductivity of the cell for the LPM, for example, can be determined by combining the material properties reported in the literature. The thermal conductivity of a battery cell, composed of many layers of different conductivities, will vary depending on the direction of heat flow [30,37]. The through-plane thermal conductivity, in the direction of the heat flow considered in the LPM, can be expressed as [37]:

$$\frac{\sum t_v}{K_{eq}} = \sum \frac{t_v}{K_v} \tag{26}$$

where K_v is the thermal conductivity of the material layer and t_v is the thickness of the material layer. The equivalent thermal conductivity of the cell, thus determined, is found to be 0.633 $\frac{W}{mK}$. The thermal conductivity and thermal expansion coefficient of the plate are 167 $\frac{W}{mK}$, and 25.2×10^{-6} $\frac{1}{K}$, respectively, from the literature.

The experimental data used for tunning of the remaining LPM parameters include those collected during CC discharging from 3.6 to 3 volts, CC fast charging from 3 to 4.2 volts, and CC discharging from 4.2 to 3 volts. The responses and predictors of the LPM are summarized in Table 3. Four LPM parameters were estimated based on the experimental data: (1) the heat capacity of the cell, C_{Cell}, (2) the heat transfer coefficient of air, h_{air}, (3) the thermal expansion coefficient of the cell, α_{Cell}, and (4) the SOC coefficient for expansion due to lithiation, ξ. The parameters were estimated such that the sum of squared errors in predicting the temperatures, $T_{s,1}$ and $T_{s,2}$, the force, F, and the displacement, $\delta^{EXP,\%}$, is minimized as follows:

$$\min \sum_i \left(T_{s,1}^{actual} - T_{s,1}^{predicted}\right)^2 + \left(T_{s,2}^{actual} - T_{s,2}^{predicted}\right)^2 + \left(F^{actual} - F^{predicted}\right)^2 + \left(\delta_{Total}^{EXP,\%,actual} - \delta_{Total}^{EXP,\%,predicted}\right)^2 \tag{27}$$

Table 3. Responses and predictors of the LPM.

Responses	Symbol
Battery surface temperature	$T_{s,1}$
Plate surface temperature	$T_{s,2}$
Total force	F
Displacement of the fixture	$\delta^{EXP,\%}$
Predictors	**Symbol**
Heat generated from the battery	Q_{gen}
Diffusion swelling	$\delta^{\%}_{Diffusion}$
Time	Time
Ambient temperature	T_{Amb}

A non-linear regression was performed using the least-squares curve fitting in MATLAB. The parameter values, thus estimated, are listed in Table 4. The estimated parameter values are in the same order of magnitude as those in the literature, e.g., in [15]. The results of model-fitting are shown in Figure 10. The experimental battery surface temperature, the plate surface temperature, and the fixture displacement are in good agreement with the LPM predictions. However, a relatively large error in prediction of force can be observed at the end of charging and during discharging.

Table 4. Estimated LPM parameters.

Parameters	Optimization Solution
$C_{Cell}\cdot$ **(J/K)**	451.62
$h_{air}\cdot$ **(W/m²K)**	95.743
$\alpha_{Cell}\cdot$ **(1/K)**	4.39×10^{-4}
$\xi\cdot$ **(m)**	1.03×10^{-4}

In order to evaluate the goodness of model-fitting, the root mean square error (RMSE) is calculated (see Table 5):

$$\text{RMSE} = \sqrt{\frac{\sum_{i=1}^{N}\left(y_i^{actual} - y_i^{predicted}\right)^2}{N}} \tag{28}$$

where y_i^{actual} is the experimental value, $y_i^{predicted}$ is the value predicted in the model, and N is the number of data points. Additionally, the RMSE divided by the range and average are also listed. It can be seen that, for each output, its RMSE is below 20% of its range indicating a good model-fit.

Table 5. Maximum error and RMSE between the experimental and predicted curves.

Responses	Maximum Error	RMSE	RMSE/ Range	RMSE/ Average
$T_{s,1}$	0.52	0.1326	0.1262	0.0066
$T_{s,2}$	0.63	0.1190	0.1142	0.0060
$\delta^{EXP,\%}$	6.06×10^{-6}	3.6364×10^{-6}	0.0968	0.2960
F	573	297.73	0.1922	0.1800

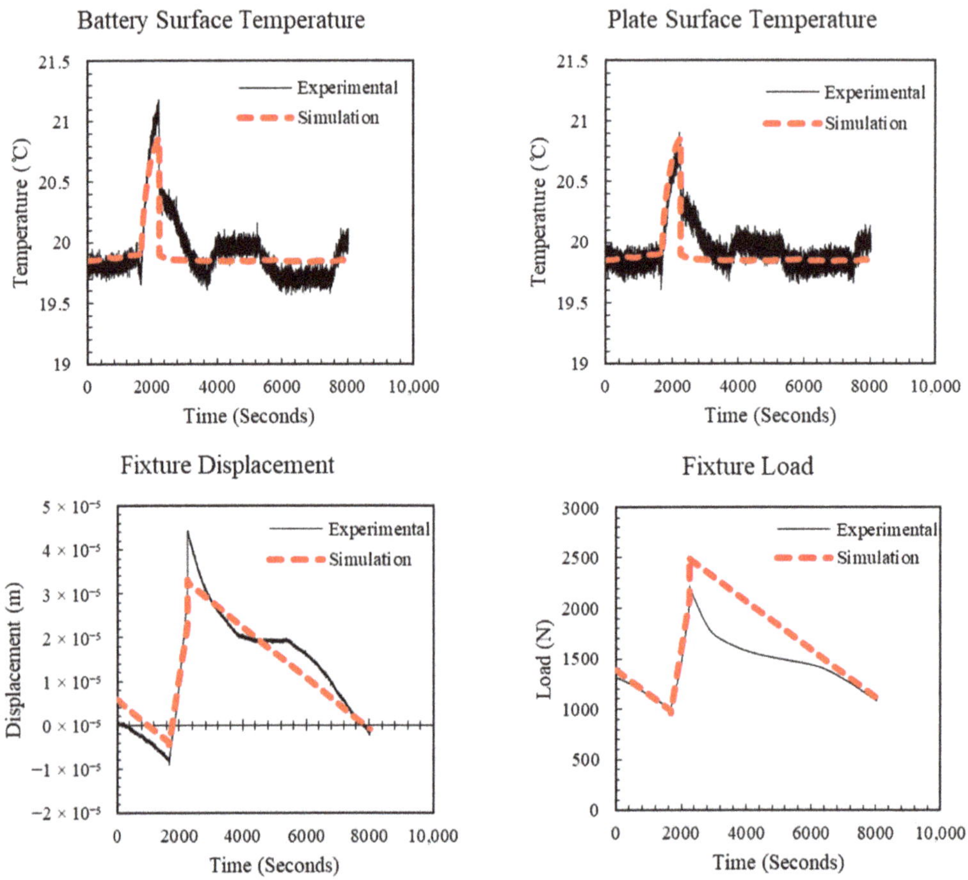

Figure 10. Comparison of the simulated output and the experimental data after model fitting.

5. Simulation Study

As mentioned in the introduction, a significant degradation in the separator's integrity can be observed after the battery cell undergoes cyclic 4C charging [8]. Therefore, the stress in the separator under 4C charging is estimated using the developed hybrid model. Moreover, the effect of charging rate, pre-torque, and ambient temperature on the stress in the separator is investigated.

5.1. Estimated Stress in the Separator

The stress in the separator under 4C charging condition of the cycling test in [8] was calculated using the hybrid model. Specifically, the core temperature, boundary pressure, and lithium concentration predicted using the electrochemical model and the LPM model were used as inputs for the solid mechanics model (See Figure 1). Figure 11a shows a homogeneous distribution of the von Mises stress in the middle section of the separator at the end of 4C charging, while Figure 11b presents the evolution of the maximum von Mises stress over time during charging. The von Mises stress is calculated as follows [38]:

$$\sigma_{\text{Mises}} = \frac{\sqrt{2}}{2}\left[(\sigma_1-\sigma_2)^2+(\sigma_2-\sigma_3)^2+(\sigma_3-\sigma_1)^2\right]^{\frac{1}{2}} \quad (29)$$

where σ_1, σ_2, and σ_3 are the principal stress components. The von Mises stress serves as a measure of the overall stress experienced by the material at a specific location, accounting for both normal and shear stresses.

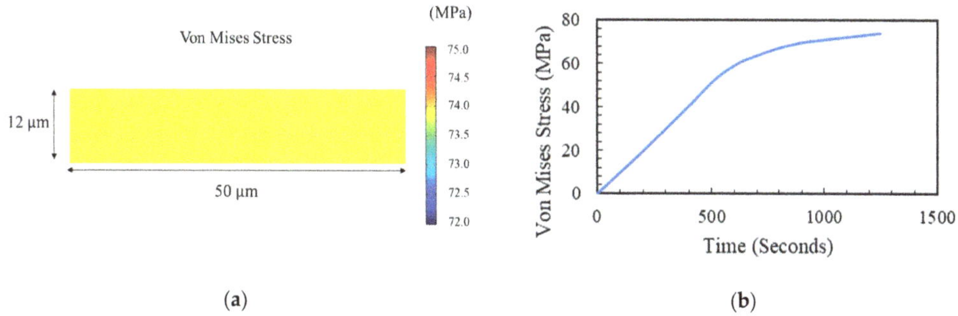

Figure 11. Von Mises stress in the separator. (**a**) Stress distribution at the end of charging. (**b**) Evolution of stress during charging.

The von Mises stress showed a homogeneous distribution within the central part of the separator during the 4C charging. The maximum von Mises stress is found to be 74.1 MPa at the end of fast charging. Creep and fatigue behavior of the separator under the estimated stress will be presented in Section 6.

5.2. Effect of Charging Rate

The hybrid model allows examination of the effect of charging rate on the separator. Evolution of the von Mises stress in the separator over time was analyzed at four different charging rates of 1C, 2C, 4C, and 5C with other conditions, such as the pre-torque and ambient temperature, fixed at those in Section 5.1. The CC–CV charging protocol was employed for each charging rate. Figure 12 shows the temperature variations during charging for different charging rates. When the charging rate is 1C, the cell temperature slightly increases from 20 °C to the maximum value of 20.25 °C at the end of the constant current (CC) period. As the charging rate increases, the maximum cell temperature increases, e.g., 24.6 °C at 5C. It can be seen that a higher charging rate results in an elevated cell temperature, which can potentially compromise battery safety and lead to accelerated degradation of cell components including the separator. Therefore, a careful consideration of the charging rate is essential for optimal battery performance and longevity.

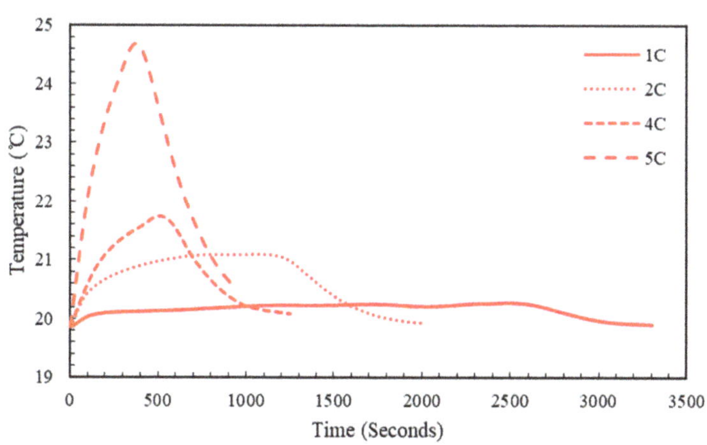

Figure 12. Evolution of cell temperature during 1C, 2C, 4C, and 5C charging.

Evolution of the von Mises stress in the separator over time at the four different charging rates is shown in Figure 13. Regardless of the charging rate, the von Mises stress exhibits a linear increase during the constant charge (CC) period followed by a slower increase during the constant voltage (CV) period. A faster increase in the von Mises stress is observed with an increase in the charging rate. It should be noted that the proposed hybrid model is capable of predicting the evolution of stress for any charging protocols. The hybrid model, therefore, opens the possibility of optimizing the charging schedule for achieving multiple objectives such as prolonging both electrode health and separator integrity.

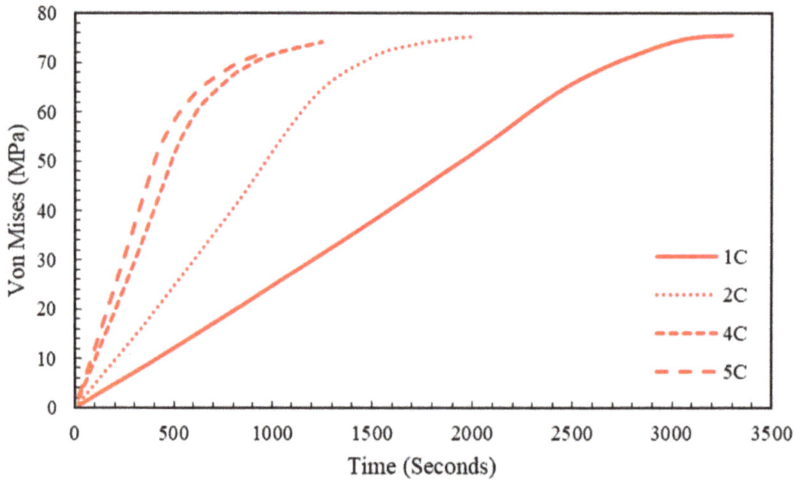

Figure 13. Von Mises stress evolution in the middle of the separator.

5.3. Effect of Pre-Torque

The effect of pre-torque on separator stress is examined by applying various pre-torque values to the hybrid model. Different pre-torques ranging from 0.5 to 12 N·m were simulated. Note, in the experiment in Section 3, a pre-torque of 1 N·m was applied to the fixture. Figure 14 shows only a minor change in the von Mises stress with increasing pre-torque. The reason why the pre-torque plays a minor role in the separator stress is as follows. The stress in the separator can be attributed to two main mechanisms. First, during charging, the anode expands in all directions due to lithiation and a temperature increase. Expansion of anode will exert a tensile load on the separator as the anode pulls the separator in the horizontal direction, as seen in Figure 5. Second, the pressure due to the pre-torque and structural constraint imposed by the fixture will compress the separator in the vertical direction. The stress due to pulling is much larger than the one due to the pressure by at least two orders in this particular configuration. Therefore, the change in pressure due to varying pre-torque will have only a minor effect on the von Mises (overall) stress.

5.4. Effect of Ambient Temperature

The effect of ambient temperature on separator stress was investigated by applying six different ambient temperatures ranging from 0 to 50 °C to the hybrid model. Figure 15 shows the relationship between the von Mises stress in the separator at the end of charging and the ambient temperature. The results show a decrease in the von Mises stress with increasing ambient temperature.

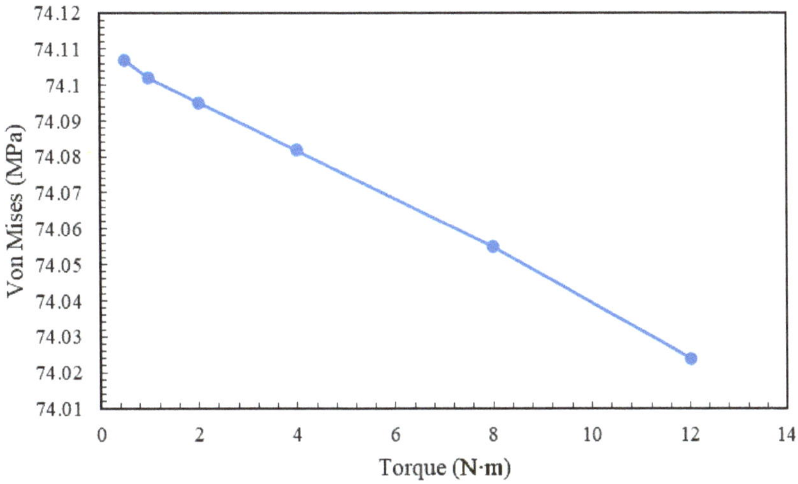

Figure 14. Von Mises stress at different torques.

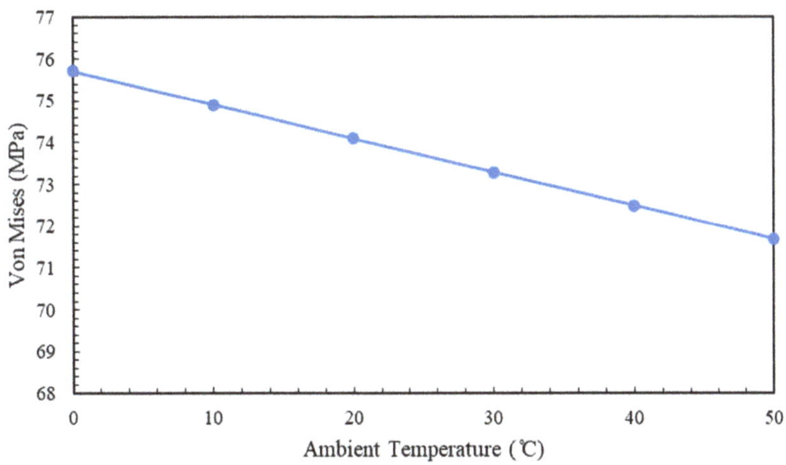

Figure 15. Von Mises stress at different ambient temperatures.

The cell temperature will increase when the ambient temperature rises. This will lead to expansion of battery components in both horizontal and vertical directions (see Figure 5). Since the thermal expansion coefficient of the separator is significantly larger than those of the other battery components, the separator will experience a pronounced length expansion in the horizontal direction, which leads to less pulling by the anode and lower stresses.

6. Creep and Fatigue Testing

As previously discussed, noticeable degradation of the separators' mechanical properties (by as much as 50%) was observed when the LIB, made of the same separator as the one in this study, was subject to repeated fast (4C) charging [8]. In Section 5, the maximum von Mises stress during 4C charging was found to be 74 MPa while only a small temperature rise was observed. Therefore, creep and fatigue tests were carried out on a pristine separator at room temperature to examine its behavior under the estimated stress during 4C charging.

Strip-shaped uniaxial tensile specimens were prepared according to ASTM D882. The samples were cut from a new separator roll using a razor. The longer axis the sample was aligned in the machine direction was with a length of 100 mm and a width of 10 mm. The gage length was chosen to be 50 mm. The creep and fatigue experiments were conducted using an ElectroPuls E1000. Foamy jaws were used to hold the specimen from each end.

Creep testing was conducted at five target stresses ranging from 50 to 90 MPa, which are lower than the yield stress of the separator (ranging from 100 to 140 MPa in the machine direction depending on the strain rate, according to the authors' previous study [8]). Note the maximum von Mises stress during 4C charging was estimated at 74.1 MPa in Section 5.1. In each creep testing, the loading increased linearly from 0 MPa to the target stress at a rate of 0.1 MPa/s. Once a target stress is reached, the loading remained at the same stress level until the displacement reaches 60 mm (except for the target stress of 50 MPa since the maximum duration of the creep test was 6 days). The creep test results are presented in Figure 16. The samples subjected to 90 and 80 MPa reached 60 mm displacement in 4 and 5.5 h, respectively. Both tests remained in the primary stage of creep, characterized by a rapid deformation. The time required to reach 60 mm displacement increases significantly to 37 h under the stress of 70 MPa, while that under the stress of 60 MPa reaches in 68 h. At 50 MPa, the test stopped after 6 days when its displacement reached 48.45 mm. The results of creep tests at 50, 60, and 70 MPa exhibit the secondary creep stage where the creep becomes more gradual after the primary creep stage. The creep test results indicate that the stress during 4C charging can cause permanent damage to the separator. The results suggest that tensile creep is a major factor contributing to the mechanical degradation of the battery separator.

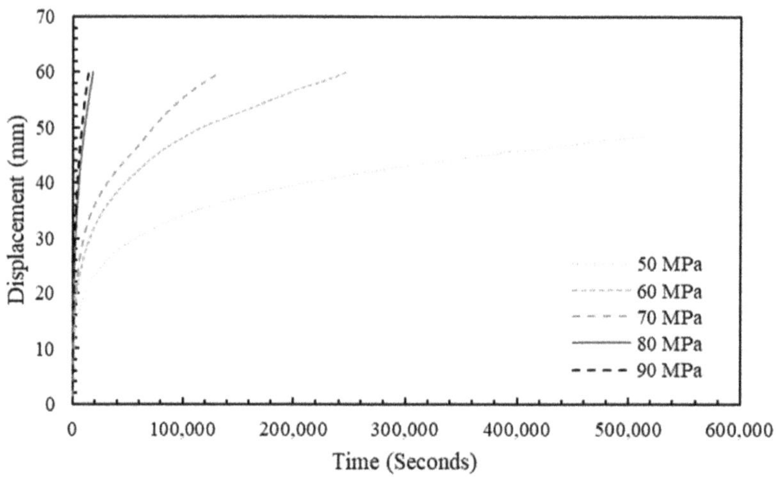

Figure 16. Displacement over time in creep tests.

The fatigue test was designed to mimic the stress evolution in the separator during cyclic fast charging for 68 cycles. Each cycle consists of (1) an increasing load along a linear ramp (0 to 70 MPa in 13 min) representing a 4C charging duration, (2) a stress hold representing a CV period for 10 min, and (3) a linear decreasing ramp (70 to 0 MPa in 90 min) representing a discharging duration. Figure 17 shows the variations of load and displacement over time during the first cycle. The displacement curve shows a rapid increase during the charging period, followed by a gradual increase in the displacement during the holing period and the beginning part of discharging. This is followed by decrease in the displacement for the remaining period of discharging. Even after the load returns to 0, a non-zero elongation of 1.3 mm remains, indicating a permanent damage in the separator.

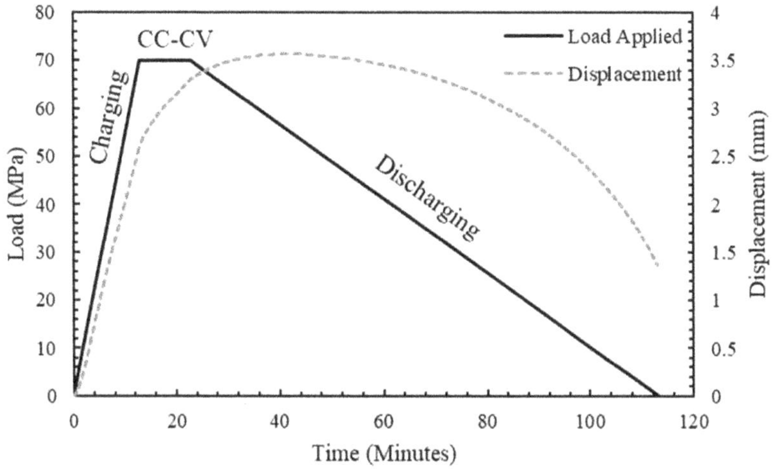

Figure 17. Load and displacement versus time.

The results of fatigue testing for 68 cycles are shown in Figure 18. The displacement over time is shown in Figure 18a while the maximum displacement of each cycle is plotted in Figure 18b. In Figure 18b, the displacement increases rapidly during the first 10 cycles and then shows a linear increase for the remaining cycles. It should be noted that the fatigue test demonstrates an extreme case of separator degradation during cycling. Once the separator elongates permanently after the first few cycles, it will be subject to stresses much lower than 70 MPa in the proceeding cycles. Nevertheless, the test results indicate that the load on the separator during 4C charging is high enough to cause damage to the separator. Moreover, the damage can be accumulated in the separator via repeated loading, explaining the separator degradation after cyclic fast charging in the previous study [8].

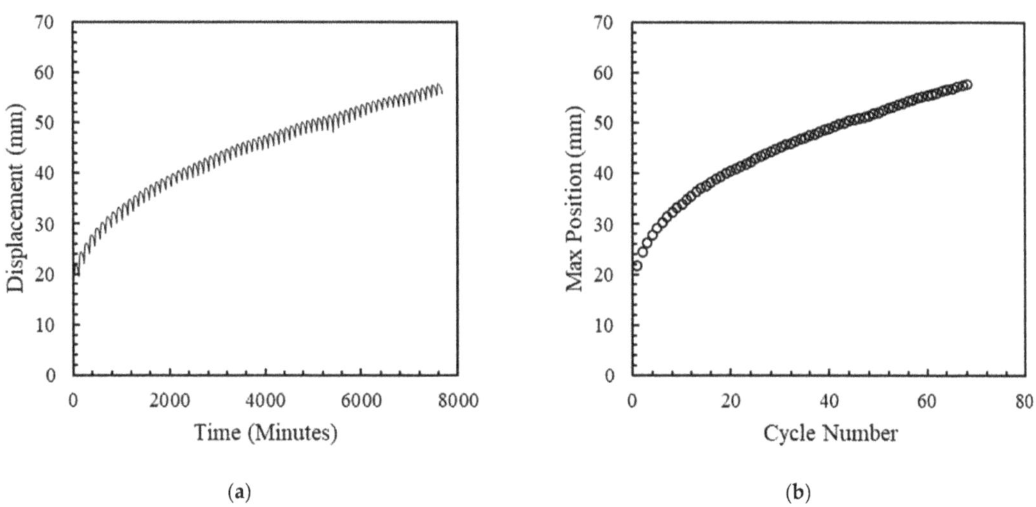

Figure 18. Results of fatigue testing. (**a**) Displacement over time. (**b**) Maximum position of each cycle.

7. Conclusions

Combining three sub-models enabled predicting the stress in the separator under various use conditions including those for battery pack assembly and thermal management.

The parameters of a lumped-parameter model were determined based on the experimental data showing good model-fit, while the material properties in other sub-models were obtained from the literature. The developed model allowed estimating the maximum von Mises stress as 74 MPa during the fast-charging condition, under which a significant separator degradation was observed after 1600 cycles in the previous study. In order to investigate the damage to the separator under this loading condition, creep and fatigue tests were conducted on a pristine separator. The high creep sensitivity and creep deformation in the test results indicate that such loading can produce and accumulate significant damage in the separator. However, care should be taken to avoid concluding that the mechanical stress is the only driving force behind separator degradation. The high temperature, which was not considered due to the low temperature rise of the setup in this study, may play a role by causing thermal degradation, which is a major mechanism for degradation of polymers in general. While the current state of the model allows for an efficient way in predicting the stresses in the cell across various charge rates, torque levels, and ambient temperature conditions, further investigations should be conducted to explore a more complex solid mechanics model which includes plasticity and damage in the constitutive equations. Additionally, the model can be further improved by incorporating concentration-dependent material properties to improve the mechanical predictions.

Author Contributions: Conceptualization, M.M., C.W.L. and G.A.; methodology, M.M., C.W.L. and G.A.; software, M.M.; validation, M.M., C.W.L. and G.A.; formal analysis, M.M.; investigation, M.M.; resources, M.M., C.W.L. and G.A.; data curation, M.M.; writing—original draft preparation, M.M.; writing—review and editing, M.M. and C.W.L.; visualization, M.M.; supervision, M.M., C.W.L. and G.A.; project administration, C.W.L.; funding acquisition, C.W.L. and G.A. All authors have read and agreed to the published version of the manuscript.

Funding: The research was partly funded by Ford Motor Company (Ford-UM Alliance: Development of a Lithium Ion Battery Separator Aging Model).

Data Availability Statement: Data will be available on request.

Acknowledgments: The authors would like to thank Tom Brackett for conducting the Li-ion batteries charging and discharging test.

Conflicts of Interest: The authors declare no conflict of interest.

Nomenclature

A	Area
a	Particle specific surface area
C_{cell}	Cell heat capacity
c	Lithium concentration
D	Diffusion coefficient
E	Young's modulus
F	Force
F_r	Faraday's constant
h_{air}	Convection heat transfer coefficient of air
I_{app}	Current density
i_{ex}	Exchange current density
J	Pore wall flux
k_i	Reaction rate constant
K	Thermal conductivity
L	Thickness
Q_{conv}	Convective heat transfer
Q_{gen}	Heat generation rate
R	Ideal gas constant

r	Particle radius coordinate
r_p	Particle radius
s	Spring constant
T	Temperature
t	Time
t_+	Transport number
U	Open circuit potential
V	Volume
α	Thermal expansion coefficient
α_a	Charge transfer coefficient for oxidation
α_c	Charge transfer coefficient for reduction
δ	Displacement
δ_{ij}	Kronecker delta
ε	Volume fraction
ϵ	Mechanical strain
η	Overpotential of the intercalation reaction
κ	Electrolyte ionic conductivity
ν	Poisson's ratio
ξ	Coefficient of expansion for state-of-charge
σ	Mechanical stress
τ_{eff}	Effective conductivity
ϕ	Potential
Δ	Displacement due to bolt
ΔL	Swelling
Ω	Partial molar volume
Superscripts	
i	Electrode under consideration (p for positive electrode and n for negative electrode)
n	Negative electrode
p	Positive electrode
sep	Separator
v	Component under consideration (p for cathode, n for anode, and sep for separator)
Subscripts	
eff	Effective
$filler$	Binder and fillers
l	Liquid phase
s	Solid phase
$surf$	On the surface

References

1. Bandhauer, T.M.; Soc, J.E.; Bandhauer, T.M.; Garimella, S.; Fuller, T.F. A Critical Review of Thermal Issues in Lithium-Ion Batteries. *J. Electrochem. Soc.* **2011**, *158*, R1. [CrossRef]
2. Lukic, S.M.; Cao, J.; Bansal, R.C.; Rodriguez, F.; Emadi, A. Energy Storage Systems for Automotive Applications. *IEEE Trans. Ind. Electron.* **2008**, *55*, 2258–2267. [CrossRef]
3. Arora, P.; Zhang, Z. Battery Separators. *Chem. Rev.* **2004**, *104*, 4419–4462. [CrossRef]
4. Lee, H.; Yanilmaz, M.; Toprakci, O.; Fu, K.; Zhang, X. A Review of Recent Developments in Membrane Separators for Rechargeable Lithium-Ion Batteries. *Energy Environ. Sci.* **2014**, *7*, 3857–3886. [CrossRef]
5. Baldwin, R.S.; Bennet, W.R.; Wong, E.K.; Lewton, M.R.; Harris, M.K. *Battery Separator Characterization and Evaluation Procedures for NASA's Advanced Lithium-Ion Batteries*; National Aeronautics and Space Administration: Cleveland, OH, USA, 2010.
6. Zhou, W. Effects of External Mechanical Loading on Stress Generation during Lithiation in Li-Ion Battery Electrodes. *Electrochim. Acta* **2015**, *185*, 28–33. [CrossRef]
7. Huang, X. Separator Technologies for Lithium-Ion Batteries. *J. Solid State Electrochem.* **2011**, *15*, 649–662. [CrossRef]
8. Makki, M.; Ayoub, G.; Lee, C.W.; Bae, C.; Colin, X. Effect of Battery Fast Cyclic Charging on the Mechanical and Fracture Behavior of the Lithium-Ion Battery Separator. *Polym. Degrad. Stab.* **2023**, *216*, 110469. [CrossRef]
9. Liu, B.; Jia, Y.; Yuan, C.; Wang, L.; Gao, X.; Yin, S.; Xu, J. Safety Issues and Mechanisms of Lithium-Ion Battery Cell upon Mechanical Abusive Loading: A Review. *Energy Storage Mater.* **2020**, *24*, 85–112. [CrossRef]
10. Christensen, J.; Newman, J. Stress Generation and Fracture in Lithium Insertion Materials. *J. Solid State Electrochem.* **2006**, *10*, 293–319. [CrossRef]

11. Gor, G.Y.; Cannarella, J.; Prévost, J.H.; Arnold, C.B. A Model for the Behavior of Battery Separators in Compression at Different Strain/Charge Rates. *J. Electrochem. Soc.* **2014**, *161*, F3065–F3071. [CrossRef]
12. Peabody, C.; Arnold, C.B. The Role of Mechanically Induced Separator Creep in Lithium-Ion Battery Capacity Fade. *J. Power Sources* **2011**, *196*, 8147–8153. [CrossRef]
13. Xiao, X.; Wu, W.; Huang, X. A Multi-Scale Approach for the Stress Analysis of Polymeric Separators in a Lithium-Ion Battery. *J. Power Sources* **2010**, *195*, 7649–7660. [CrossRef]
14. Shi, D.; Xiao, X.; Huang, X.; Kia, H. Modeling Stresses in the Separator of a Pouch Lithium-Ion Cell. *J. Power Sources* **2011**, *196*, 8129–8139. [CrossRef]
15. Oh, K.Y.; Epureanu, B.I. A Novel Thermal Swelling Model for a Rechargeable Lithium-Ion Battery Cell. *J. Power Sources* **2016**, *303*, 86–96. [CrossRef]
16. Oh, K.Y.; Epureanu, B.I.; Siegel, J.B.; Stefanopoulou, A.G. Phenomenological Force and Swelling Models for Rechargeable Lithium-Ion Battery Cells. *J. Power Sources* **2016**, *310*, 118–129. [CrossRef]
17. Xu, J.; Wang, L.; Guan, J.; Yin, S. Coupled Effect of Strain Rate and Solvent on Dynamic Mechanical Behaviors of Separators in Lithium Ion Batteries. *Mater. Des.* **2016**, *95*, 319–328. [CrossRef]
18. Xu, J.; Liu, B.; Wang, X.; Hu, D. Computational Model of 18650 Lithium-Ion Battery with Coupled Strain Rate and SOC Dependencies. *Appl. Energy* **2016**, *172*, 180–189. [CrossRef]
19. Liu, B.; Zhao, H.; Yu, H.; Li, J.; Xu, J. Multiphysics Computational Framework for Cylindrical Lithium-Ion Batteries under Mechanical Abusive Loading. *Electrochim. Acta* **2017**, *256*, 172–184. [CrossRef]
20. Duan, X.; Wang, H.; Jia, Y.; Wang, L.; Liu, B.; Xu, J. A Multiphysics Understanding of Internal Short Circuit Mechanisms in Lithium-Ion Batteries upon Mechanical Stress Abuse. *Energy Storage Mater.* **2022**, *45*, 667–679. [CrossRef]
21. Yuan, C.; Wang, L.; Yin, S.; Xu, J. Generalized Separator Failure Criteria for Internal Short Circuit of Lithium-Ion Battery. *J. Power Sources* **2020**, *467*, 228360. [CrossRef]
22. Doyle, M.; Fuller, T.F.; Newman, J. Modeling of Galvanostatic Charge and Discharge of the Lithium / Polymer / Insertion Cell. *J. Electrochem. Soc.* **1993**, *140*, 1526–1533. [CrossRef]
23. Doyle, M.; Newman, J. Comparison of Modeling Predictions with Experimental Data from Plastic Lithium Ion Cells Comparison of Modeling Predictions with Experimental Data from Plastic Lithium Ion Cells. *J. Electrochem. Soc.* **1996**, *143*, 1890. [CrossRef]
24. Smith, K.A.; Rahn, C.D.; Wang, C. Model Order Reduction of 1D. *J. Dyn. Syst. Meas. Control* **2008**, *130*, 1–8. [CrossRef]
25. Smith, K.A.; Rahn, C.D.; Wang, C. Control Oriented 1D Electrochemical Model of Lithium Ion Battery. *Energy Convers. Manag.* **2007**, *48*, 2565–2578. [CrossRef]
26. Lin, X.; Perez, H.E.; Siegel, J.B.; Stefanopoulou, A.G.; Li, Y.; Anderson, R.D.; Ding, Y.; Castanier, M.P. On-Line Parameterization of Lumped Thermal Dynamics in Cylindrical Lithium Ion Batteries for Core Temperature Estimation and Health Monitoring. *IEEE Trans. Control Syst. Technol.* **2012**, *21*, 1745–1755.
27. Lin, X.; Perez, H.E.; Mohan, S.; Siegel, J.B.; Stefanopoulou, A.G.; Ding, Y.; Castanier, M.P. A Lumped-Parameter Electro-Thermal Model for Cylindrical Batteries. *J. Power Sources* **2014**, *257*, 1–11. [CrossRef]
28. Kim, Y.; Siegel, J.B.; Stefanopoulou, A.G. A Computationally Efficient Thermal Model of Cylindrical Battery Cells for the Estimation of Radially Distributed Temperatures. In Proceedings of the 2013 American Control Conference, Washington, DC, USA, 17–19 June 2013; IEEE: Piscataway, NJ, USA; pp. 698–703.
29. Bernardi, D.; Pawlikowski, E.; Newman, J. A General Energy Balance for Battery Systems. *J. Electrochem. Soc.* **1985**, *132*, 5. [CrossRef]
30. Chen, S.C.; Wan, C.C.; Wang, Y.Y. Thermal Analysis of Lithium-Ion Batteries. *J. Power Sources* **2005**, *140*, 111–124. [CrossRef]
31. Zhang, X.; Shyy, W.; Marie Sastry, A. Numerical Simulation of Intercalation-Induced Stress in Li-Ion Battery Electrode Particles. *J. Electrochem. Soc.* **2007**, *154*, A910. [CrossRef]
32. Zhang, X.; Sastry, A.M.; Shyy, W. Intercalation-Induced Stress and Heat Generation within Single Lithium-Ion Battery Cathode Particles. *J. Electrochem. Soc.* **2008**, *155*, A542. [CrossRef]
33. Bulla, M.; Kolling, S.; Sahraei, E. A Material Model for the Orthotropic and Viscous Behavior of Separators in Lithium-Ion Batteries under High Mechanical Loads. *Energies* **2021**, *14*, 4585. [CrossRef]
34. Bulla, M.; Kolling, S.; Sahraei, E. An Experimental and Computational Study on the Orthotropic Failure of Separators for Lithium-Ion Batteries. *Energies* **2020**, *13*, 4399. [CrossRef]
35. Makki, M.; Ayoub, G.; Lee, C.W. Modeling the Anisotropic Behavior of Highly Orthotropic Lithium-Ion Batteries Polymer Separators. *Int. J. Solids Struct.* **2023**, *264*, 112102. [CrossRef]
36. Zhou, W.; Hao, F.; Fang, D. The Effects of Elastic Stiffening on the Evolution of the Stress Field within a Spherical Electrode Particle of Lithium-Ion Batteries. *Int. J. Appl. Mech.* **2013**, *5*, 1350040. [CrossRef]
37. Drake, S.J.; Wetz, D.A.; Ostanek, J.K.; Miller, S.P.; Heinzel, J.M.; Jain, A. Measurement of Anisotropic Thermophysical Properties of Cylindrical Li-Ion Cells. *J. Power Sources* **2014**, *252*, 298–304. [CrossRef]
38. Dieter, D.E. *Mechanical Metallurgy*, 3rd ed.; McGraw-Hill Publishing Company: New York, NY, USA, 1986.

batteries

MDPI

Article

Influence of the Arrangement of the Cells/Modules of a Traction Battery on the Spread of Fire in Case of Thermal Runaway

Ana Olona [1] and Luis Castejón [2,*]

1 Research Department, Electric Vehicle and Mobility Area, Instituto de Investigación Sobre Vehículos, S.A., Ctra. N232, km 273, 50690 Pedrola, Spain; a.olona@centro-zaragoza.com
2 Department of Mechanical Engineering, University of Zaragoza, C/María de Luna s/n, 50018 Zaragoza, Spain
* Correspondence: luiscast@unizar.es; Tel.: +34-976762556

Abstract: When designing the battery of an electric vehicle, different parameters must be considered to obtain the safest arrangement of the battery/modules/cells from the mechanical and thermal points of view. In this study, the thermal runaway propagation mechanism of lithium-ion cells is analyzed as a function of their arrangement within a battery pack in case of a fire propagation of a battery pack in which a thermal runaway has occurred. The objective is to identify which cell/module arrangement is most critical within the battery pack, using microscopic analysis of the structure and chemical composition of the most damaged cells, both horizontally and vertically, of a battery belonging to a burnt vehicle. And their final condition was compared with the condition of new cells of the same type. In this way, the structure and chemical composition of the cathode, anode, and separator after thermal runaway were compared. This research was carried out to obtain information to understand the mechanical properties of lithium-ion cells and their behavior after thermal runaway heating leading to the propagation of a fire. Through the analysis carried out, it is concluded that cells placed in a vertical arrangement have worse behavior than cells in a horizontal arrangement. Regarding the safety of the battery, the results of this study will allow us to determine which arrangement and structure of the cells in the battery pack is safer against thermal runaway due to thermal failure.

Keywords: thermal runaway; pouch cell; chemical composition analysis; structural analysis; thermal failure test

1. Introduction

Battery fire safety remains an important concern due to high-temperature instability [1]. Over the past 30 years, there have been many accidents involving lithium-ion batteries, indicating that those batteries are not fire-safe. In addition, several serious fire accidents have occurred in Tesla Model S electric vehicles [2–7]. Some caught fire while being driven, and others during the charging and discharging process.

In general, lithium-ion batteries are most vulnerable when subjected to high temperature, electrical failure, or mechanical impact [8]. Lithium-ion batteries have limited temperature and voltage thresholds, and their components are unstable. Namely, the electrolyte is flammable, and the electrodes are made of a material with a high specific energy. All these characteristics imply this type of cell is prone to thermal runaway (TR) in some cases of abuse [9,10].

A lithium-ion battery may fail because of mechanical, electrical, or thermal failures that occur at the cell, module, or battery pack level (Figure 1). These failures occur in successive series, leading to the phenomenon of TR. Thermal failure due to overheating can occur, which may be due to incorrect contact connections or excessive heat near the battery pack, such as in a fire caused by combustion in a car accident. In addition, overheating may occur during fast charging as it uses high voltage and current levels. Thermal failure can also be due to an internal short circuit, which occurs due to cell separator failures, manufacturing defects, or lithium build-up (dendrite formation) on the anode. This failure may also be due to defects originating during cell manufacture, which induce an internal short circuit.

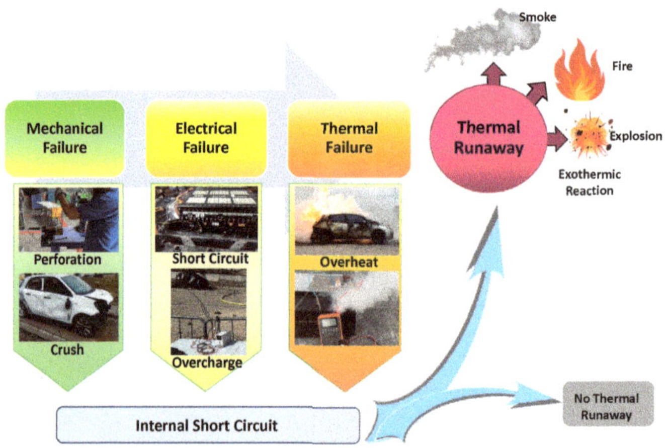

Figure 1. Failure mechanism related to lithium-ion battery failure and subsequent thermal runaway mechanism for electric vehicles (EVs) [11].

Thermal abuse failure due to external heat or flame exposure, which is also due to heat build-up inside from the incorrect removal of reaction heat or the initiation of unwanted internal exothermic reactions, can lead to thermal runaway. Controlled fire tests carried out on a complete electric vehicle with an installed high-voltage battery, such as scale tests with different cells and different cell arrangements, will allow characterization of the behavior of the cells in a fire due to electrical failure, high temperature, or significant deformation. The thermal runaway reaction is influenced by the stranded energy, which is the energy that remains stored inside the lithium-ion battery even after it has been discharged to a certain extent. In the context of a fire involving a lithium-ion battery, the stranded energy can represent a significant risk. To handle safely a lithium-ion battery involved in a fire, it is very important to be aware of the potential for trapped energy and to ensure that the battery is fully discharged and cooled before attempting to handle or transport it.

When a high-voltage battery is deformed or damaged, energy remains stored inside any undamaged battery cells and modules, and there is no way to discharge it. That stranded energy can cause a high-voltage battery fire in an electric vehicle to ignite several times after firefighters extinguish the fire. Emergency responders cannot measure how much energy remains inside a damaged battery, and they have no way of draining that energy. Engineers or other specialists can check the remaining voltage using the battery management system if the system is operational, and some high-voltage batteries have incorporated discharge ports. However, when an accident occurs, the high-voltage battery system can be damaged and deformed, making it impossible to access the battery management system or discharge ports. On the one hand, the duration of the post-crash fire and, on the other hand, the multiple reignitions of the battery could indicate that the battery contained stranded energy. Several researchers [12] from Oak Ridge National Laboratory conducted a study in which they concluded that lithium-ion cells with SoC > 80% subjected to mechanical damage were more prone to thermal runaway, while cells with a low SoC would not suffer thermal runaway but would only experience a slight temperature rise.

Several studies have been carried out to analyze the behavior of lithium-ion cells under external heating conditions. Huang et al. [13] heated three 50 Ah lithium-ion cells at different SoCs (States of Charge) with an electric heater and concluded that the cells only turned on when the cell surface reached between 112 and 121 °C. The cell response and thermal runaway characteristics depend on the SoC. If the SoC is increased, then the time elapsed from heating to ignition is gradually reduced [13–15], and the reactions become more dangerous. Feng et al. [16] divided the internal chemical reactions occurring in lithium-ion batteries into different stages depending on the temperature reached. When lithium-ion batteries are overheated, their temperature gradually increases. The SEI (solid electrolyte interphase) film begins to break up when the cell reaches 90–130 °C [17]. As the SEI breaks up, the electrolyte reaches the graphite surface more easily and readily reacts with the intercalated lithium, releasing oxygen and generating more heat [18]. On the other hand, polyethylene (PE) can melt at 130–140 °C, causing an internal micro-short circuit and, thus, a continuous increase in battery temperature [19]. When the temperature reaches around 200 °C, the cathode materials start to decompose and release oxygen [20]. In addition, the decomposition reaction of the SEI film and the chemical reaction between the cathode metal oxide materials and the electrolyte is accelerated, as well as the reaction between the negative active substances and the electrolyte, which will produce more heat and gases [15] such as oxygen, HF, CO and NO [21–25]. As the temperature increases, the internal chemical reaction of the cell is much more complex and intense, generating a large amount of combustible gases and heat [26], followed by a thermal runaway reaction, causing subsequently the lithium-ion cell to burn or even at times explode.

It is important to understand the risk/hazard for the design of single cells and battery packs and modules to minimize or avoid the impact of thermal runaway. However, there are few studies on the methods of assessing the safety of cells or battery modules. Several studies have analyzed the influence of cell arrangement in a lithium-ion battery pack on thermal runaway (TR) propagation in both open and enclosed spaces [27]. Experimental results show that in a closed environment, when the battery's SoC is 100% and the cell spacing is more than 8 mm vertically and 4 mm horizontally, the possibility of thermal propagation for the battery pack is higher than that in an open environment. It is concluded that the TR propagation possibility of the battery pack in a vertical arrangement is higher than in a horizontal arrangement. Studies indicate that TR only occurs when the State of Charge (SoC) is higher than 50%. On the other hand, the critical spacing that triggers TR is 4 mm for a battery with an SoC at 80% and 6 mm for a battery with an SoC at 100% [28]. Feng et al. [29] conducted a mechanical failure (penetration) inducing TR on a module of six cells in series contact, concluded that the cell casing transfers the most heat, and developed a thermally resistant layer placed between the cells in order to prevent the propagation of TR in the module [30]. Lopez et al. [31] analyzed the influence of inter-cell spacing on TR and concluded that cell spacing in the battery pack is a critical factor as it greatly affects TR propagation. They analyzed different battery modules connected in series and concluded that the solution to TR propagation is good heat conduction. Therefore, there are several studies on the propagation of TR in series-connected battery modules [29–31], but there are few studies to analyze the propagation of TR in parallel-connected battery modules. Wike et al. [32] stated that the temperature increase after TR in a parallel-connected battery cell module is higher than if the cells in the module are connected in series. Lamb et al. [33] also found that the temperature increase in a module with parallel connected cells was higher, and as a result, the TR propagation was accelerated. Gao et al. [34] carried out a study of the propagation of TR in a large format battery module with parallel connections, 18 pouch cells in parallel 24 Ah. It is observed that the undamaged cells transfer the current to the cell experiencing TR, which causes it to experience a 10 °C higher temperature compared to a module with the cells connected in series. Therefore, a module with parallel connections is less safe.

In this study, overheating tests of a lithium-ion battery placed in an electric vehicle have been carried out using a fossil fuel heater to analyze how the arrangement of the cells influences the propagation of thermal runaway. The authors hope that this research can provide useful conclusions for the improvement of the Li-ion cell arrangement in electric vehicles with the aim of reducing the damage caused by thermal runaway in lithium-ion batteries in automotive applications.

2. Materials and Methods

2.1. Methodology

The correctness of the results obtained will depend on the ability of the cell tests to reproduce the real behavior of a battery pack. As a starting point, a controlled fire test with a vehicle fitted with a battery pack was carried out to characterize its behavior.

2.1.1. Controlled Fire Test of an Electric Vehicle

Firstly, a controlled fire test was carried out on a complete electric vehicle, specifically a Nissan Leaf, in which the temperature progression and the behavior of the battery were analyzed. The fire was started by positioning a fuel burner on the bottom of the vehicle battery. Time and the battery temperature were monitored through the test with the help of a thermographic camera. The following Table 1 shows the main specifications of the battery analyzed.

Table 1. Specifications of the battery of the burnt Nissan Leaf vehicle.

Vehicle	Battery Electric Vehicle
Date of first registration	17 December 2015
Range	121 km (EPA test)
Battery specifications	
Capacity	24 kWh
Battery Voltage	360 V
Battery	Lithium-ion battery
Cell type	Laminate-type, pouch cells
Cathode Active Material	LMO ($LiMn_2O_4$) with NCA (lithium nickel-cobalt-aluminium oxide $Ni_{0.8}Co_{0.15}Al_{0.05}O_2$)
Anode Active Material	Graphite
Capacity	32.5 Ah
Nominal Voltage	3.75 V
Battery Modules	48
Cells per module	4
Energy Density	157 Wh/kg
Battery Weight	~180 kg
Battery Price	7000 €
State of Charge (SoC)	68.0%

The test started, and the vehicle was becoming hotter and hotter as the combustion process progressed. Within 3 min from the beginning of the test, the battery pack reached a temperature close to 805 °C. It was not until about eight minutes from the beginning that small explosions from the vehicle airbag gas generators began to be heard, followed by the phenomenon of thermal runaway in which the battery cells entered a self-heating and uncontrollable state.

Initially, the fire was extinguished using a fire blanket (Figure 2). After just over ten minutes from the beginning of the test, the temperature in the battery area exceeded 1000 °C. It was at this point that the firefighters proceeded to cover the vehicle with the blanket. Immediately after the vehicle had been covered by the blanket, the surface temperature of the vehicle dropped from approximately 850 °C to about 350 °C. From this moment on, the blanket insulates the vehicle from the radiation emitted by the fire, preventing any element from spreading into the environment and, at the same time, cooling the vehicle. In addition, the blanket prevents the oxygen supply, preventing the fire from continuing when all the oxygen confined under the blanket is consumed.

Figure 2. Zaragoza City Council firefighters covering the burnt vehicle with the fire blanket.

As the minutes passed, the temperature recorded by the thermal imaging camera dropped to around 100 °C. However, electric vehicle fires have a high risk of fire reignition because the battery continues to produce all the elements necessary for the fire to start—namely heat, fuel, and oxygen. If the vehicle is covered by the blanket, there is no risk, as it is only a matter of time before the battery temperature drops. But if the vehicle is uncovered, the risk of a restart is very high. This is why 9 min after, the vehicle was completely covered, it was decided to remove the blanket to show how the process of restarting the fire is reproduced. Just after removing the blanket, it can be seen how the fire restarts from the battery area and quickly spreads again throughout the vehicle.

As the test continued, the fire increased again the temperature during the following two and a half minutes, reaching a peak temperature of around 600 °C. Once this temperature is reached, the vehicle is covered again to insulate and extinguish the vehicle fire again, given that the blanket is reusable.

The following shows the state of the battery pack (Figure 3a) and a detailed view of the battery modules (Figure 3b) before the fire test:

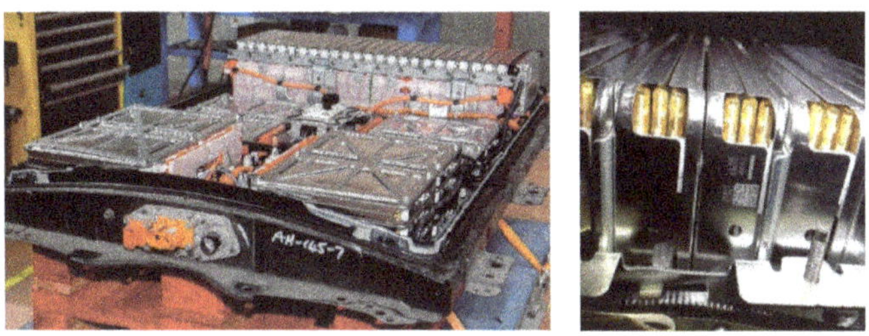

(**a**) (**b**)

Figure 3. (**a**) Battery pack before fire test. (**b**) Detail of a battery module constructed with a 4-cell configuration).

To finalize the test, once the blanket was removed, the battery was cooled with water and the vehicle was left in quarantine to check that the fire did not restart.

The following table (Table 2) gives details of the fire test specifications:

Table 2. Characteristics of the controlled fire test carried out on the Nissan Leaf vehicle.

Fire test specifications	
Place and date	Zaragoza, 5 December 2019
Vehicle under test	Battery Electric Vehicle
TEST	
Vehicle	Vehicle only with voltage battery
Fire	Ignition Burners (fossil fuel)
Method to extinguish the fire	Bridgehill Car Fire Blanket
Temperature monitoring	Thermal Imaging Infrared Camera: FLIR T640
Thermal Imaging Infrared camera distance	13.7 m
Ambient Temperature	7.1 °C
Average Wind Speed	2.5 m/s
Maximum Wind Speed	8.9 m/s
Maximum Temperature reached	~1000 °C
Time to start Battery Thermal Runaway	~8 min
After removing the blanket	
Amount of water to extinguish the fire	400 L

Images of the electric vehicle fire test explained above are shown below (Figure 4):

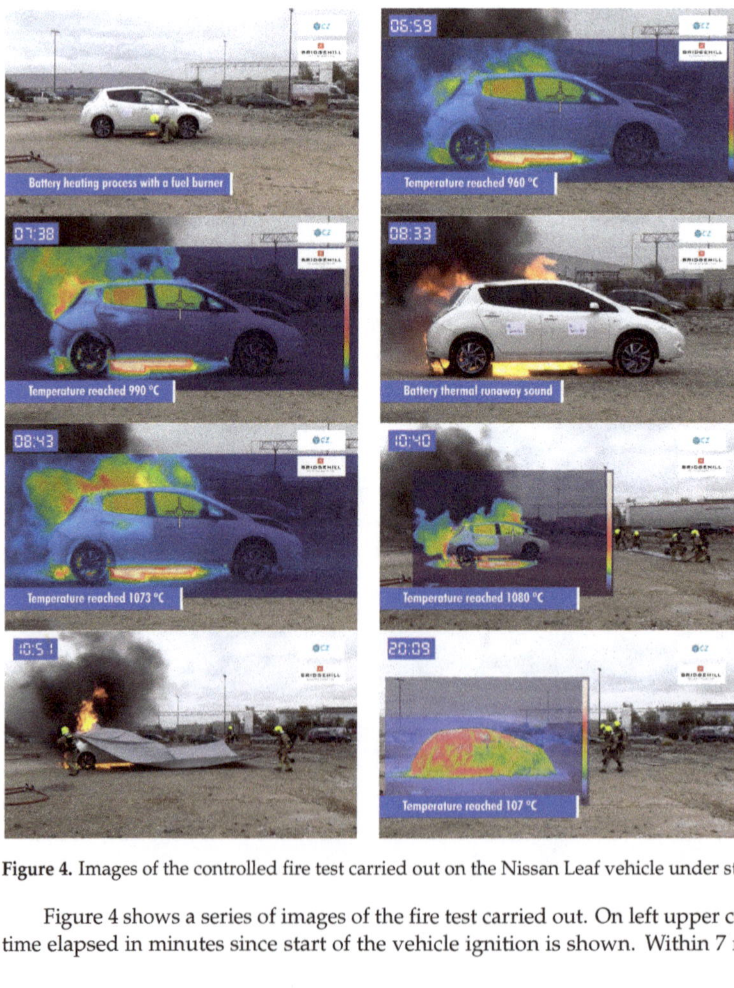

Figure 4. Images of the controlled fire test carried out on the Nissan Leaf vehicle under study.

Figure 4 shows a series of images of the fire test carried out. On left upper corner, the time elapsed in minutes since start of the vehicle ignition is shown. Within 7 min from the start of the test (second image upper right), the battery pack reached a temperature close to 960 °C. It was not until about eight minutes from the begging that small explosions from the vehicle airbag gas generators began to be heard, followed by the phenomenon of thermal runaway in the battery cells (fourth image). The range of the thermal event onset temperature is 80–150 °C.

2.1.2. Identification of the Most Damaged Cells in the Analyzed Battery

Once the test had been carried out and the necessary quarantine time had elapsed, the cells were dismantled (Figure 5) to see how they behaved and their condition according to the position they were in. The aim was to analyze how the enclosure was damaged according to the position of each cell within the enclosure. Moreover, deformation suffered by each of the cells was studied. Finally, the condition of these cells has been perfectly documented.

Figure 5. Pictures of the state of the battery pack, the module, and the lithium-ion cells inside the battery pack.

Once the burnt battery of the Nissan Leaf had been disassembled, the most damaged modules were visually identified, being those modules that showed the greatest deformation and breakage, both vertically (module 11) and horizontally (module 30), and all modules were disassembled and visually analyzed.

In Figure 6 a number has been assigned to each of the battery modules in order to identify them.

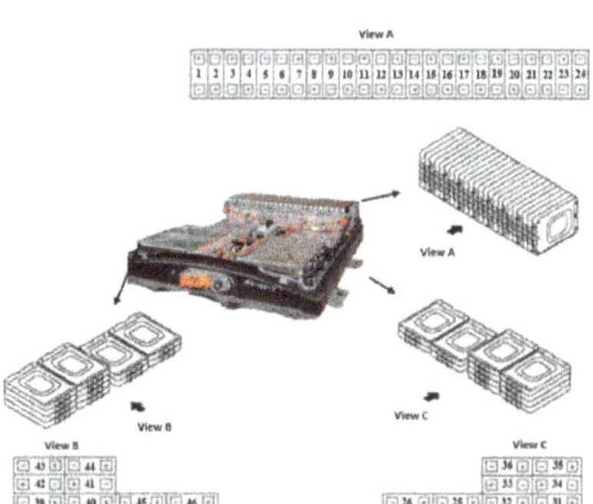

Figure 6. Exploded view and module identification of the traction battery of a first-generation Nissan Leaf.

Figure 7 shows burnt Nissan Leaf battery disassembled in the arrangement in which it is in the vehicle.

Figure 7. Burnt Nissan Leaf battery disassembled in the arrangement in which it is in the vehicle.

The following pictures show the state of the most damaged modules, firstly, Figure 8 shows the state of the burnt Nissan Leaf vehicle battery module 30 and secondly, Figure 9 shows the state of the burnt Nissan Leaf vehicle battery module 11:

Figure 8. Images of the condition of module 30 (horizontal arrangement) of the battery pack of the burnt Nissan Leaf vehicle. This module has been identified as the most damaged among the modules in the horizontal arrangement.

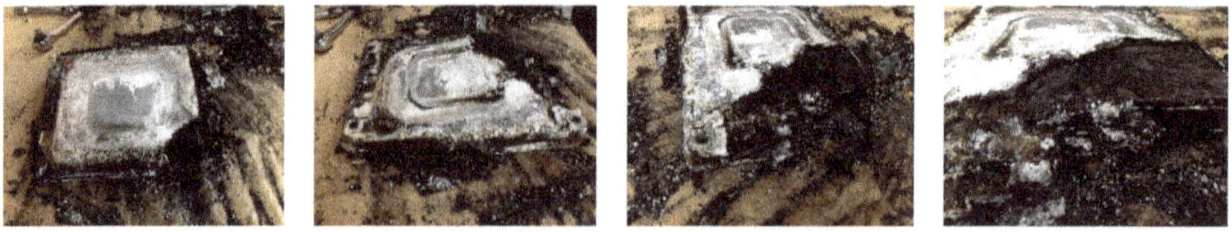

Figure 9. Images of the condition of module 11 (vertical arrangement) of the battery pack of the burnt Nissan Leaf vehicle. This module has been identified as the most damaged among the vertically arranged modules.

The following Figure 10 identifies the position of the most damaged modules in the battery pack analyzed:

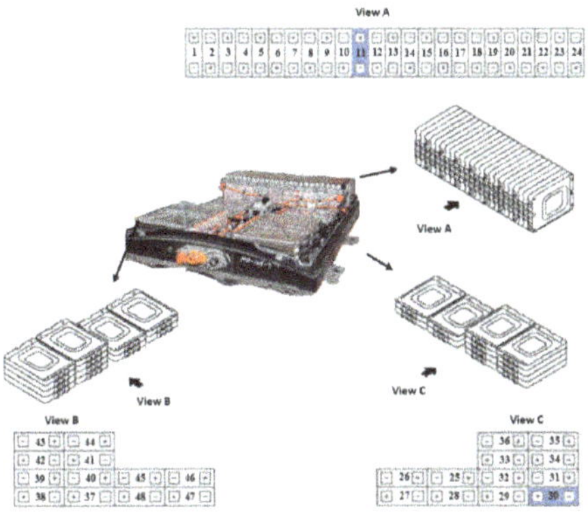

Figure 10. Exploded view of the first-generation Nissan Leaf battery, with identification of the most damaged modules, shaded in blue.

2.1.3. Generation of Samples for Structural and Cross-Sectional Analysis

A study of the structure was carried out by taking samples and analyzing them transversally (there are several layers of anodes and cathodes, as well as separators; it is a stack sample of materials) (see Figure 11), and on the other hand, a surface analysis of the chemical composition was carried out. To perform that, anode, cathode, and separator were separated (see Figure 12).

For the structure analysis test, three transversal section samples were taken from a cell of module 11, three transversal section samples from a cell of module 30, and two transversal section samples from a new cell. For chemical composition and morphology analysis, one sample was taken from the cathode, one sample from the anode, and one sample from the separator for module 11, module 30, and the new cell, respectively. The samples for the tests were obtained at Centro Zaragoza facilities, and once the samples had been obtained, the tests were carried out at the Material Science Department of the University of Zaragoza. Sampling has been carried out considering the indications of a previous study [35]. Images from sample preparation to Sibe tested for the new cell are shown below.

(a) (b) (c)

Figure 11. Sample and obtaining the sample for structural analysis. (**a**) Location of the points from where the samples have been obtained for the structural analysis of the stacked layers: 1. Anode tab (copper); 2. Cathode tab (aluminum); 3. (**b**) Sample generated in position 3. (**c**) Section of the pouch cell once the sample has been extracted.

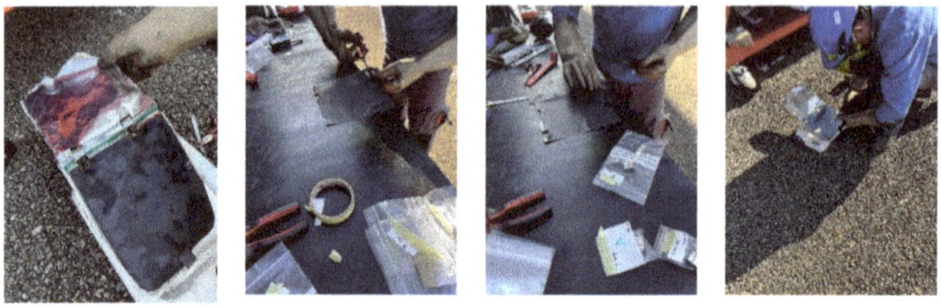

(a) (b) (c) (d)

Figure 12. Obtaining the sample for surface analysis. (**a**) Image of the inside of a pouch cell from where the samples for surface analysis of the stacked layers (anode, cathode, and separator in the case of the intact cell) have been obtained. (**b**) Sample generated from the anode (copper tab). (**c**) Sample generated from the cathode (aluminum tab). (**d**) Sample generated from the separator.

Figure 13 below shows the position from which samples have been taken from a calcined cell of module 30:

Figure 13. Obtaining samples from a module 30 calcined cell. The letters A–D identify the cell within the module and the numbers 1–3 indicate the position from which the sample has been taken.

For the preparation of the surface sections of the individual layers, after obtaining the samples, with a surface area of approximately 5 mm × 5 mm, of each of the different layers of the cell, anode, cathode, and separator (in the case of the new cell), they were coated with a layer of high purity carbon of 10.40-nanometer thick carbon layer to make them conductive, using a Leica EM ACE600 sputtering coater, (Leica Microsystems, Spain) and placed on an aluminum support with a double-sided carbon adhesive tape so that they do not move from the support during the analysis. A small sample of cobalt is placed to calibrate the equipment.

After obtaining the samples, with a surface area of approximately 10 mm × 10 mm, from the electrode stack, the samples are embedded in resin to improve the preparation result and then polished so that the surface is scratch- and deformation-free, flat, and highly reflective. A prepared sample embedded in resin is shown below (Figure 14).

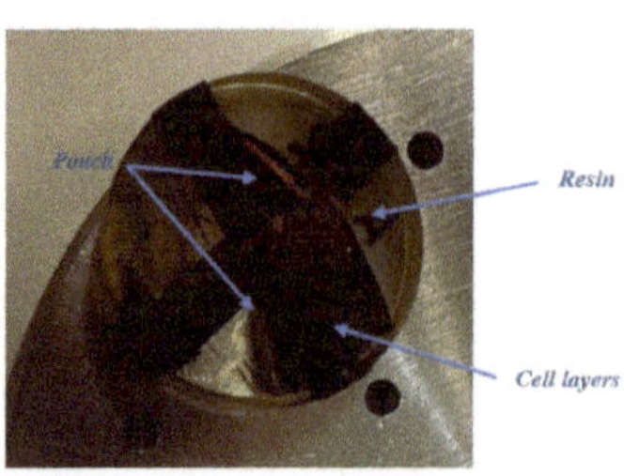

Figure 14. Layer structure of a resin-embedded and polished cell, used to identify cell parameters and cell structure.

2.1.4. Method for Carrying Out the Surface and Structural Analysis

A Carl Zeiss MERLINTM (Carl Zeiss SMT AG, Germany) Field Emission Scanning Electron Microscope (FESEM) was used to perform the analyses. It has a hot-field emission electron emission cannon and allows observations up to 0.8 nm spatial resolution and acceleration voltages between 0.02 and 30 kV. It has secondary and backscattered electron detectors in the chamber and in the column (in-lens). It also has an EDS detector for analysis of the energy of X-Max scattered X-rays (20 mm^2) with SDD (Silicon Drift Detector) from Oxford Instruments (United Kingdom). For the generation of images in scanning electron microscopy, the interaction products most frequently used are secondary electrons (SEs) and backscattered electrons (BSEs). For the distinction and detection of SEs and BSEs, it must consider two parameters: energy and angle distribution. For that mission, an energy-selective backscattered detector (EsB) and an angle-selective backscattered detector (AsB®) have been used. The standard detectors are the in-lens detector (annular SE detector), which detects SE signals and is used to analyze surface structure, and the SE2 detector (Everhart-Thornley type), which detects SE2 signals and is used to analyze topography.

Morphology can be analyzed with SE2 and in lens, backscattered and low-energy backscattered EsB (different gray level due to different chemical composition), and high-energy backscattered AsB.

The following Table 3 show information on different detectors used by the FESEM:

Table 3. Information on the different detectors used by the FSEM.

Standard Detectors	Detected Signals	Typical Application
In-lens detector (annular SE detector)	SE	Surface Structure
SE detector (Everhart-Thornley type)	SE2	Topography
Optional Detectors	**Detected Signals**	**Typical Application**
EsB® detector with filtering grid (in-column detector)	BSE	Pure material contrast
AsB® detector, integrated	BSE	Channeling contrast (cristal orientation), compositional contrast

The SE2 detector detects SEs as well as BSEs. Electrons on the way to the detector are grouped by the collector and directed to the scintillator. The collector voltage (suction voltage), with a range between −250 V and +400 V, generates an electrical field located in front of the detector, thus directing the low-energy SEs toward the scintillator. The collector bias should be set to +300 V for all standard applications. The whole AsB® angular selective backscattered electron detector is built within the GEMINI® lens's pole piece. This eliminates the need to align the AsB® detector with the optical axis in order to image BSEs at extremely small working distances. It is possible to distinguish between low-angle and high-angle BSEs with this detector arrangement. The AsB® detector is equipped with four diodes, and each can be individually adjusted via a menu. While the topography mode displays surface features, the compositional mode generates images that display the specimen's atomic contrast. Secondary electrons have an energy <50 eV and are inelastically scattered primary electrons. They are produced from the adjacent surface of the incoming primary electron beam, providing the best lateral resolution when it comes to several nanometres. The backscattered electrons disperse elastically in the field of the atomic nucleus. The energy range is from >50 eV to the excitation energy. When the atomic number or atomic weight of a phase or region is higher, a greater number of electrons will be backscattered from this area of the sample (material contrast).

The Figure 15 below shows two views of the surface tests carried out.

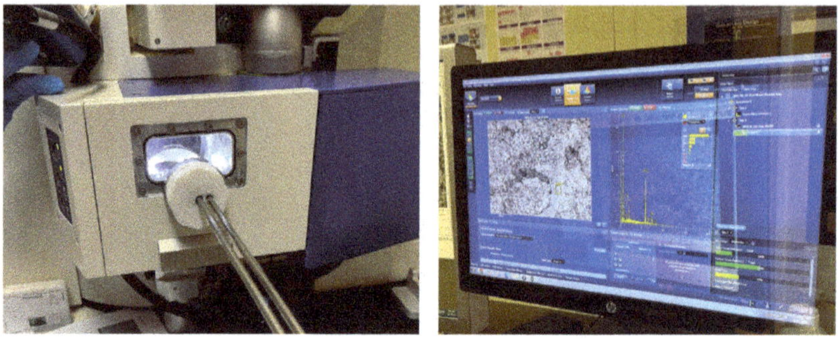

Figure 15. Views of the surface tests carried out.

2.1.5. Method for Carrying Out the Chemical Analysis

To obtain information on the chemical composition of the analyzed surface, electron energy loss spectroscopy combined with electron microscopy was used. To do this, high-energy electrons were directed toward the samples, which caused the electrons in the inner shell to ionize, creating a vacancy in the inner layer. This unfilled position is occupied by an electron of a higher energy level, releasing energy that might be observed as an X-ray quantum or Auger electron (electron that is forced to leave the atom). This energy, known as X-ray radiation, is unique to each element. An EDAX Super Octane (EDAX, Pleasanton, CA, USA) silicon drift detection system (energy resolution of about 123 eV @ Mn Ka) with a silicon nitride window for optimal sensitivity in the low-energy region was used to capture the EDX spectra.

3. Results and Discussion

The conclusions obtained from the microscopic analysis of lithium-ion cells are the basis for research into the operation, safety, and degradation of lithium-ion batteries. There are studies [35] in which analyses have been carried out using images taken under a microscope to obtain information that allows us to understand the mechanical properties of lithium-ion cells. Parameters such as layer thickness, material composition, and surface properties are important in the analysis and further development of lithium-ion batteries. The purpose of this section is to compare the structure and chemical composition of the anode, cathode, and separator in cells that have undergone thermal runaway compared to an intact (new) cell.

In this section, not only the methods used to procure the surface properties and chemical composition of all the batteries but also the results of SEM images of transverse sections of zone 1 (Anode) of an original cell and a cell from module 11 and module 30 of the burnt Nissan Leaf vehicle battery are presented. The surface properties and chemical composition of the different components of an original cell and a burnt cell from module 11 and module 30 are also analyzed.

3.1. Surface Test Results: Surface Properties and Chemical Composition Analysis

To carry out the surface analysis tests, three aluminum supports are prepared: support A, support B, and support C. Two cathode samples, one from the upper cathode and one from the lower cathode, and two anode samples, one from the upper anode and one from the lower anode, are prepared in support A from cell B of module 30. Two cathode samples, one from the upper cathode and one from the lower cathode, and two anode samples, one from the upper anode and one from the lower anode, are prepared in support B from cell C of module 11. And in support C, a sample of the cathode, anode, and separator of a new Nissan Leaf cell is prepared.

The following images (Figure 16) compare different pictures obtained with FESEM of the upper cathode of module 30, module 11, and the new cell.

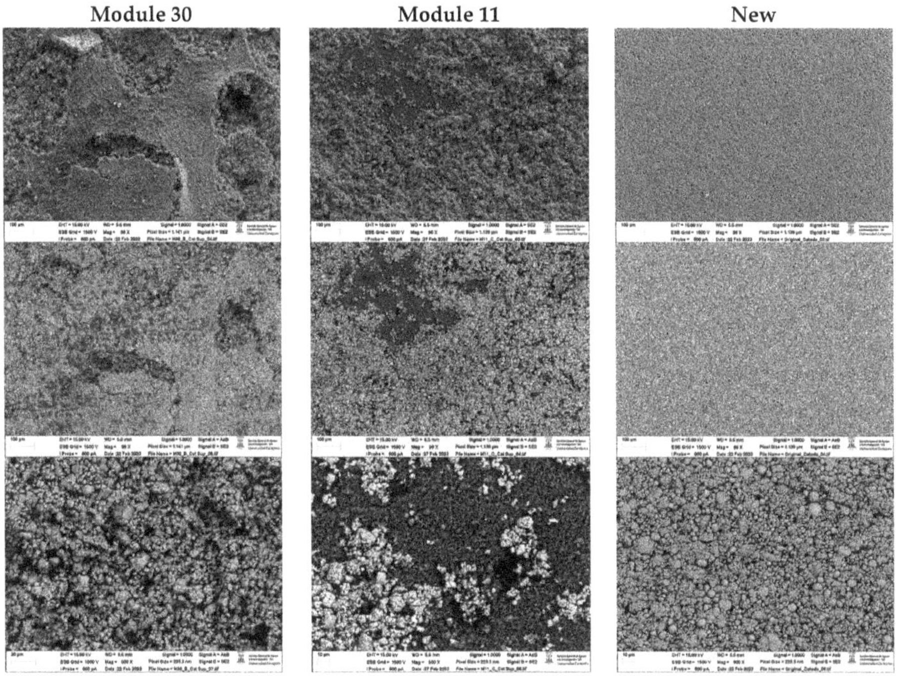

Figure 16. Comparison of images obtained with the FESEM of different areas of the upper cathode of module 30, module 11, and the new cell.

The following images compare different images obtained with FESEM of the lower cathode of module 30, module 11, and the new cell.

With respect to the cathode electrodes in the figure above, the cathode of an original cell has been compared with two cathodes (one upper and one lower) of module 30 and module 11 of the battery of the burning Nissan Leaf after experiencing the thermal runaway. It should be remarked that rather than a single particle size, most materials, especially the active materials used in battery electrodes, contain a variety of particle sizes.

It is noted in Figure 17 that after experiencing the thermal runaway phenomenon, the cathode surface is covered with off-white floccules. The distribution of these floccules on the surface of the cathode electrodes is not uniform, being more concentrated in one area and more dispersed in another area.

Due to the morphology of these substances, it can be deduced that they are decomposed separators due to the high temperature since, at 150 °C, the separator shrinks and groups together. In other words, after TR, the separator decomposes as the temperature increases.

The temperature of decomposition of a PE-PP-based polymer is more than 300 °C. Therefore, it is appropriate to cover the surfaces of the cathode electrodes with unevenly distributed separator particles.

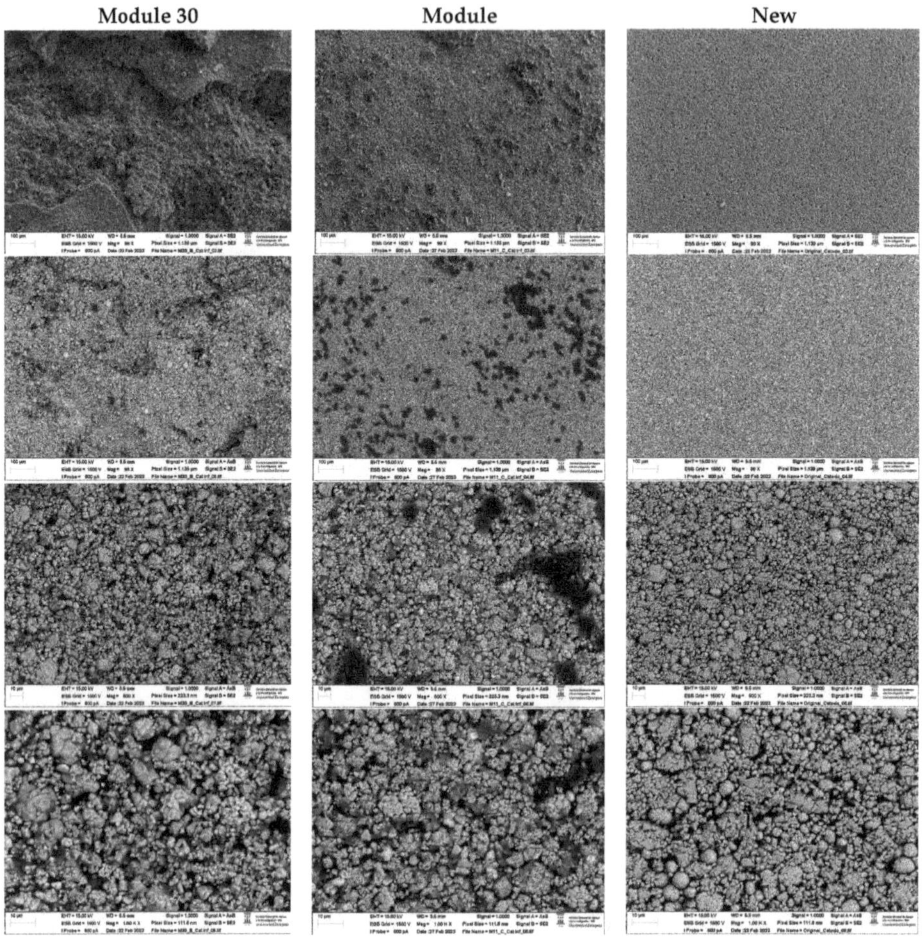

Figure 17. Comparison of images obtained with the FESEM of different areas of the lower cathode of module 30, module 11, and the new cell.

It should be noted that the burned modules being analyzed had a 68% SoC (State of Charge). As the SoC increases, the number of flocs increases greatly, and darker-colored and spherical flocs appear, indicating that having an elevated SoC during TR, the reaction temperature inside the cell is higher, and the agglomeration and decomposition reactions of the separators are more intense [36].

After TR, the spherical cluster structure of the cathode active materials and the flake structure of the graphite were destroyed. The cathode and anode are joined together, and it is difficult to identify them (Figures 5 and 13). There are also irregularly shaped impurities/debris, some rectangular and some spherical, and they are chaotically distributed. When the temperature is increased due to the TR, the aluminum current collector of the chaotic electrode had oxidized and adhered to the anode side.

Therefore, it is normal to observe in the cathode traces of fragments of cathode materials after TR, ash from the separators and cathode material, and products of exothermic reactions and graphite peels (Figure 17, images of the last row).

It is observed that in the case of module 11, there are more dark-colored flocs than in the case of module 30, and there are more in the upper cathode than in the lower cathode. Therefore, it is concluded that module 11, in a vertical arrangement, experiences a higher temperature in the TR at the same SoC as module 30, which is in a horizontal arrangement. If the structural transformation (morphology analysis with InLens) of the cathode materials after the TR is analyzed (Figures 18 and 19), it is observed that in the case of modules 30 and 11 with a SoC of 68% after the TR, the layered structure of the material was destroyed, the resulting particles dispersed out of the original layered structure and subsequent adhesion occurred. The cathode (positive electrode) material reacted at high temperatures and decomposed. On the other hand, carbon particles from the anode (negative electrode) were doped into the cathode (positive electrode) through the damaged diaphragm.

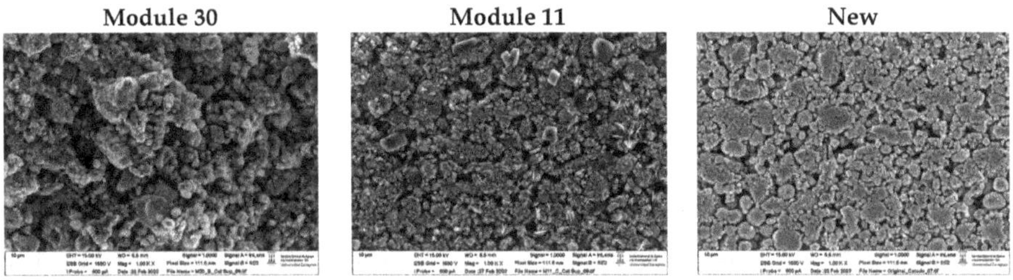

Figure 18. Morphological analysis of the transformation of the upper cathode of Module 30 and Module 11 after the thermal runaway compared with the initial cathode structure of a new cell.

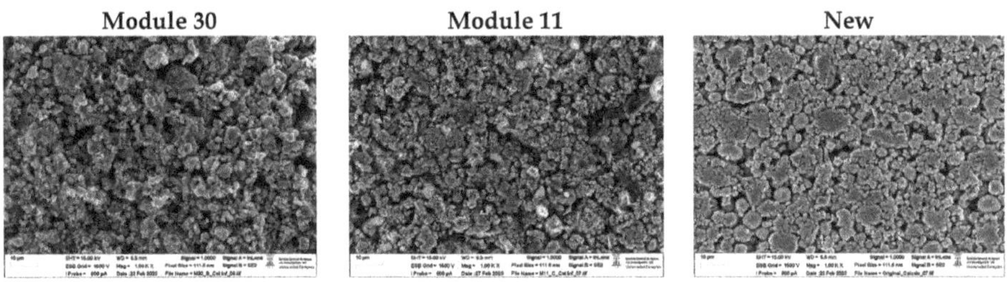

Figure 19. Morphological analysis of the transformation of the lower cathode of Module 30 and Module 11 after the thermal runaway compared with the initial cathode structure of a new cell.

It is observed that in the case of the lower cathode of module 30 (horizontal arrangement), the particles are smaller compared to the upper cathode of module 30; this may be due to the higher temperature as it is more exposed.

In the case of module 11 (vertical arrangement), there is no difference between the particle size of the upper cathode and the lower cathode. Hereafter, the chemical composition analysis of the upper cathode of module 30 and module 11 will be presented in comparison to the composition of the original cathode. The cathode showed a granular surface structure, as can be seen in the following image (Figure 20). The largest particles were about 6 μm in size. The following chemical elements were identified by EDXS: nickel (Ni), manganese (Mn), cobalt (Co), oxygen (O), and carbon (C). This led to the conclusion that this compound has the chemistry of a $LiNiMnCoO_2$ (NMC) cathode. Two different types of compounds were observed on the cathode surface, marked in Figure 21 by two red boxes. The results of a detailed analysis of the formed compounds show that the compound marked with "1" had the NMC chemistry mentioned above, while the second compound was determined to be a manganese oxide compound (LMO), some fluorine was also observed that may be from the electrolyte (LiPF6) when in contact with the cathode and anode. Considering these issues, cathode layer chemistry could be concluded as a mixture of NMC and LMO.

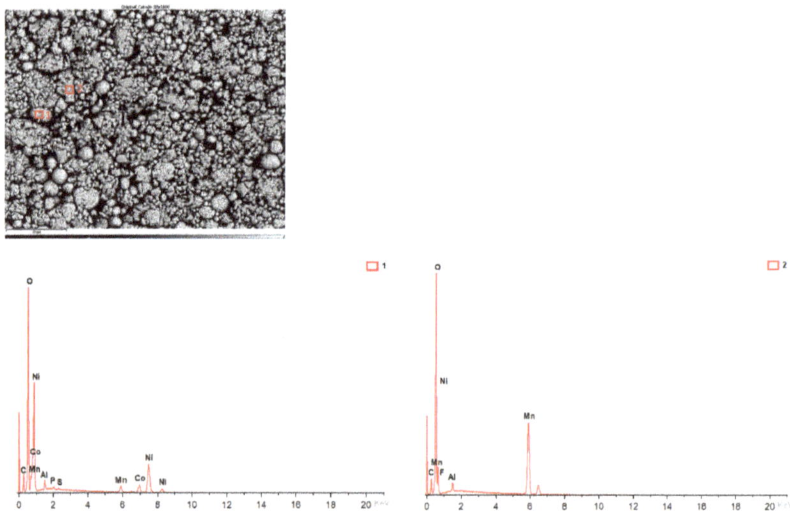

Figure 20. Chemical composition of the original cathode.

During the EDXS measurement, different areas were focused on, and the corresponding peaks are shown in Figure 20. Details of the two spectra of the component values measured in atomic and weight % are listed in Table 4.

Table 4. EDXS weight ratio of components in original cathode using two spectra focused on two distinct areas.

		C		O		F		Al		Mn		Co		Ni	
		Weight (%)	Atomic (%)	Weight (%)	Atomic (%)	Weight (%)	Atomic (%)	Weight (%)	Atomic (%)	Weight (%)	Atomic (%)	Weight (%)	Atomic (%)	Weight (%)	Atomic (%)
Compound 1: $LiNiMnCoO_2$	Spectrum 1	8.46	19.32	29.5	50.6			1.42	1.44	3.89	1.94	9.57	4.45	46.69	21.82
Compound 2; LMO + F	Spectrum 2	5.19	11.84	31.29	53.52	2.54	3.66	1.17	1.19	59.8	29.79				

Next, the chemical composition of the upper cathode of the ignited module 30 is analyzed. The upper cathode of module 30 no longer shows a defined granular surface, as in the case of the original cathode; the structure was broken. The largest particles are about 5 µm in size. The following chemical elements were identified by EDXS: nickel (Ni), manganese (Mn), cobalt (Co), oxygen (O), fluorine (F), phosphorus (P), carbon (C), and aluminum (Al). Two different compound types were observed on the cathode surface of module 30, marked in Figure 21 by two red crosses. The results obtained from an exhaustive analysis of the formed compounds show that the compound marked with "1" had the chemistry of a compound with manganese and oxygen and some fluorine and phosphorus, while in the rectangular-shaped compound "3" fluorine and nickel were observed. This last compound is due to the reactions that have taken place in the TR, and that is why there are leftovers of the electrolyte.

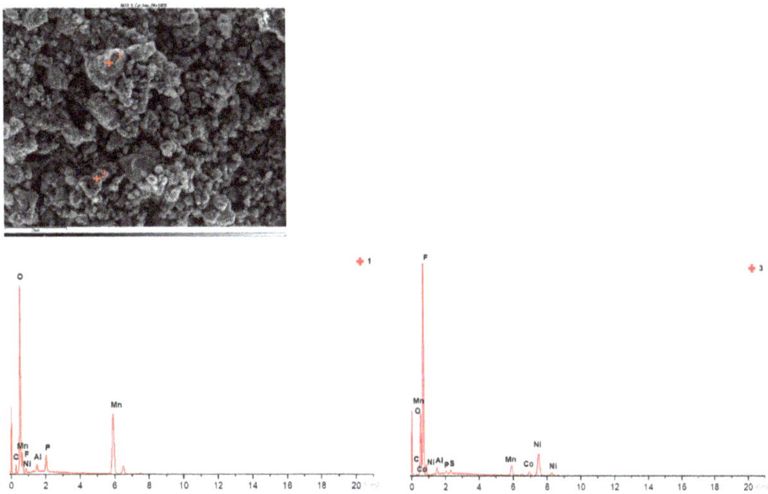

Figure 21. Chemical composition of the upper cathode of module 30.

During the EDXS measurement, different areas were focused on, and the corresponding peaks are shown in Figure 21. Details of the two spectra of the component values measured in atomic and weight % are listed in Table 5.

Table 5. EDXS weight ratio of components in the upper cathode of the ignited module 30 using two spectra focused on two distinct areas.

		O		F		Al		P		Mn		Ni	
		Weight (%)	Atomic (%)	Weight (%)	Atomic (%)	Weight (%)	Atomic (%)	Weight (%)	Atomic (%)	Weight (%)	Atomic (%)	Weight (%)	Atomic (%)
Compound 1: Mn and O, with some F and P	Spectrum 1	26.91	50.88	6.7	10.66	1.26	1.41	2.94	2.87	60.38	33.24	1.81	0.93
Compound 2: F and Ni	Spectrum 2	30.55	47.88	20.86	27.54	4.49	4.17	1.47	1.19	34.57	15.78	7.19	3.07

Next, the chemical composition of the lower cathode of the ignited module 30 is analyzed. The lower cathode of module 30 also does not show a defined granular surface, as in the case of the original cathode; the structure was broken. The larger particles are about 4 µm in size, so the particles are smaller than in the case of the upper cathode of the same module, probably because they have been exposed to higher temperature values. Two different compound types were observed on the surface of the lower cathode of module 30, marked in Figure 22 with two red crosses. A detailed analysis of the compounds obtained shows that the compound marked with "1" had the chemistry of a compound with oxygen and manganese, while it was determined that the second structure was also the chemistry of a nickel and oxygen compound; some manganese, fluorine, cobalt, and aluminum were also observed.

During the EDXS measurement, different areas were focused on, and the corresponding peaks are shown in Figure 22. Details of the two spectra of the component values measured in atomic and weight % are listed in Table 6.

The distribution of elements in each area of the upper cathode of module 30 is also mapped. This information is also collected with the EDS detector and then processed with the AZtec software (Figure 23).

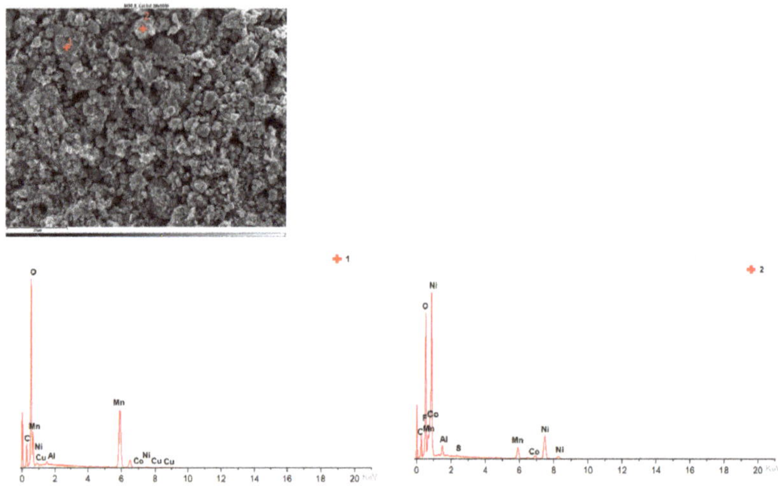

Figure 22. Chemical composition of the lower cathode of module 30.

Table 6. EDXS weight ratio of components in the lower cathode of the ignited module 30 using two spectra focused on two distinct areas.

		O		F		Al		Mn		Co		Ni	
		Weight (%)	Atomic (%)	Weight (%)	Atomic (%)	Weight (%)	Atomic (%)	Weight (%)	Atomic (%)	Weight (%)	Atomic (%)	Weight (%)	Atomic (%)
Compound 1: O and Mn	Spectrum 1	29.71	59.05			0.59	0.7	67.4	39.01	0.74	0.4	1.56	0.85
Compound 2: O and Ni, with some Mn, F, Co and Al	Spectrum 2	24.77	49.88	6.17	10.46	2.04	2.43	10.63	1.19	5.68	3.1	50.56	27.74

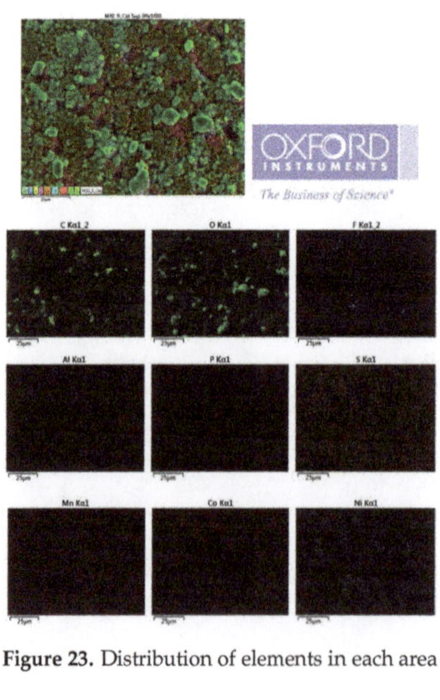

Figure 23. Distribution of elements in each area of the upper cathode of module 30.

Below (Figure 24) is a map of the distribution of elements in each area of the lower cathode of module 30.

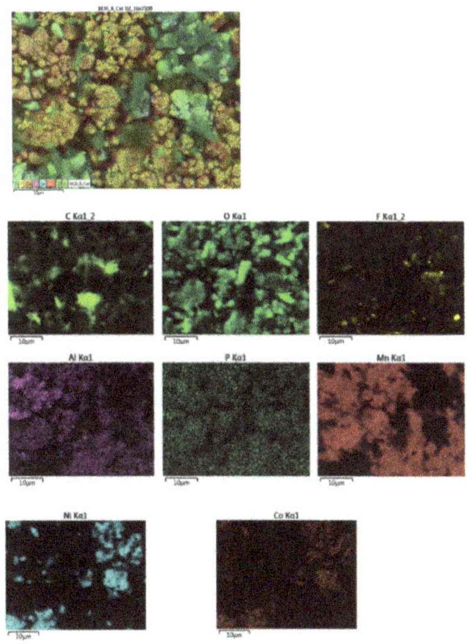

Figure 24. Distribution of elements in each area of the lower cathode of module 30.

In the lower cathode, there is a higher amount of fluorine, aluminum, manganese, phosphorus, nickel, and cobalt than in the upper cathode. The following figure shows the chemical composition of the upper cathode of modulo 11. The upper cathode of module 11 also does not show a defined granular surface, as in the case of the original cathode; the structure was broken. The larger particles are about 5 μm in size, so the particles are smaller than in the case of the original cathode. Three different types of structures are observed on the surface of the upper cathode of modulo 11, marked in Figure 25 with three red squares.

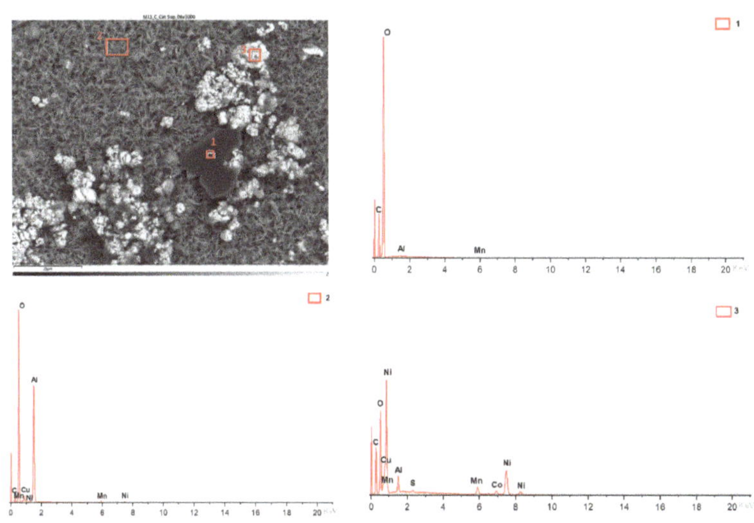

Figure 25. Chemical composition of the upper cathode of module 11.

Details of the three spectra of the component values measured in atomic and weight % in the upper cathode of module 11 (Figure 25) are listed in Table 7.

Table 7. EDXS weight ratio of components in the upper cathode of the ignited module 11 using three spectra focused on three distinct areas.

		C		O		Al		Mn		Co		Ni	
		Weight (%)	Atomic (%)	Weight (%)	Atomic (%)	Weight (%)	Atomic (%)	Weight (%)	Atomic (%)	Weight (%)	Atomic (%)	Weight (%)	Atomic (%)
Compound 1: C and O	**Spectrum 1**	17.92	22.61	81.51	77.2	0.12	0.07	0.44	0.12				
Compound 2: O and Al	**Spectrum 2**	6.49	9.97	60.91	70.18	25.88	17.68	2.79	0.94			3.93	1.23
Compound 3: C, O and Ni, with some Al, Mn and Co	**Spectrum 3**	23.4	47.63	17.05	26.05	2.59	2.35	5.46	2.43	5.51	2.29	45.69	19.03

A thorough analysis of the chemicals that were created reveals that the compound marked with "1" had the chemistry of a carbon and oxygen compound, while the second structure was determined to be an aluminum and oxygen compound, formed from the fusion of the cathode aluminum collector, and the third structure was a chemical compound containing nickel, carbon, oxygen, cobalt, manganese, and aluminum. They are all elements that are products of the cathode and collector coating. Regarding module 11, a higher amount of aluminum is observed; this may be since, in this case, the aluminum collector has been more damaged than in the case of module 30.

The following Figure 26, shows the chemical composition of the lower cathode of module 11. The lower cathode of module 11 also does not show a defined granular surface, as in the case of the original cathode; the structure was broken. Five different types of compounds are observed on the surface of the lower cathode of module 11, shown in Figure 26. A detailed analysis of the compounds formed reveals that the compound marked with "1" had peaks of nickel, manganese, oxygen, copper, carbon, cobalt, and fluorine, so it presents all the elements of the cathode coating; there is also copper that can come from the anode collector. The second structure was a compound of nickel, manganese, cobalt, oxygen, and carbon, and the third structure was a compound of copper and oxygen, probably because the anode collector had melted. If compound 4 is analyzed, it is a nickel, oxygen, manganese, and fluoride-containing chemical compound. And finally, compound 5 is a manganese and oxygen compound formed from the active material of the cathode, and there is also aluminum coming from the aluminum collector of the cathode. In this area, an aluminum peak appears; this is the result of a thermal failure that leaves the aluminum current collector exposed.

During the EDXS measurement, different areas were focused on, and the corresponding peaks are shown in Figure 26. Details of the five spectra of the component values measured in atomic and weight % are listed in Table 8.

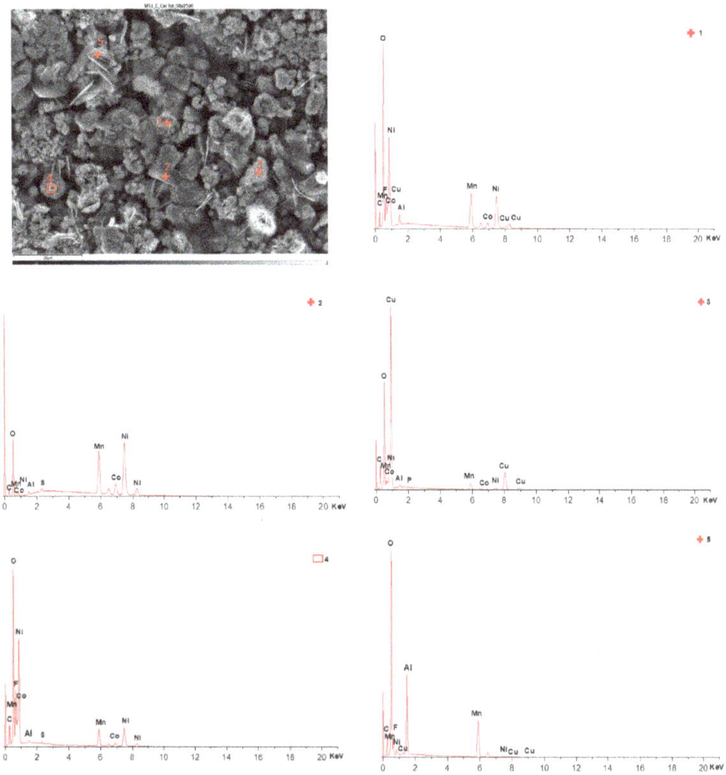

Figure 26. Chemical composition of the lower cathode of module 11.

Table 8. EDXS weight ratio of components in the lower cathode of the ignited module 11 using five spectra focused on five distinct areas.

	C		O		F		Al		Mn		Co		Ni		Cu	
	Weight (%)	Atomic (%)	Weight (%)	Atomic (%)	Weight (%)	Atomic (%)	Weight (%)	Atomic (%)	Weight (%)	Atomic (%)	Weight (%)	Atomic (%)	Weight (%)	Atomic (%)	Weight (%)	Atomic (%)
Spectrum 1 Compound 1: Ni, Mn, O, Cu, C, Co with F	5.73	15.49	17.97	36.46	4.28	7.32	1.08	1.3	18.06	10.67	4.4	2.42	37.3	20.62	11.19	5.72
Spectrum 2 Compound 2: Ni, Mn, Co, O, C	2.02	7.83	5.95	17.34			0.36	0.61	21.87	18.56	9.06	7.17	60.4	47.98		
Spectrum 3 Compound 3: Cu and O	14.26	33.54	20.81	36.75			0.45	0.47	5.56	2.86	0.49	0.23	2.75	1.32	55.56	24.71
Spectrum 4 Compound 4: Ni, O, Mn with F	8.5	18.07	24.26	38.71	14.7	19.75	0.29	0.28	13.23	6.15	4.44	1.92	34.36	14.94		
Spectrum 5 Compound 5: Mn and O with Al	9.2	17.35	37.15	52.58	3.81	4.54	11.71	9.83	34.42	14.19			2.14	0.82	1.3.6	0.48

A comparison of the systematic analysis of the EDSX data shown in Figures 20–22, 25 and 26 (pristine cathode, upper and lower cathode of module 30, and upper and lower cathode of module 11) is shown below (Table 9).

Table 9. Comparison of the systematic analysis of the EDX data of the different cathodes analyzed.

CATHODE	Compound 1	Compound 2	Compound 3	Compound 4	Compound 5	Comments
Pristine Cathode	$LiNiMnCoO_2$	LMO + F				1. Compound 1 is the chemistry of a NMC cathode 2. Compound 2 is an LMO, and the F may be from the electrolyte (LiPF6) in contact with the cathode and anode.
Upper Cathode of Module 30	Mn and O, with some F and P	F and Ni				1. Compound 2 is due to the reactions in the TR, and is why there are leftovers of the electrolyte.
Lower Cathode of Module 30	O and Mn	O and Ni with some Mn, F, Co, and Al				1. Aluminum of compound 2 comes from the fusion of the cathode aluminum collector.
Upper Cathode of Module 11	C and O	O and Al	C, O, Mn, Co, and Ni, with some Al			1. Compound 2 is formed from the fusion of the cathode aluminum collector. 2. The aluminum of compound 3 comes from the fusion of the cathode aluminum collector.
Lower Cathode of Module 11	Ni, Mn, O, C, Co with Al and Cu	Ni, Mn, C, Co and O	Cu and O	Ni, O, Mn with F	Mn and O with Al	1. Compound 1 presents all the elements of the cathode coating; the copper may come from the anode collector. 2. Compound 3 has probably been formed because the anode collector has melted. 3. Compound 4 has F due to the reactions in the TR and why there are leftovers of the electrolyte. 4. Compound 5 is formed from the active material of the cathode, and there is also aluminum from the aluminum collector of the cathode.

After carrying out the comparative analysis presented in Table 9, it can be concluded that the cathodes of the ignited modules present compounds containing the elements of the cathode coating. However, they also present in some areas aluminum from the melting of the aluminum collector of the cathode, as well as copper from the melting of the copper collector of the anode. There are also traces of fluorine from the electrolyte (LiPF6). Copper only appears on the lower cathode of module 11, and more compounds appear, so it can be concluded that this is the most damaged cathode analyzed.

The following images (Figure 27) compare different images obtained with FESEM of the upper anode of module 30, module 11, and the new cell. And in Figure 28, different images obtained with FESEM of the lower anode of module 30, module 11, and the new cell are compared.

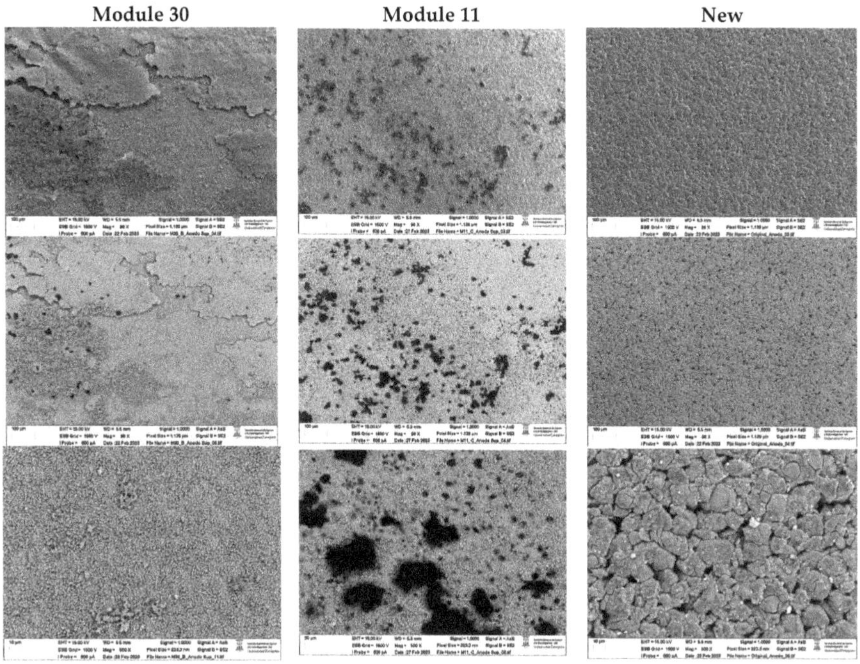

Figure 27. Comparison of images obtained with the FESEM of different areas of the upper anode of module 30, module 11, and the new cell.

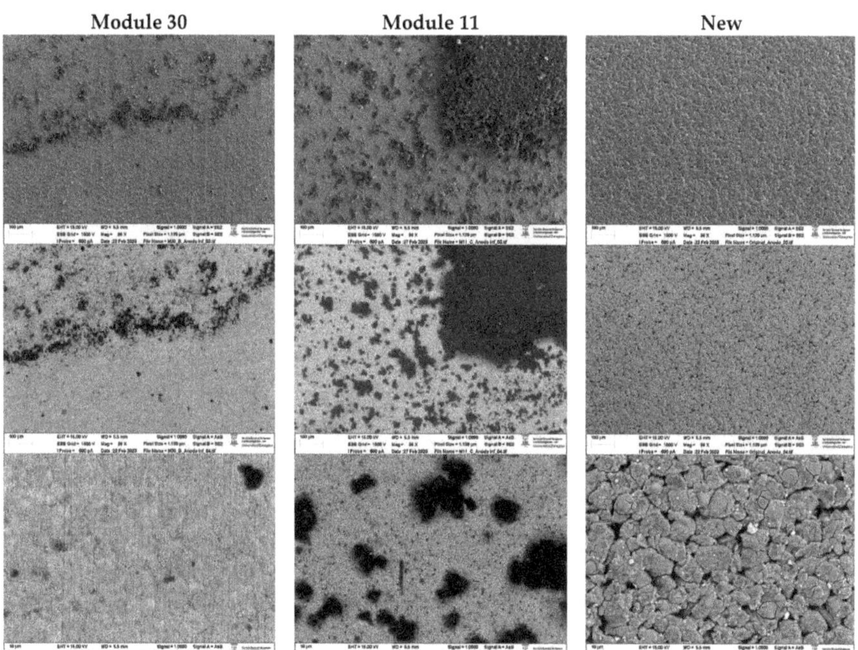

Figure 28. Comparison of images obtained with the FESEM of different areas of the lower node of module 30, module 11, and the new cell.

With respect to the anode electrodes in the previous figure, as it has been done with the cathode electrodes, the anode of an original cell with two anodes (one upper and one lower) was compared with module 30 and module 11 of the battery of the burning Nissan Leaf, i.e., after experiencing the thermal runaway. It is observed that after experiencing the thermal runaway, spherical particles exist on the graphite anode surface. These spherical surfaces are covered with smaller stereoscopic particles, as can be seen in Figures 28 and 29.

The original graphite anode has a porous structure that disappears when it undergoes a TR because the pores are occupied by spherical particles. When TR happens, the exothermic reaction on the anode side is linked to the creation of stereoscopic particles. However, at higher temperatures, Li_2CO_3 can be formed by the reaction of the intercalated carbon with the electrolytes [31], since at a temperature above 120 °C, the SEI decomposes. In addition, as the temperature increases, at approximately 290 °C, the lithited carbon could react with fluorine to form LiF [31].

After the analysis of the cathode and anode states after experiencing the TR, it is concluded that the spherical structure of the cathode active materials and the flake structure of the graphite are destroyed. It was quite difficult to obtain the sample from the ignited modules to identify them due to the fact that the cathode and anode were joined together.

It is observed that there are many impurities, not regular in shape, on the surface of the electrode. Sometimes it is rectangular, and sometimes it is spherical. The surface distribution is chaotic, as can be seen in Figure 29b,c. Because the temperature at which TR is reached in NMC pouch cells is high, the aluminum current collector of the cathode is oxidized and bonded to the anode. Therefore, it is normal to identify debris in the image observed with the SEM. Studies [30] indicate that this debris can be fragments of cathode materials, separator and cathode ash, products of exothermic reactions, as well as graphite flaking. In some areas, a dense and brittle layer has been found on the surface of the electrode. In other areas, flake graphite particles were found under this layer. Thus, it follows that this layer was an oxidized aluminum current collector. Unlike in the cathode, the pores of the anode have been clogged by materials from TR reactions. In some areas, a dense, brittle layer covering the surface of the electrode is also observed. In other areas, flake graphite particles were observed under this layer. Therefore, the authors contemplated that this layer was an oxidized Al current collector. Judging by the structure of the base layer, it was the layer of active material on the anode. Unlike the active material coating of the cathode of new cells, the pores of the electrode that suffered thermal runaway had been completely clogged by the products of the thermal runaway reactions. The active material of the cathode had completely ceased to function. If the structural transformation (morphology analysis with InLens) of the anode materials after the TR is analyzed (Figures 30 and 31), it is observed that in the case of modules 30 and 11 with a SoC at 68% after the TR, the layered structure of the material was destroyed, the particles dispersed outside the original layered structure and adhesion occurred. The cathode (positive electrode) material reacted at high temperatures and decomposed. On the other hand, carbon particles from the anode (negative electrode) went into the positive electrode (cathode) structure through the damaged diaphragm.

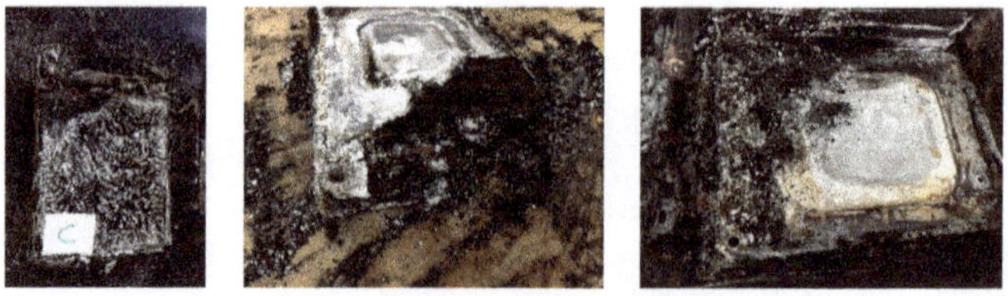

(a) (b) (c)

Figure 29. (**a**) Cell c of module 30; (**b**) Module 11 set on fire; (**c**) Module 30 burned down.

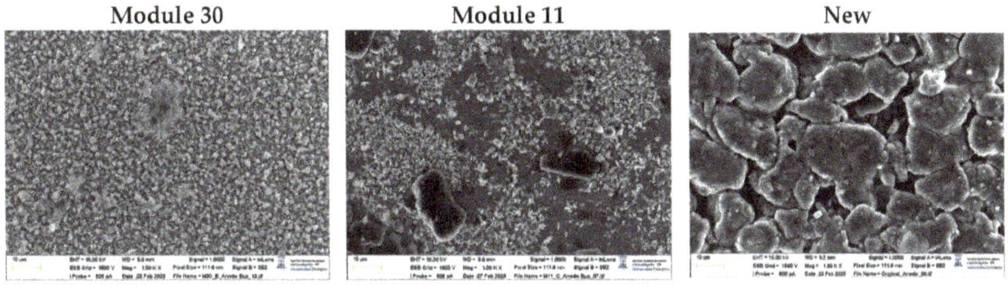

Figure 30. Morphological analysis of the transformation of the upper anode of module 30 and module 11 after the thermal runaway compared with the initial anode structure of a new cell.

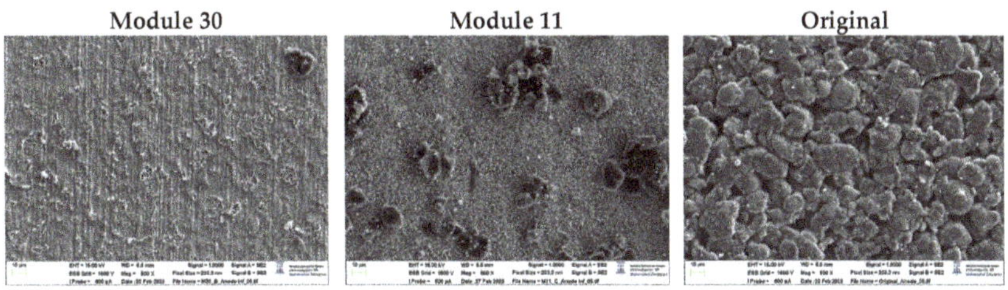

Figure 31. Morphological analysis of the transformation of the lower anode of module 30 and module 11 after the thermal runaway compared with the initial anode structure of a new cell.

The chemical composition of the new cell anode is then analyzed. The anode has a granular structure. The diameter of the largest particles was 14 μm. Four types of zones are identified, where the chemical elements will be analyzed. The chemical analysis revealed a high content in zone 2 of carbon (C), which allows the active material of the anode to be identified as graphite. There are traces of phosphorus (P) that can be identified as remnants of the electrolyte (assumed that it is LiPF6 dissolved in a carbonate mixture solvent). Also, in zone 1, a fluorine peak is identified; it can be seen on the surface of the graphite particles and can be attributed to the decomposition of the electrolyte that is part of the SEI. The small peaks of Cu can be attributed to the preparation of the sample and can be neglected. On the other hand, in zones three and four, peaks of calcium and iron are identified due to the shape of which impurities can be considered to have appeared due to the preparation of the sample. Manganese also appears.

Details of the four spectra of the component values measured in atomic and weight % in the original anode (Figure 32) are listed in Tables 10 and 11.

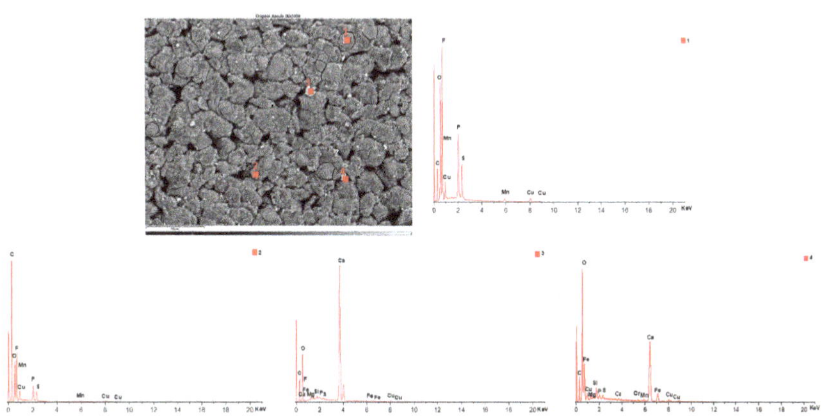

Figure 32. Chemical composition of the new cell anode.

Table 10. EDXS weight ratio of components in the original anode using the first two spectra focused on two distinct areas.

	C		O		F		P		S		Mn		Cu	
	Weight (%)	Atomic (%)	Weight (%)	Atomic (%)	Weight (%)	Atomic (%)	Weight (%)	Atomic (%)	Weight (%)	Atomic (%)	Weight (%)	Atomic (%)	Weight (%)	Atomic (%)
Spectrum 1 Compound 1: C, O, F and P	17.9	27.54	21.36	24.67	37.97	36.93	8.55	5.1	5.31	3.06	2.14	0.72	6.77	1.97
Spectrum 2 Compound 2: C, O, F with P	56.04	67.86	16.53	15.03	18.19	13.92	2.64	1.24	1.84	0.83	0.86	0.23	3.91	0.89

Table 11. EDXS weight ratio of components in the original anode using the last two spectra focused on two distinct areas.

	C		O		F		Mg		Si		Ca		Fe		Cu	
	Weight (%)	Atomic (%)	Weight (%)	Atomic (%)	Weight (%)	Atomic (%)	Weight (%)	Atomic (%)	Weight (%)	Atomic (%)	Weight (%)	Atomic (%)	Weight (%)	Atomic (%)	Weight (%)	Atomic (%)
Spectrum 3 Compound 3: O, Ca and C	7.4	14.33	35.64	51.81	0.89	1.08	0.56	0.53	0.46	0.38	53.72	31.17	0.94	0.39		
Spectrum 4 Compound 4: C, O, Fe with F	10.61	25.37	20.01	35.94	1.62	2.44	0.58	0.69	1.35	1.38	0.42	0.3	60.39	31.07	2.8	1.27

The following Figure 33, shows the chemical composition of other compounds identified in the original anode. Zone 1 identifies a chemical compound of manganese and oxygen, and zone 2 identifies a chemical compound of carbon and oxygen with fluorine and phosphorus, which may be due to electrolyte, copper, and sulfur (the latter two elements may be due to impurities from sample preparation). The manganese in structure 1 may come from the manganese-based cathode because of the possibility of manganese ions intercalating along with lithium ions into graphite in the initial stage of the lithium-ion intercalation process during storage [37].

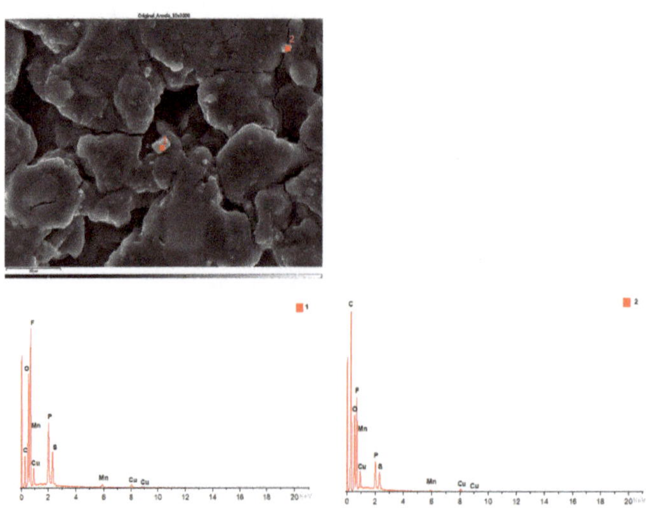

Figure 33. Chemical composition of the new cell anode 2.

Details of the two spectra of the component values measured in atomic and weight % in the original anode (Figure 33) are listed in Tables 12 and 13, respectively.

Table 12. EDXS weight ratio of components in the original anode using first spectrum focused area one.

	C		O		F		Al		P		Mn		Cu	
	Weight (%)	Atomic (%)	Weight (%)	Atomic (%)	Weight (%)	Atomic (%)	Weight (%)	Atomic (%)	Weight (%)	Atomic (%)	Weight (%)	Atomic (%)	Weight (%)	Atomic (%)
Spectrum 1 Compound 5: Mn and O	6.69	14.5	31.1	50.57	5.27	7.22	1.15	1.11	0.47	0.39	53.59	25.38	1.45	0.59

Table 13. EDXS weight ratio of components in the original anode using second spectrum focused area two.

	C		O		F		Si		P		S		Mn		Cu	
	Weight (%)	Atomic (%)	Weight (%)	Atomic (%)	Weight (%)	Atomic (%)	Weight (%)	Atomic (%)	Weight (%)	Atomic (%)	Weight (%)	Atomic (%)	Weight (%)	Atomic (%)	Weight (%)	Atomic (%)
Spectrum 2 Compound 6: C and O with F and P, Cu and S	37.83	51.52	23.87	24.41	19.17	16.51	1.78	1.04	3.4	1.8	2.98	1.52	0.92	0.27	8.8	2.27

Afterward, the chemical composition of the upper anode of the ignited module 30 is analyzed. The upper anode of module 30 also does not show a defined granular surface, as in the case of the original anode; the structure was broken. Two different types of compounds were observed on the surface of the upper anode of module 30, shown in Figure 34. A detailed analysis of the compounds formed shows that the compound marked with "2" had the chemical of a copper and oxygen compound, which may be due to the anode collector having melted due to the elevated temperatures; moreover, while zone 3 was a carbon and oxygen compound, it may be due to the coating of the graphite anode.

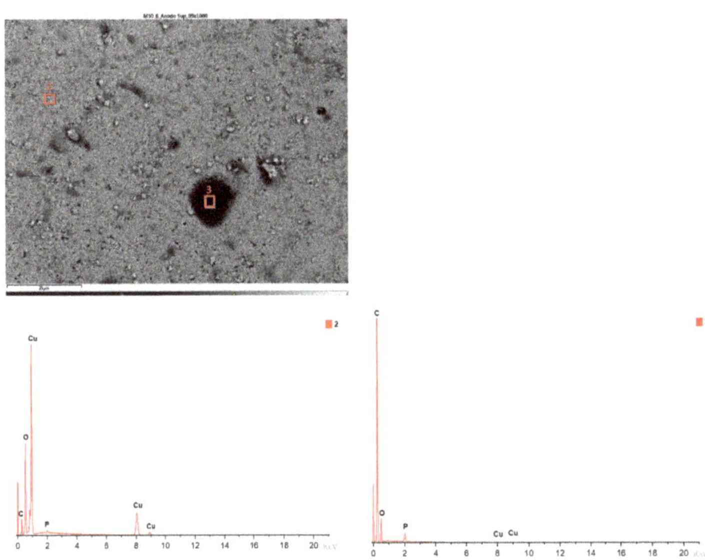

Figure 34. Chemical composition of the upper anode of module 30.

As has been performed previously, details of the two spectra of the component values measured in atomic and weight % in the upper anode of module 30 (Figure 34) are listed in Table 14.

Table 14. EDXS weight ratio of components in the upper anode of the ignited module 30 using two spectra focused on two distinct areas.

		C		O		F		P		Cu	
		Weight (%)	Atomic (%)	Weight (%)	Atomic (%)	Weight (%)	Atomic (%)	Weight (%)	Atomic (%)	Weight (%)	Atomic (%)
Compound 1: Cu and O	**Spectrum 2**	12.44	29.63	22.9	40.97			0.58	0.54	64.08	28.86
Compound 2: C and O	**Spectrum 3**	80.31	85.45	16.39	13.09	0.82	0.55	1.93	0.8	0.55	0.11

In Figure 35 below, the chemical composition in another area of the upper anode of the ignited module 30 is analyzed. A single type of compound was observed on the surface of the upper anode of module 30. This compound features copper, carbon, and oxygen.

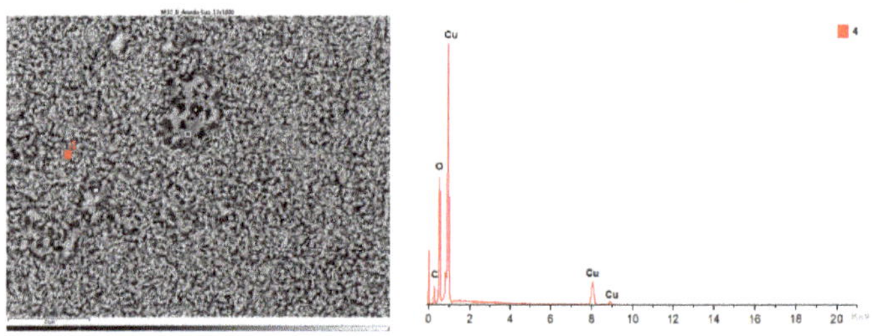

Figure 35. Chemical composition of the upper anode of module 30, second area analyzed.

Details of the one spectrum of the component values measured in atomic and weight % in the upper anode of module 30 (Figure 35) are listed in Table 15.

Table 15. EDXS weight ratio of components in the upper anode of the ignited module 30 using one spectrum-focused area four.

		C		O		Cu	
		Weight (%)	Atomic (%)	Weight (%)	Atomic (%)	Weight (%)	Atomic (%)
Compound 3: Cu, C and O	**Spectrum 4**	10.15	25.18	23.45	43.67	66.4	31.14

In Figure 36 below, the chemical composition in another area of the upper anode of the ignited module 30 is analyzed. Four different types of compounds were observed. In the first two ones, a compound of copper, carbon, and oxygen was observed. In zone 3, fluoride, copper, and oxygen were identified; the existence of fluoride is due to the electrolyte, and the existence of copper is due to copper from the anode collector. In zone 4, cooper and oxygen were identified.

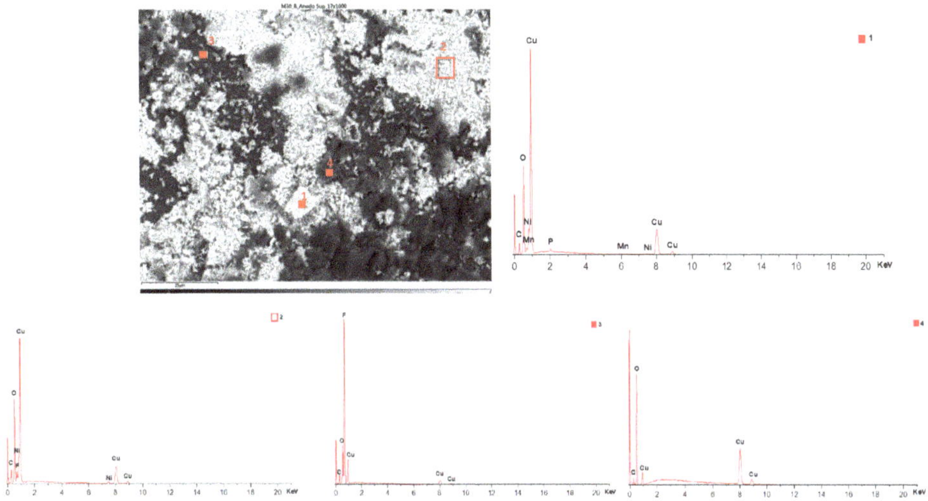

Figure 36. Chemical composition of the upper anode of module 30, third area analyzed.

As has been performed previously, details of the four spectra of the component values measured in atomic and weight % in the upper anode of module 30 (Figure 36) are listed in Table 16.

Table 16. EDXS weight ratio of components in the upper anode of the ignited module 30 using five spectra focused on five distinct areas.

	C		O		F		P		Mn		Ni		Cu	
	Weight (%)	Atomic (%)	Weight (%)	Atomic (%)	Weight (%)	Atomic (%)	Weight (%)	Atomic (%)	Weight (%)	Atomic (%)	Weight (%)	Atomic (%)	Weight (%)	Atomic (%)
Spectrum 1 Compound 3: Cu, C and O	8.8	23.11	21.12	41.67			0.63	0.65	0.36	0.21	1.1	0.59	67.98	33.77
Spectrum 2 Compound 3: Cu, C and O	11.2	26.72	22.2	39.69	3.28	4.95					4.02	1.96	59.27	26.68
Spectrum 3 Compound 4: F, Cu and O	5.77	9.93	7.7	9.96	68.03	74.09							18.51	6.03
Spectrum 4 Compound 1: Cu and O	5.03	7.99	71.19	84.87									23.78	7.14

Next, the chemical composition of the lower anode of the ignited module 30 is analyzed. The lower anode of module 30 also does not show a defined granular surface, as in the case of the original anode; the structure was broken. Two different types of compounds were observed on the surface of the lower anode of module 30, shown in Figure 37. A detailed analysis of the compounds formed shows that the compound marked with "1" had the chemistry of a Carbon and oxygen compound, which may be due to the anode coating, and the second structure features copper, carbon, and oxygen.

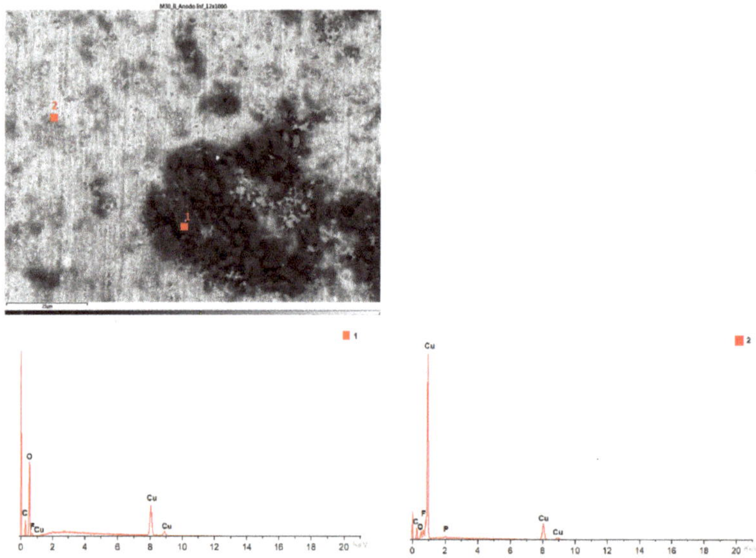

Figure 37. Chemical composition of the lower anode of module 30.

As has been performed previously, details of the two spectra of the component values measured in atomic and weight % in the lower anode of module 30 (Figure 37) are listed in Table 17.

Table 17. EDXS weight ratio of components in the lower anode of the ignited module 30 using two spectra focused on two distinct areas.

		C		O		F		P		Cu	
		Weight (%)	Atomic (%)	Weight (%)	Atomic (%)	Weight (%)	Atomic (%)	Weight (%)	Atomic (%)	Weight (%)	Atomic (%)
Compound 1: C and O with F	Spectrum 1	19.23	24.44	70.73	67.49	10.04	8.07				
Compound 2: Cu, C and O	Spectrum 2	15.55	45.38	2.89	6.34	2.29	4.22	0.56	0.63	78.71	43.42

Figure 38 below shows the chemical composition in another area of the upper anode of the ignited module 11. Six different types of compounds were observed on the surface of the upper anode of module 11. The compound marked with "1" is carbon, coming from the graphite of the anode coating; the compound marked with "2" is a compound of fluorine, copper, and oxygen; fluorine comes from the electrolyte; and the rest of the compounds have copper, carbon, and oxygen.

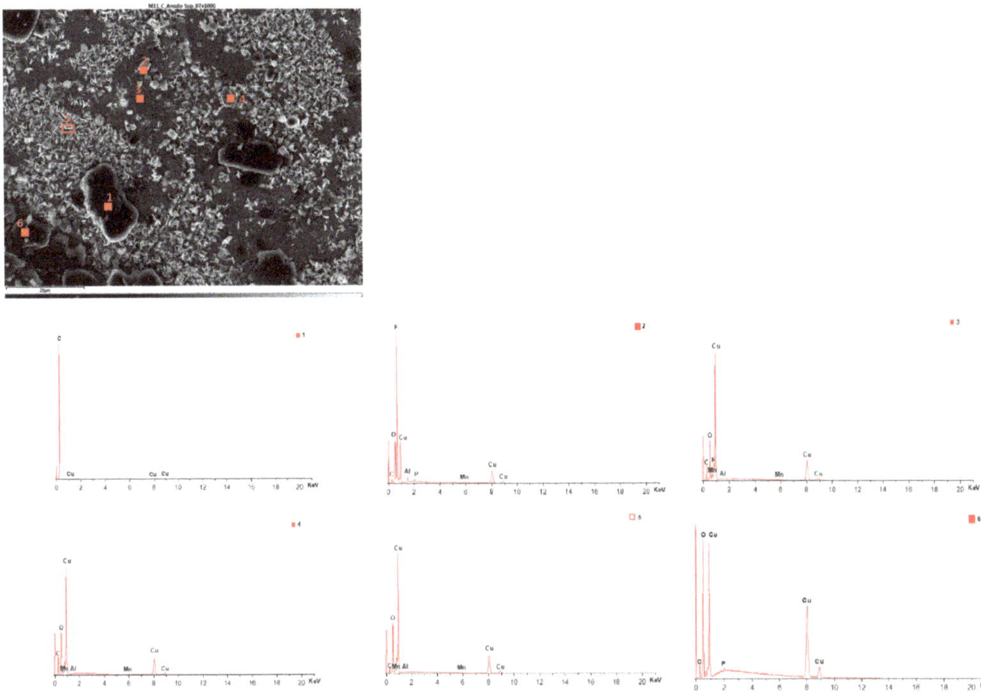

Figure 38. Chemical composition of the upper anode of module 11.

Details of the six spectra of the component values measured in atomic and weight % in the upper anode of module 11 (Figure 38) are listed in Table 18.

Table 18. EDXS weight ratio of components in the upper anode of ignited module 11 using six spectra focused on six distinct areas.

	C		O		F		Al		P		Mn		Cu	
	Weight (%)	Atomic (%)	Weight (%)	Atomic (%)	Weight (%)	Atomic (%)	Weight (%)	Atomic (%)	Weight (%)	Atomic (%)	Weight (%)	Atomic (%)	Weight (%)	Atomic (%)
Spectrum 1 Compound 1: C	100	100												
Spectrum 2 Compound 2: F, Cu and O	3.16	6.28	11.72	17.47	48.95	61.48	1.45	1.29	1.02	0.78	0.77	0.33	32.93	12.37
Spectrum 3 Compound 3: Cu, C and O	17.7	40.76	16.52	28.58	1.73	2.52	0.35	0.36			0.68	0.34	63.01	27.44
Spectrum 4 Compound 4: Cu, C and O	22.43	46.75	19.34	30.25							0.92	0.42	57.31	22.57
Spectrum 5 Compound 5: Cu, C and O	8.13	21.76	21.06	42.31							1.55	0.91	69.25	35.03
Spectrum 6 Compound 3: Cu, C and O	9.09	19.54	34.02	54.92	2.19	2.98			0.76	0.63			53.93	21.92

And finally, in the following Figure 39, the chemical composition in another area of the lower anode of the burning module 11 is analyzed. Six different types of compounds were observed on the surface of the lower anode of module 11. The first structure was a compound of fluorine, copper, and oxygen; in compound 2, copper and oxygen were identified; compound 3 is like the first, which is a compound of fluorine, copper and oxygen. In compound 4 there is more copper than fluorine and oxygen, and in structures 5 and 6, there is copper oxide and carbon oxide.

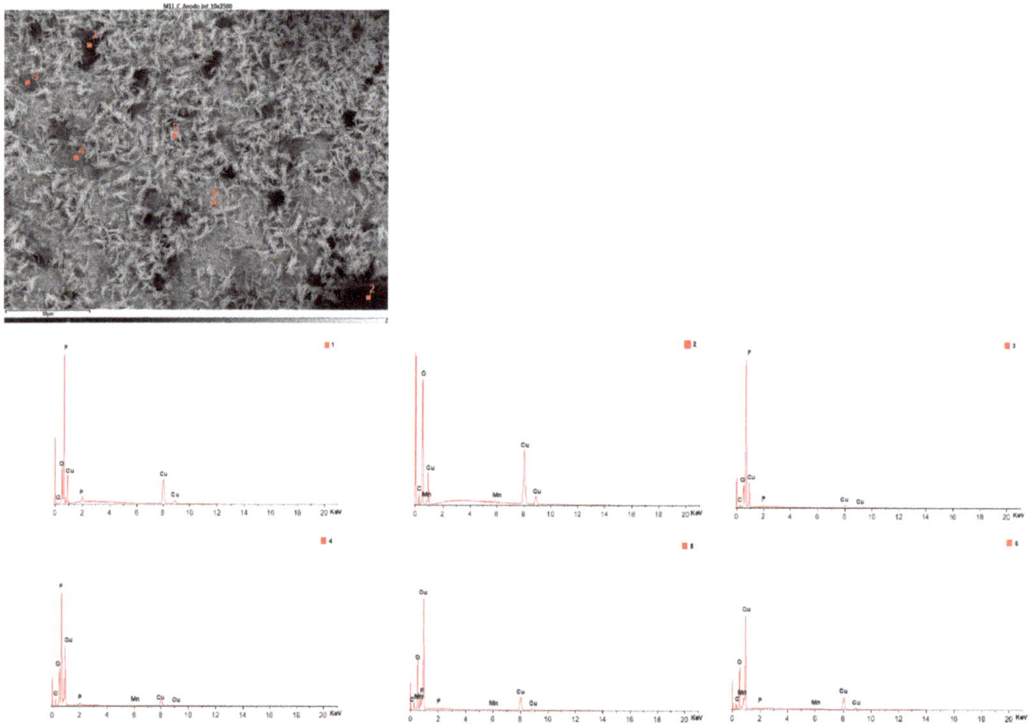

Figure 39. Chemical composition of the lower anode of module 11.

Details of the six spectra of the component values measured in atomic and weight % in the lower anode of module 11 (Figure 39) are listed in Table 19.

Table 19. EDXS weight ratio of components in the lower anode of ignited module 11 using six spectra focused on six distinct areas.

		C		O		F		P		Mn		Cu	
		Weight (%)	Atomic (%)	Weight (%)	Atomic (%)	Weight (%)	Atomic (%)	Weight (%)	Atomic (%)	Weight (%)	Atomic (%)	Weight (%)	Atomic (%)
Compound 1: F, Cu and O	Spectrum 1			11.27	16.76	56.01	70.15	2.12	1.63			30.6	11.46
Compound 2: Cu and O	Spectrum 2	9.43	16.41	54.94	71.73					2.98	1.13	32.65	10.73
Compound 1: F, Cu and O	Spectrum 3	3.02	5.83	8.14	11.78	58.17	70.9	0.83	0.62			29.84	10.87
Compound 1: F, Cu and O	Spectrum 4	7.66	15.87	11.24	17.48	37.6	49.28	0.71	0.57	0.41	0.18	42.39	16.61
Compound 3: Cu, C and O	Spectrum 5	10.03	25.01	20.16	37.73	3.81	6	0.2	0.2	0.79	0.43	65.01	30.63
Compound 3: Cu, C and O	Spectrum 6	10.49	26.3	22.19	41.76					0.56	0.31	66.76	31.63

The analysis of the cell anodes after the fire test shows that the anode has peaks of oxygen and peaks of carbon, which indicates the formation of lithium carbonate (Li_2CO_3).

A comparison of the systematic analysis of the EDSX data shown in Figures 32–39 (pristine anode, upper and lower anode of module 30, and upper and lower anode of module 11) is shown below (Table 20).

Table 20. Comparison of the systematic analysis of the EDX data of the different anodes analyzed.

ANODE	Compound 1	Compound 2	Compound 3	Compound 4	Compound 5	Compound 6	Comments
Pristine Anode	C, O, F with P	C, O, F with P	O, Ca, and C	C, O, Fe, with F	Mn and O	C and O with F and P, Cu, and S	1. The F of compound 1 can be attributed to the decomposition of the electrolyte. 2. The C of compound 2 allows the active material of the anode (graphite). The traces of P are remnants of the electrolyte (LiPF6). 3. In compounds 3 and 4, there are Ca and Fe due to the preparation of the sample (impurities). 4. The Mn in compound 5 comes from the manganese-based cathode. 5. In compound 6, the traces of F and P may be due to electrolytes, and the traces of Cu and S may be due to the sample preparation (impurities).
Upper Anode of Module 30	Cu and O	C and O	Cu, C, and O	F, Cu, and O			1. Compound 1 may be due to the anode collector having melted due to elevated temperatures. 2. The compound 2 may be due to the coating of the graphite anode. 3. In the case of compound 3, it can come, on the one hand, from Cu due to the melting of the copper collector of the anode and, on the other hand, from C due to the coating of the graphite anode. 4. The existence of F in compound 4 is due to the electrolyte.
Lower Anode of Module 30	C and O with F	Cu, C, and O					1. The compound 1 may be due to the coating of the graphite anode. And the existence of F is due to the electrolyte. 2. In the case of compound 2, its elements can come, on the one hand, from Cu due to the melting of the copper collector of the anode, and on the other hand, C from the coating of the graphite anode.
Upper Anode of Module 11	C	F, Cu, and O	Cu, C, and O				1. Compound 1 is C, coming from graphite. 2. The existence of F in compound 2 is due to the electrolyte. And Cu comes from the melting of the copper collector of the anode. 3. In the case of compound 3, its elements can come, on the one hand, from Cu due to the melting of the copper collector of the anode, and on the other hand, C from the coating of the graphite anode.
Lower Anode of Module 11	F, Cu, and O	Cu and O	Cu, C, and O				1. The existence of F in compound 2 is due to electrolyte. And Cu comes from the melting of the copper collector of the anode. 2. In the case of compound 3, its elements can come, on the one hand, from Cu due to the melting of the copper collector of the anode and, on the other hand, from C due to the coating of the graphite anode.

After carrying out the comparative analysis presented in Table 20, it can be concluded that the anodes of the ignited modules present compounds containing the elements of the anode coating. However, they also present in some areas cooper from the melting of the cooper collector of the anode but no aluminum from the melting of the cathode collector. There are also traces of fluorine from the electrolyte (LiPF6).

A morphological analysis of the separator of the new cell is then performed since the separator cannot be recovered in the burned modules because it has melted in the thermal runaway reaction. The surface structure of the separator was investigated to determine the thickness of the separator fibers and the size of the pores, as well as to determine whether an alumina layer had been deposited on the separator membrane, as described above. Figure 40 shows the results of the SEM image. The structure of the fibers of the polypropylene membrane $(C_3H_6)_n$ is shown on the right. The separator fibers were aligned perpendicular to the cell tabs.

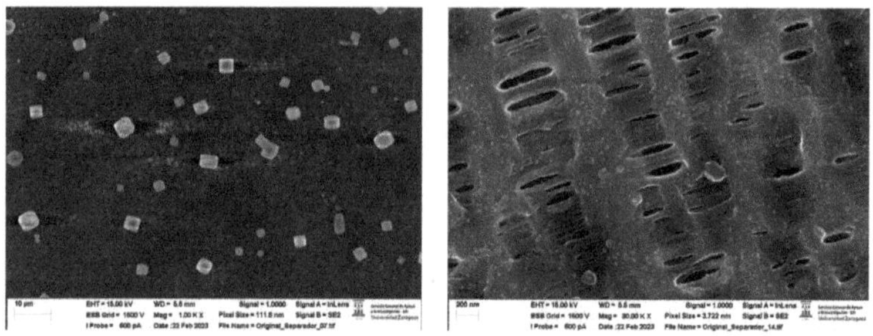

Figure 40. Morphological analysis of the separator. The first image has been obtained with a magnification of 1.00 K X, and the second image with a magnification of 30.00 K X.

The total thickness of the separating fiber was about 289 nm, as shown in Figure 41. The analyzed structure shows pores of different diameters, the smallest of about 41.61 nm and the largest of about 88.78 nm. The pore size in lithium-ion cells must be in the submicron range; it is a general requirement to prevent dendritic penetration of lithium from occurring over the life of the battery [38].

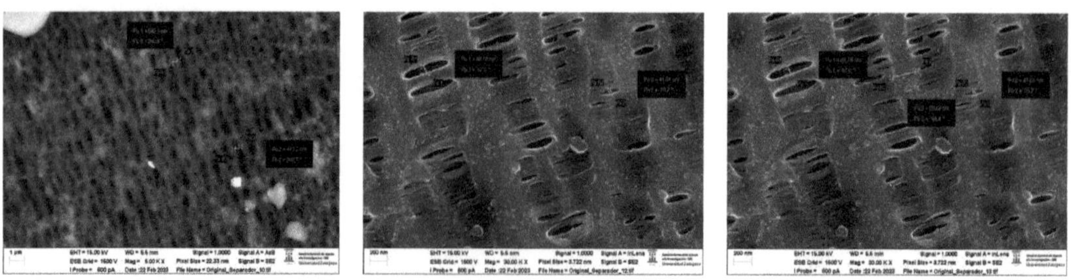

Figure 41. Morphological analysis of the separator. Measurement.

After that, the analysis of the separator's chemical composition will be performed. There are three types of compounds, the first being carbon with fluorine and phosphorus. The carbon is from the structure of the polypropylene separator itself; the fluorine and the phosphorus come from the electrolyte ($LiPF_6$) with which it is soaked. Compounds 2 and 3 have the same composition as the first compound.

As has been performed previously, details of the three spectra of the component values measured in atomic and weight % in the separator (Figure 42) are listed in Table 19.

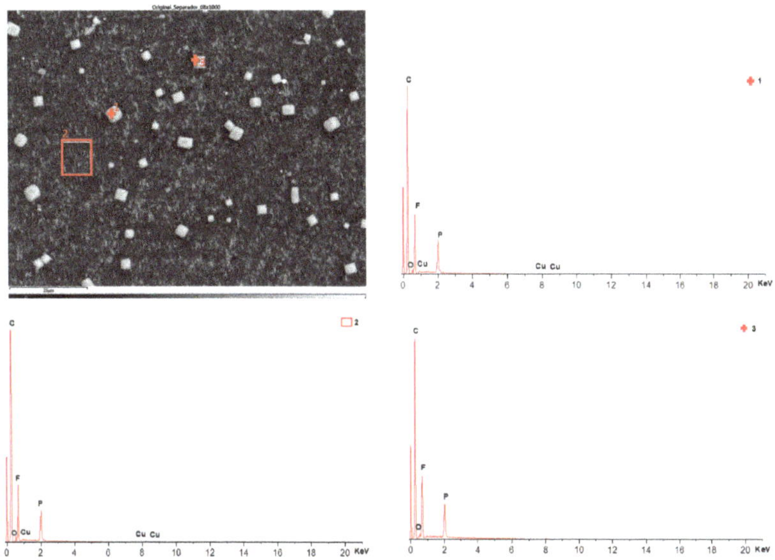

Figure 42. Chemical composition of the separator.

Details of the three spectra of the component values measured in atomic and weight % in the separator (Figure 42) are listed in Table 21.

Table 21. EDXS weight ratio of components in the separator using three spectra focused on three distinct areas.

	C		O		F		P		Cu	
	Weight (%)	Atomic (%)	Weight (%)	Atomic (%)	Weight (%)	Atomic (%)	Weight (%)	Atomic (%)	Weight (%)	Atomic (%)
Spectrum 1	81.14	71.43	1.57	1.84	14.57	20.29	2.61	5.94	0.11	0.51
Spectrum 2	83.58	74.96	1.87	2.24	12.39	17.59	2.07	4.79	0.09	0.42
Spectrum 3	81.16	71.74	1.73	2.03	14.49	20.26	2.61	5.96		

The layered structure was destroyed when the thermal runaway occurred. In the battery sample with a 68% SoC, there were characteristic fluoride peaks, indicating that after decomposition, the cathode material reacts with the electrolyte [39]. Characteristic carbon peaks are also shown; this is because, in the thermal runaway reaction, some of the carbon dust went into the positive electrode (cathode) material through the broken diaphragm [40]. It is observed that in the samples of module 11 and module 30, in which a thermal runaway reaction has occurred, the structure of the positive electrode (cathode) sample is destroyed, the particles were dispersed outside the original stratified structure, and the adhesion phenomenon occurred, indicating fragmentation. Particles are not found, and external impurities are placed in the structure [41].

The battery that caught fire in the test had a 68% SoC. Some studies [36] indicate that in the case of NMC batteries, the increase in SoC makes the thermal runaway reaction more intense, the temperature at which the thermal runaway starts is decreased, and the structural damage to the electrode materials is strong. In high-temperature conditions, the internal diaphragm of the battery is damaged, causing the battery to short circuit and worsen the thermal runaway reaction. During the thermal runaway reaction, the structure of the positive electrode (anode) material is destroyed, and the particles are dispersed outside the structure and adhere to the negative electrode (cathode).

If the particle size distribution (PSD) is compared for the case of the new cell and the burnt cells, it is observed that in the case of the latter, the particles of the positive electrode or cathode are smaller; they are further apart to facilitate the passage of gases and this gap will be occupied by another particle. In the case of burning cells, the particles are much smaller than in the case of new cells. Heat generation depends on the surface area and particle size of the positive electrode or cathode. There are studies [42] that show that the microstructure of particles can influence exothermic reactions of exothermic degradation.

3.2. Structural Test Results: Layer Properties of the Cell and Identification of Cell Details

In this section, the transversal-section investigation of the cells will be carried out, which will allow information about the structure and manufacture of the cells to be obtained, and the thickness measurement of each of the layers of the cell will be carried out.

The stacking of the different layers of the cell will be analyzed in the transversal section, which will allow us to identify interesting fabrication details of the pouch cells. First, the sample taken from zone 3 (Figure 12a) is analyzed, the central area on the opposite side to the area where the battery tabs of the cell are (Figure 43).

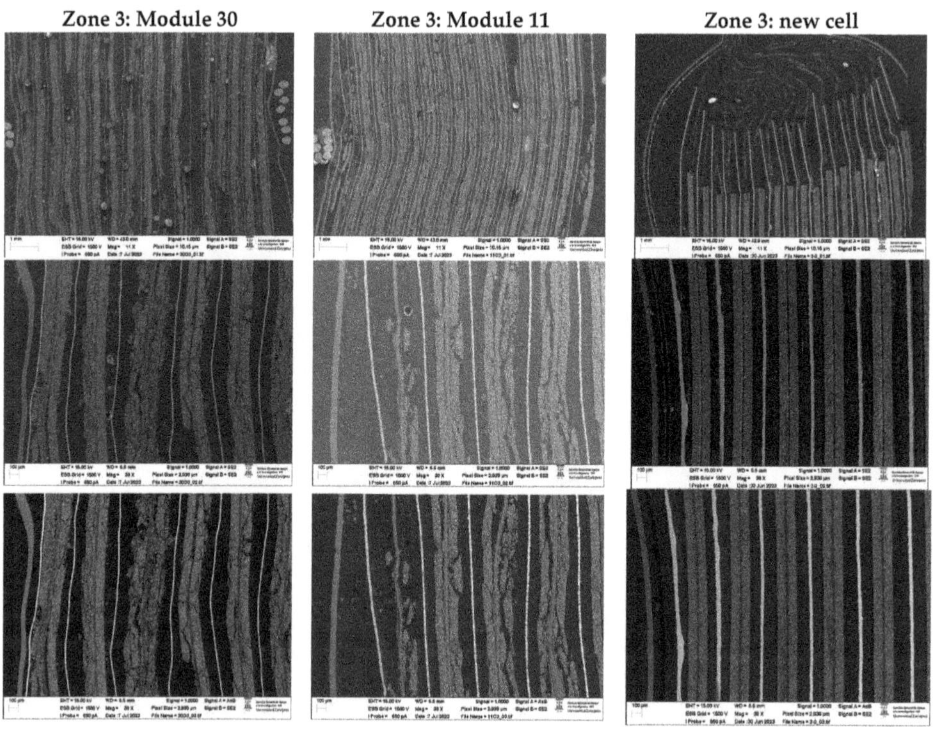

Figure 43. Images of the section of the pouch cell in zone 3 (central area on the opposite side to the one where the battery tabs of the cell are located Figure 12a).

It is observed that all the separating membranes are welded to the bag in a sealing point near the edge. Regarding the mechanical behavior of the cell and its possible failure, this fact indicates that when mechanical forces are applied to the battery near this axis, the separator may experience high tensile loads, and mechanical rupture of the separator membrane may occur. This would cause the battery electrodes to come into contact, which is the prerequisite for experiencing an electrical failure and thermal runaway. It is noticed that near the battery edge, the distances between layers become smaller due to the hot-welded area of the bag. Since there are no layers in this part of the battery, a short circuit will not occur even if some of the battery layers are too close to each other.

Another purpose of an anode layer that is longer than the cathode is to increase the battery rigidity near the edge, which smooths the distribution of charge in the event of charging at the edge. The investigated cell contained 18 anodic layers, 17 cathode layers, and 34 separator sheets. The electrodes on both sides of the battery were anodes. This observation indicates that the active material deposited outside the last copper foil remained electrochemically inactive and, therefore, did not participate in the energy storage or charge transfer process.

In lithium-ion batteries, a layer of aluminum oxide (Al_2O_3) is usually deposited on the separator membrane to improve its thermal resistance as well as its thermal properties.

If the morphology is analyzed, it can be concluded that the outer area of zone three is more damaged in the case of module 11, while in the case of module 30, the central area is more damaged. It is also observed that the thickness of the copper collector decreases in case of thermal runaway, and it is observed to be narrower for module 30. To know the chemical composition of zone 3 in the original cell, a map of the distribution of elements was made (Figure 44). The copper collector (yellow) and the aluminum collector (blue) are identified. The coating of the anode is also observed, which is made of graphite (carbon, red color) and the coating of the cathode (composed of oxygen, manganese, and nickel). Electrolyte fluoride is also observed.

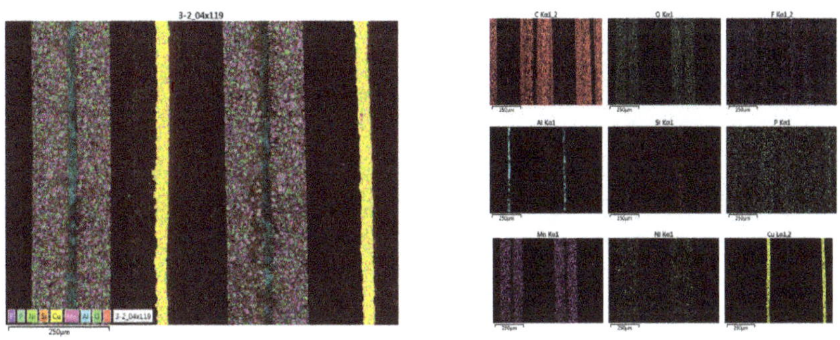

Figure 44. Distribution of the elements of a new pouch cell in zone 3 (central area on the opposite side to the one where the battery tabs of the cell are located).

After that, the chemical composition of zone 3 was analyzed for the case of the cell of the burned module 30.

Figure 45d shows the structure of the pouch cell of the burned module 30 in zone 3, three zones have been marked with a green box in which the composition of each of the layers of this pouch cell will be analyzed (Figure 45a–c). Figure 45a identifies the pouch of the pouch cell, which is made of aluminum (zones 2 and 3). In zone 5, there is carbon that may be graphite from the anode coating. Figure 46c identifies the cathode (manganese nickel oxide), and there are also carbon particles. Around the aluminum manifold, alumina (Al_2O_3) has been deposited on the separator in the cathode area. Alumina (aluminum oxide) has been deposited. In Figure 45b, the anode area has been detected, as well as graphite (carbon) coating and a layer with copper, oxygen, and the anode collector shell.

Moreover, the chemical composition of zone 3 (Figure 12a) is analyzed in the case of the cell of the burning module 11.

Figure 46d shows the structure of the pouch cell of the burned module 11 in zone 3, three zones have been marked with a green box in which the composition of each of the layers of this pouch cell will be analyzed (Figure 46a–c).

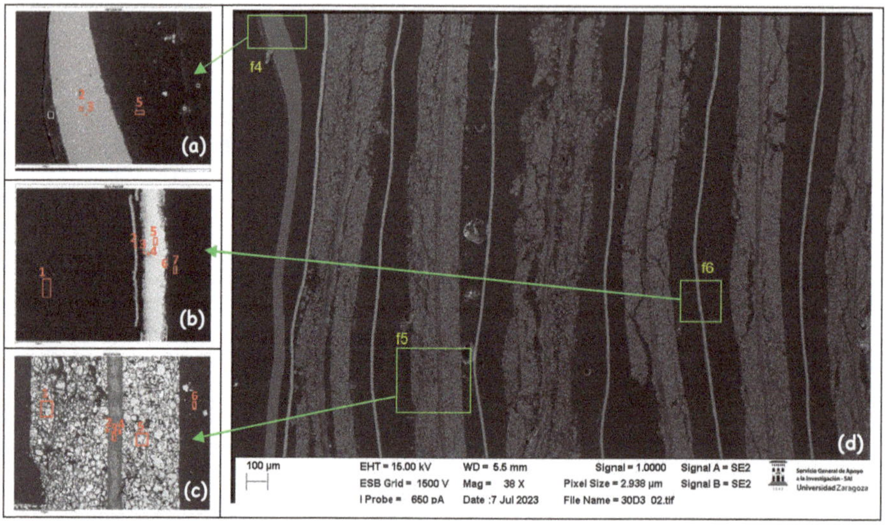

Figure 45. Chemical composition of each of the layers of the pouch cell of module 30 in zone 3 (central zone on the opposite side to that of the battery tabs of the cell Figure 12a).

Figure 46a identifies the pouch of the pouch cell, which is made of aluminum (zones 1 and 2); it appears more damaged than the bag of module 30. In zones 6 and 8, there is a compound of carbon and oxygen with some fluorine from the electrolyte, and in zone 7, there is a compound of carbon and oxygen with aluminum from the cathode collector. Figure 46c identifies the cathode (manganese oxide), and there are also carbon particles. In this case, aluminum is not observed. In Figure 46b, the anode zone has been detected. Then, there is a copper layer and a copper collector. Lastly, there are silicon particles, and a graphite coating layer is detected. The silicon particles could be due to traces of a sealant material.

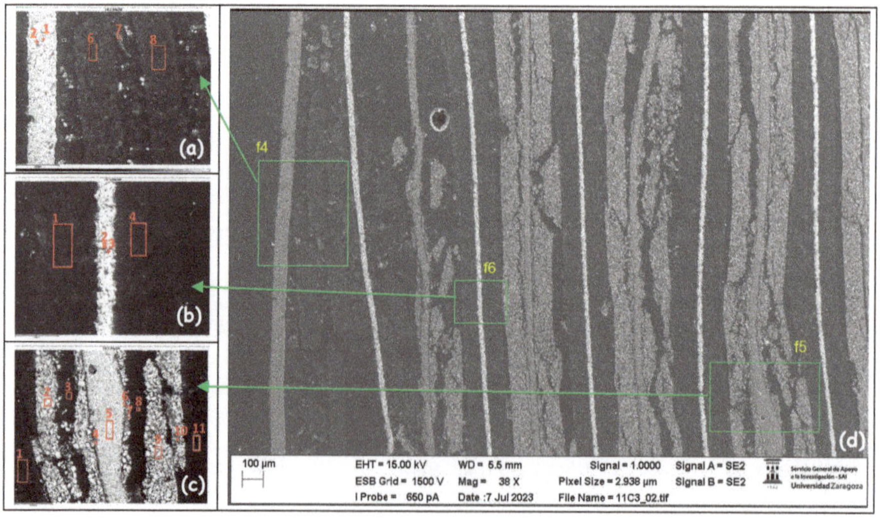

Figure 46. Chemical composition of each of the layers of the pouch cell of module 11 in zone 3 (central zone on the opposite side to that of the battery tabs of the cell Figure 12a).

The sample obtained in zone 1 (anode battery tabs) (Figure 12a) is analyzed below (Figure 47):

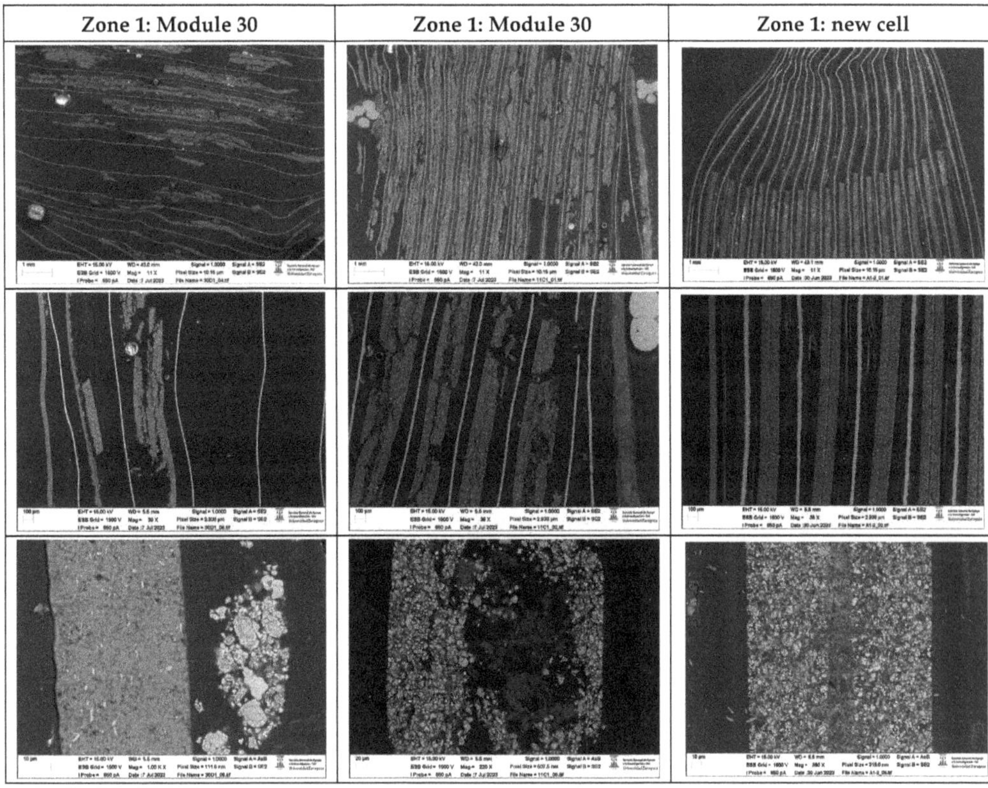

Figure 47. Images of the pouch cell section in zone one (anode zone) of the cell (Figure 12a).

If the morphology is analyzed, it can be observed that in zone one (Figure 12a), module 30 is more damaged than module 11. It is also observed that the thickness of the copper collector decreases in the event that thermal runaway has occurred, and it is observed to be narrower in the case of module 30.

Subsequently, the chemical composition of zone 1 (Figure 12a) is analyzed in the case of the cell of module 30 that caught fire (Figure 48).

Figure 48e shows the structure of the pouch cell of the burned module 30 in zone 1, four zones have been marked with a green box in which the composition of each of the layers of this pouch cell will be analyzed (Figure 48a–d).

Figure 48a identifies a compound of nickel and oxygen in zone 1 with traces of carbon and manganese, a compound of manganese and oxygen in zone 2, and a compound of carbon and oxygen in zone 3. Figure 48b identifies a compound of copper, carbon, and oxygen in zone 1 and zone 3; copper in zone 2, which is the anode collector; and a compound of carbon and oxygen in zone 4. Figure 48c identifies carbon, anode coating in zone 1, a compound of manganese, oxygen, and carbon in zone 2, cathode coating in zone 2, cathode aluminum collector in zone 3, fluorine from the electrolyte in zone 4, and a compound of aluminum and oxygen in zone 5. Finally, Figure 48d identifies a compound of copper, carbon, and oxygen in zones 1 and 3, copper in zone 2, which is the copper collector of the anode, and a compound of carbon and oxygen in zone 4.

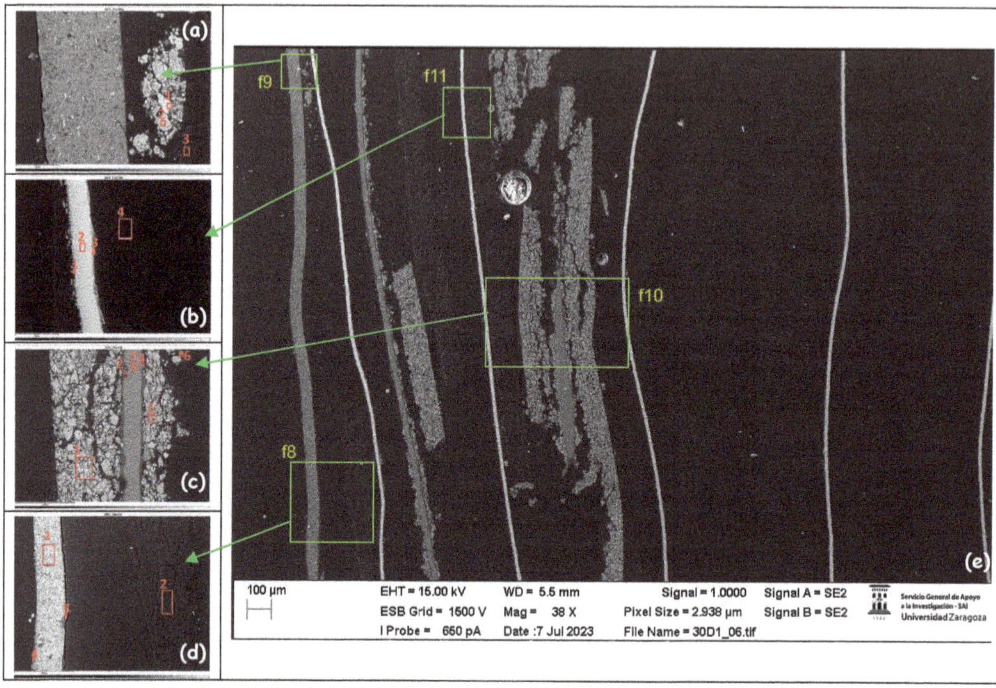

Figure 48. Chemical composition of each of the layers of the pouch cell of module 30 in zone 1 (anode zone) (Figure 12a).

Next, the chemical composition of zone 1 (Figure 12a) is analyzed in the case of the cell of the burned module 11.

Figure 49c shows the structure of the pouch cell of the burned module 11 in zone 1, two zones have been marked with a green box in which the composition of each of the layers of this pouch cell will be analyzed (Figure 49a,b).

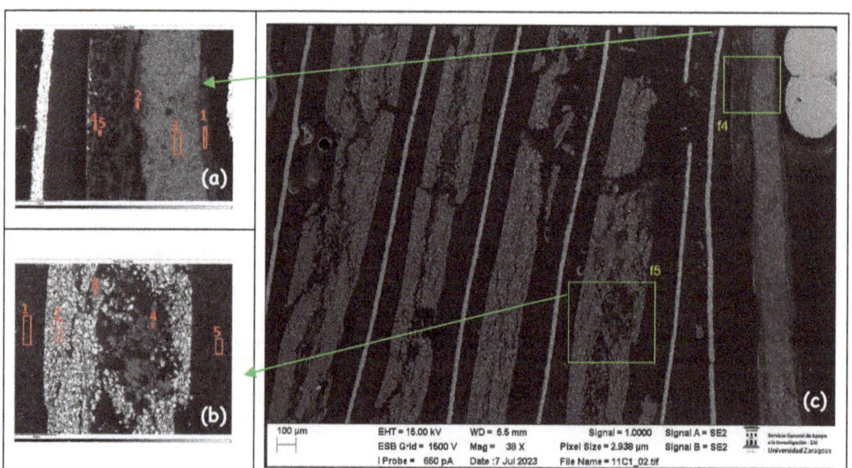

Figure 49. Chemical composition of each of the layers of the pouch cell of module 11 in zone 1 (anode zone) (Figure 12a).

Figure 49a identifies a compound of carbon, oxygen, and aluminum in zones 1 and 2; aluminum and carbon in zone 3, which is the pouch bag; and in zone 4, there are carbon, oxygen, aluminum, and some copper. And in zone 5, there is carbon, which is the coating of the anode. In Figure 49b, a compound of carbon and oxygen is identified in zones 1, 4, and 5; in zone 4, there is also some aluminum. In zone 2, there is a compound of manganese, carbon, and oxygen cathode coating, and in the center, zone 3, there is a compound of aluminum and oxygen from the aluminum collector that has melted down and disappeared.

The chemical composition of zone 1 is then analyzed in the case of the new cell (Figure 50).

Figure 50e shows the structure of a new pouch cell in zone 1, four zones have been marked with a green box in which the composition of each of the layers of this pouch cell will be analysed (Figure 50a–d).

Figure 50a identifies a compound of carbon and oxygen in the three zones analyzed, and Figure 50b identifies a compound of carbon and oxygen in zones 1, 2, 5, and 6, a compound of manganese, carbon, oxygen, and fluoride in zone 3, and carbon and aluminum in zone 4. Figure 50c identifies the bag, which is aluminum in zone 3, coated on both sides by a compound of carbon and oxygen. Figure 50d identifies the copper collector of the anode in zone 4, which is coated by a compound of carbon and oxygen.

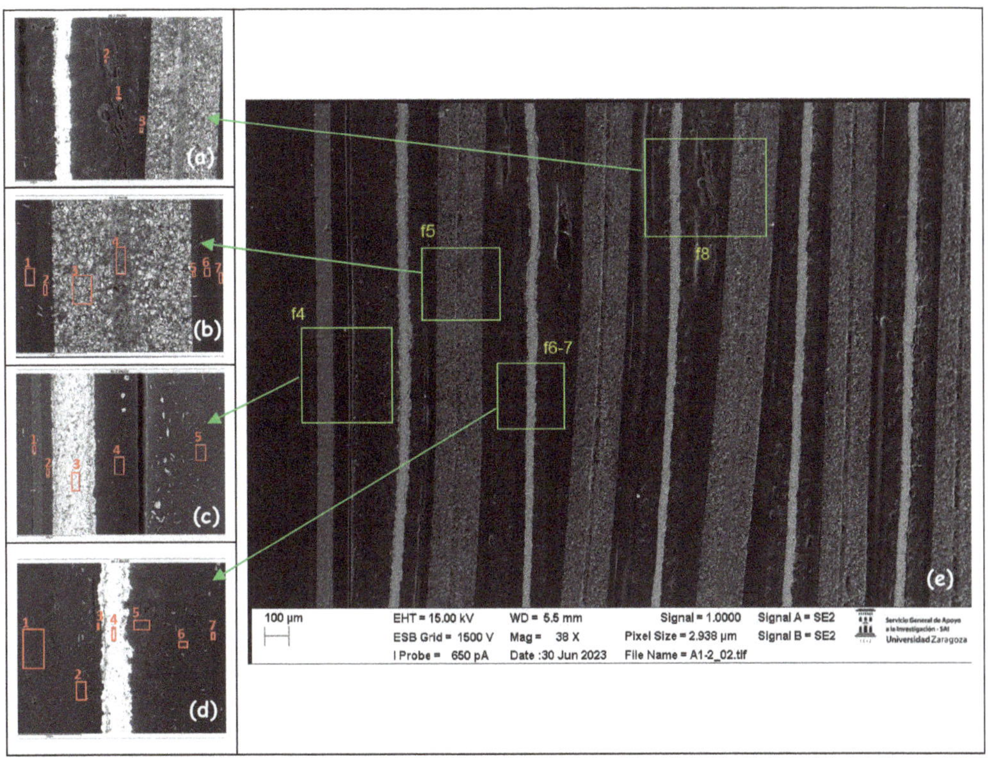

Figure 50. Chemical composition of each of the layers of the new pouch cell in zone 1 (anode zone).

The sample obtained in zone 2 (cathode battery tabs) is analyzed below (Figure 51):

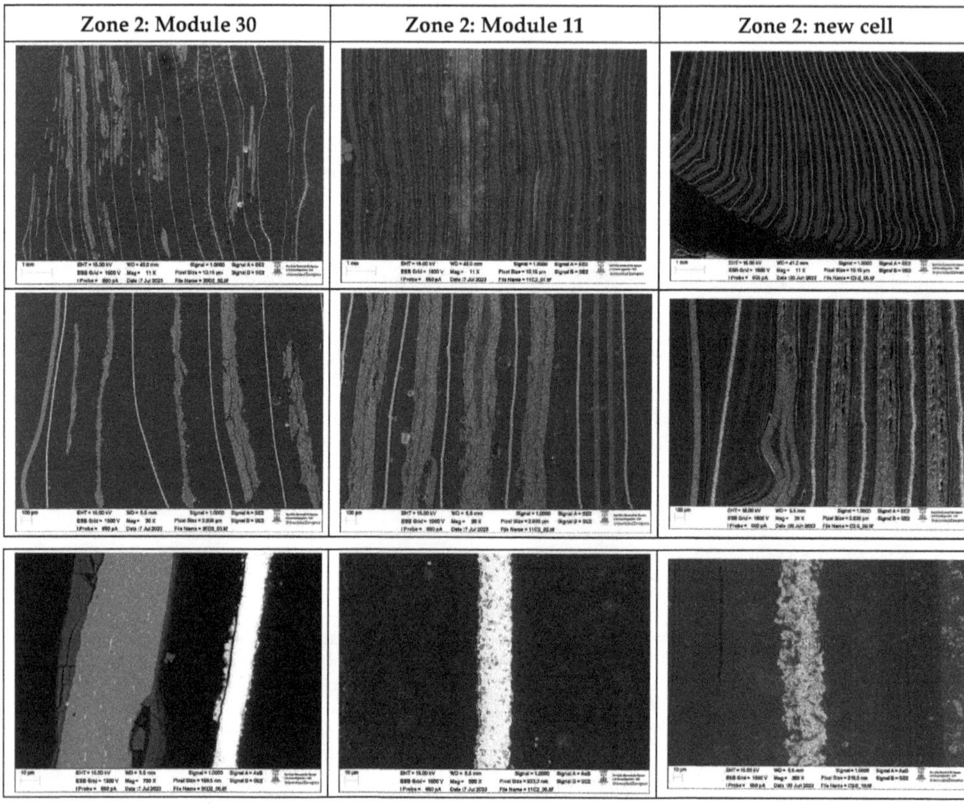

Figure 51. Images of the pouch cell section in zone two (cathode zone) of the cell.

If the morphology is analyzed, module 30 is more damaged than module 11 in zone 2. It is also observed that the thickness of the copper collector decreases in the case of the cell experiencing thermal runaway, and it is observed to be narrower in the case of module 30.

Next, the chemical composition of zone 2 is analyzed in the case of the cell of the burning module 30 (Figure 52).

Figure 52d shows the structure of the module 30 pouch cell in zone 2, three zones have been marked with a green box in which the composition of each of the layers of this pouch cell will be analyzed (Figure 52a–c).

Figure 52a identifies a layer of a compound of carbon and oxygen (zone 1), a layer of a compound of aluminum and oxygen (zone 2) and the aluminum pouch (zone 3), a layer of an aluminum and oxygen compound (zone 5) and a layer of a carbon and oxygen compound (zone 6), anode copper collector (zone 8), and a layer of a compound of cooper and oxygen (zone 7 and zone 9). Figure 52b identifies a compound of manganese, carbon, and oxygen in zone 1 and 5, a layer of an aluminum and oxygen compound in zones 2 and 4, and aluminum and carbon in zone 3, the cathode collector. Figure 52c identifies a compound of copper, carbon, and oxygen in zones 1 and 3 and copper and carbon in zone 2, which is the collector and anode.

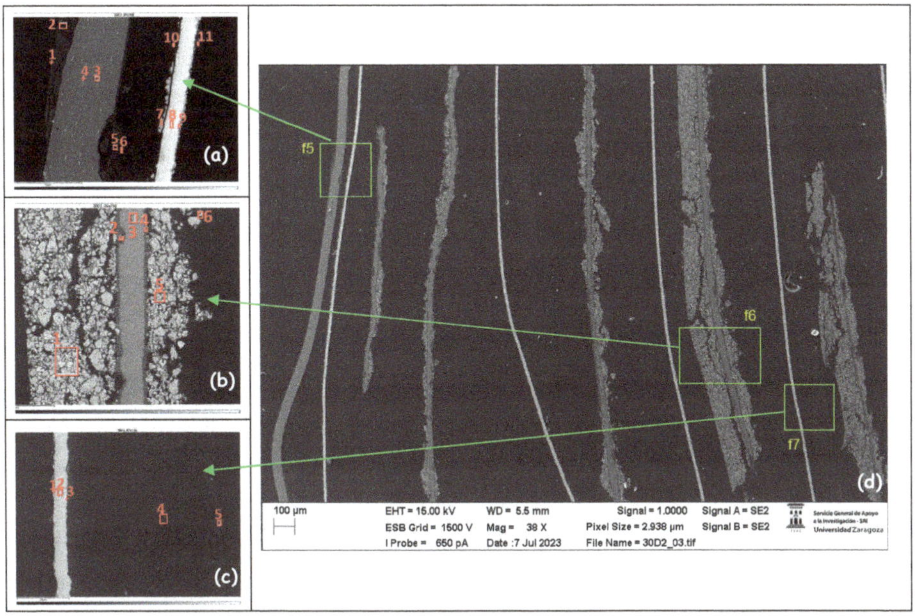

Figure 52. Chemical composition of each of the layers of the module 30 pouch cell in zone 2 (cathode zone).

Next, the chemical composition of zone 2 is analyzed in the case of the cell of the burning module 11 (Figure 53). Figure 53d shows the structure of the module 11 pouch cell in zone 2, three zones have been marked with a green box in which the composition of each of the layers of this pouch cell will be analyzed (Figure 53a–c).

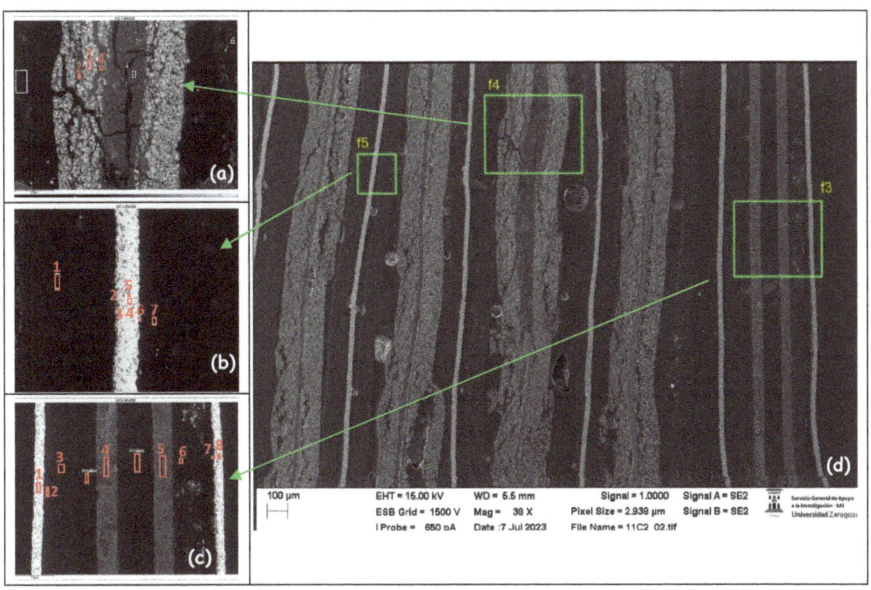

Figure 53. Chemical composition of each of the layers of the module 11 pouch cell in zone 2 (cathode zone).

Figure 53a shows the cracking of the collector and cathode coating, identifying a compound of aluminum and oxygen (zone 2 and zone 4) within a layer of manganese, carbon, nickel, and oxygen (zone 3). Figure 53b identifies a layer of carbon (zone 1 and zone 7); a layer with a compound of copper, carbon, and oxygen (zones 2 and 3); a zone with a compound of copper and oxygen that is the collector of the anode; and a zone that has silicon (zone 5 and zone 6). Figure 53c shows a copper zone (zone 1 and zone 8) with two collectors; zones 4 and 5 are carbon and aluminum; resin layers are also identified; and in zone 7, carbon with traces of silicon and copper is observed. The silicon particles could be due to traces of a sealant material.

The chemical composition of zone 2 is then analyzed in the case of the new cell (Figure 54).

Figure 54d shows the structure of the new pouch cell in zone 2, three zones have been marked with a green box in which the composition of each of the layers of this pouch cell will be analyzed (Figure 54a–c).

Figure 54a shows the copper collector with carbon and some silicon (zone 4), a layer of a compound of carbon and oxygen on either side of the collector that comes from the graphite in the anode. Figure 54b shows the aluminum pouch with carbon and a compound of carbon and oxygen layers. Figure 54c shows layers of a carbon and oxygen compound, zone 3 shows aluminum and carbon, zone 4 shows a compound of manganese and oxygen, and zone 5 shows carbon and silicon. As stated above, the silicon particles could be due to traces of a sealant material.

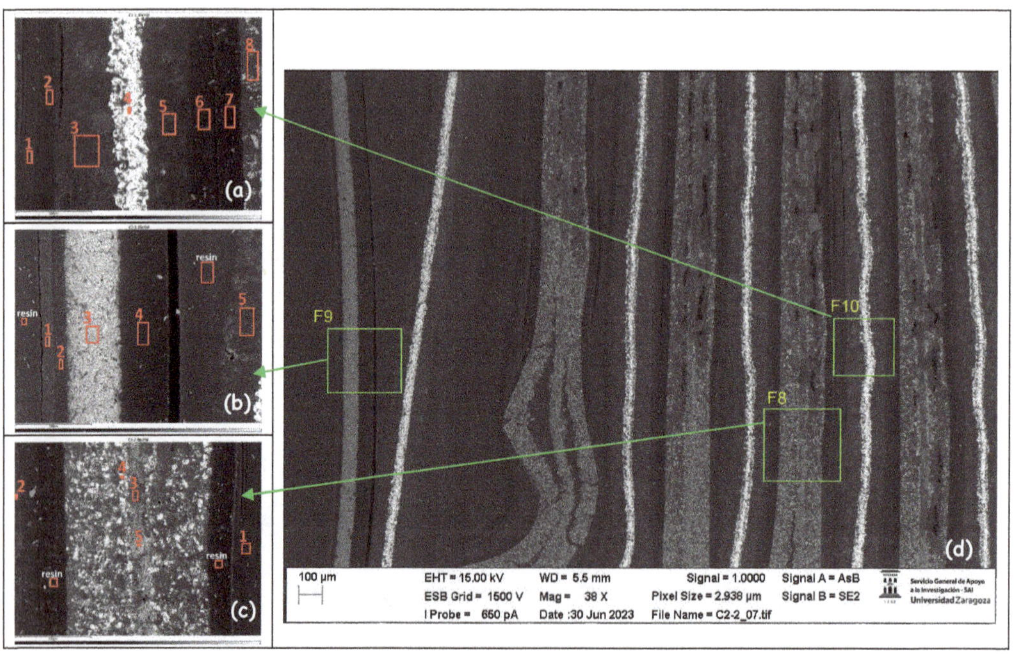

Figure 54. Chemical composition of each of the layers of the new pouch cell in zone 2 (cathode zone).

4. Conclusions

After analyzing the results obtained on the state of the cells subjected to thermal runaway in each of the analyzed arrangements, both vertical and horizontal arrangement, and compared with the state of a new cell, the following conclusions are reached:

- After thermal runaway, the cathode surface is covered with off-white floccules, fragment debris from cathode materials, ash from cathode material and separators, and products of exothermal reactions and traces of anode graphite. It is observed that in the case of module 11, there are more dark-colored floccules than in the case of module 30, and there are more in the upper cathode than in the lower cathode. Therefore, it is concluded that module 11, in a vertical arrangement, experiences higher temperatures in the thermal runaway at the same SoC as module 30, which is in a horizontal arrangement.

- Regarding the morphology, it is observed that in the case of the lower cathode of module 30 (horizontal arrangement), the particles are smaller compared to the upper cathode of module 30. This may be due to the higher temperature since it is more exposed. In the case of module 11 (vertical arrangement), there is no difference between the particle size of the upper cathode and the lower cathode.
- It is observed that in the case of modules 30 and 11 with a SoC at 68% after thermal runaway, the pouch cell layered structure was destroyed, and adhesion of the particles dispersed outside the original layered structure occurred. The positive electrode material (cathode) reacted at high temperatures and decomposed. On the other hand, carbon particles from the anode (negative electrode) went into the cathode (positive electrode) structure through the damaged diaphragm.
- In the lower cathode of module 30, there are higher amounts of fluorine, aluminum, manganese, phosphorus, nickel, and cobalt than in the upper cathode.
- Regarding module 11, a higher amount of the aluminum and oxygen compound is observed. This may be because, in this case, the aluminum collector has been more damaged than in the case of module 30.
- The analysis of the cell anodes after the fire test shows that the anode has peaks of oxygen and peaks of carbon, which could indicate the formation of lithium carbonate (Li_2CO_3). Nevertheless, we have to note that the EDS data can only identify elemental composition and its weight/atomic percentage without offering insights into the chemical structure of components.
- After carrying out the comparative analysis of the different cathodes analyzed, it can be concluded that the cathodes of the ignited modules present compounds containing the elements of the cathode coating, but also present in some areas aluminum from the melting of the aluminum collector of the cathode and copper from the melting of the copper collector of the anode. There are also traces of fluorine from the electrolyte (LiPF6). Copper only appears on the lower cathode of module 11, yet more compounds appear, so it can be concluded that this is the most damaged cathode analyzed.
- It is observed that when the cell fails, the pressure caused by the swelling due to outgassing leaves visible fractures in the cathode.
- Analysis of the stacked layer samples allowed us to understand interesting aspects of the interior of the investigated battery, including in the tab area of the battery an additional separator layer, and it showed that on the edge of the battery, the separator and bag were welded together.
- When analyzing the structure, it is observed that in zone 3, the cell of module 11 on the outside and the cell of module 30 on the inside are more damaged.
- In both zone 1 (anode) and zone 2 (cathode), the cell of module 30 is more damaged than that of module 11 and in the case of zone 2.

The results of this study indicate that a more thorough structural analysis of the different layers of the cell (visualization of coatings deposited due to the exothermic reaction, study of the separator pore size, among other characteristics) should be carried out to improve the identification of the material composition of each layer. On the other hand, the technique for obtaining the test samples is also important so that no impurities contaminate the sample.

Regarding the safety of the battery, the results of this study will allow us to determine which arrangement and structure of the cells within the battery pack is safer against thermal runaway due to thermal failure. On the other hand, the identification of the chemical composition of each of the cell layers and the compounds generated as a residue of a lithium-ion battery after a thermal abuse test is very useful for the intervention of first responders and for developing safe procedures to implement in response to accidents involving lithium-ion batteries.

Author Contributions: Conceptualization, A.O. and L.C.; data curation, A.O. and L.C.; formal analysis, A.O. and L.C.; funding acquisition, A.O. and L.C.; investigation, A.O. and L.C.; methodology, A.O. and L.C.; project administration, L.C.; resources, A.O. and L.C.; supervision, L.C.; writing—original draft, A.O. and L.C.; writing—review and editing, A.O. All authors have read and agreed to the published version of the manuscript.

Funding: University of Zaragoza: Industrial Doctorate (DI 4/2020).

Data Availability Statement: The data presented in this study are available on request from the corresponding author. The data are not publicly available due to privacy restrictions.

Acknowledgments: The authors are grateful for the support received through the University of Zaragoza and Instituto de Investigación sobre Vehículos, S.A. (Centro Zaragoza), in which the work presented in this article was framed.

Conflicts of Interest: The authors declare no conflicts of interest.

References

1. Noh, H.J.; Youn, S.; Yoon, C.S.; Sun, Y.K. Comparison of the structural and electrochemical properties of layered $Li[Ni_xCo_yMn_z]O_2$ (x = 1/3, 0.5, 0.6, 0.7, 0.8 and 0.85) cathode material for lithium-ion batteries. *J. Power Sources* **2013**, *233*, 121–130. [CrossRef]
2. Electrek 2018. Available online: https://electrek.co/2018/06/16/tesla-model-s-battery-fire-investigating/ (accessed on 28 November 2023).
3. Yahoo!News. 2019. Available online: https://sg.news.yahoo.com/tesla-car-catches-fire-hong-kong-parking-lot-050418281--finance.html?guccounter=1&guce_referrer=aHR0cHM6Ly93d3cuZ29vZ2xlLmNvbS8&guce_referrer_sig=AQAAAClpkmgR9urlu5KzzVB-y0uh0DW8ghCIrL9EdFrWSrIT8vBTETyCHsDA7mD5_1traEOgmIBOpuOZXf4YnfLD95g-yLPIlILgysRGQE48TSCJ4CdwpZKHvgTWfNP0r98GLV4CHFC9g-qRRZo2c0GcQPZZ3X4LK_4Y66TUg90JdNlN (accessed on 28 November 2023).
4. CNN Business. 2019. Available online: https://edition.cnn.com/2019/05/16/business/tesla-fire-battery-software-update/index.html#:~:text=Tesla%20is%20upgrading%20the%20battery,and%20improve%20their%20overall%20longevity (accessed on 28 November 2023).
5. Electrek. 2023. Available online: https://electrek.co/2023/04/27/tesla-fire-police-believed-battery-arson/ (accessed on 28 November 2023).
6. Carnewschina. 2023. Available online: https://carnewschina.com/2023/08/09/tesla-model-s-fire-in-sichuan/ (accessed on 28 November 2023).
7. Cbsnews. 2023. Available online: https://www.cbsnews.com/sacramento/news/tesla-spontaneously-catches-fire-in-rancho-cordova-high-end-vehicle-scrap-yard/ (accessed on 28 November 2023).
8. Sun, P.; Bisschop, R.; Niu, H.; Huang, X. A Review of Battery Fires in Electric Vehicles. *Fire Technol.* **2020**, *56*, 1361–1410. [CrossRef]
9. Balakrishnan, P.G.; Ramesh, R.; Kumar, T.P. Safety mechanisms in lithium-ion batteries. *J. Power Sources* **2006**, *155*, 401–414. [CrossRef]
10. Chen, M.; Yuen, R.; Wang, J. An experimental study about the effect of arrangement on the fire behaviors of lithium-ion batteries. *J. Therm. Anal. Calor* **2017**, *129*, 181–188. [CrossRef]
11. Feng, X.; Ouyang, M.; Liu, X.; Lu, L.; Xia, Y.; He, X. Thermal Runaway mechanism of lithium ion battery for electric vehicles: A review. *Energy Storage Mater.* **2018**, *10*, 246–267. [CrossRef]
12. Wang, H.; Lara-Curzio, E.; Rule, E.T.; Winchester, C.S. Mechanical abuse simulation and thermal runaway risks of large-format Li-ion batteries. *J. Power Sources* **2017**, *342*, 913–920. [CrossRef]
13. Huang, P.; Wang, Q.; Li, K.; Ping, P.; Sun, J. The combustion behavior of large scale lithium titanate battery. *Sci. Rep.* **2015**, *5*, 7788. [CrossRef] [PubMed]
14. Fu, Y.; Lu, S.; Li, K.; Liu, C.; Cheng, X.; Zhang, H. An experimental study on burning behaviors of 18650 lithium ion batteries using a cone calorimeter. *J. Power Sources* **2015**, *273*, 216–222. [CrossRef]
15. Chen, M.; Zhou, D.; Chen, X.; Zhang, W.; Liu, J.; Yuen, R.; Wang, J. Investigation on the Thermal Hazards of 18650 Lithium IonBatteries by Fire Calorimeter. *J. Therm. Anal. Calorim.* **2015**, *122*, 755–763. [CrossRef]
16. Feng, X.; Fang, M.; He, X.; Ouyang, M.; Lu, L.; Wang, H.; Zhang, M. TR features of large format prismatic lithium ion battery using extended volume accelerating rate calorimetry. *J. Power Sources* **2014**, *255*, 294–301. [CrossRef]
17. Roth, E.P.; Doughty, D.H. Thermal abuse performance of high-power 18650 Li-ion cells. *J. Power Sources* **2004**, *128*, 308–318. [CrossRef]
18. Wang, Q.; Sun, J.; Yao, X.; Chen, C. Thermal behavior of lithiated graphite with electrolyte in lithium-ion batteries. *J. Electrochem. Soc.* **2006**, *153*, A329–A333. [CrossRef]

19. Ma, S.; Jiang, M.; Tao, P.; Song, C.; Wu, J.; Wang, J.; Deng, T.; Shang, W. Temperature effect and thermal impact in lithium-ion batteries: A review. *Prog. Nat. Sci.* **2018**, *28*, 653–666. [CrossRef]
20. Zhang, S. Problems and their origins of Ni-rich layered oxide cathode materials. *Energy Storage Mater.* **2020**, *24*, 247–254. [CrossRef]
21. Peng, Y.; Yang, L.; Ju, X.; Liao, B.; Ye, K.; Li, L.; Cao, B.; Ni, Y. A comprehensive investigation on the thermal and toxic hazards of large format lithium-ion batteries with $LiFePO_4$ cathode. *J. Hazard. Mater.* **2020**, *381*, 120916. [CrossRef]
22. Wang, Q.; Ping, P.; Zhao, X.; Chu, G.; Sun, J.; Chen, C. Thermal Runaway Caused Fire and Explosion of Lithium Ion Battery. *J. Power Sources* **2012**, *208*, 210–224. [CrossRef]
23. Bai, F.; Chen, M.; Song, W.; Feng, Z.; Li, Y.; Ding, Y. Thermal management performances of PCM/water cooling-plate using for lithium-ion battery module based on non-uniform internal heat source. *Appl. Therm. Eng.* **2017**, *126*, 17–27. [CrossRef]
24. Garg, M.; Tanim, T.R.; Rahn, C.D.; Bryngelsson, H.; Legnedahl, N. Elevated temperature for life extension for lithium ion power cells. *Energy* **2018**, *159*, 716–723. [CrossRef]
25. Galushkin, N.E.; Yazvinskaya, N.N.; Galushkin, D.N. Mechanism of gases generation during lithium-ion batteries cycling. *J. Electrochem. Soc.* **2019**, *166*, A897–A908. [CrossRef]
26. Wang, G.; Kong, D.; Ping, P.; He, X.; Lv, H.; Zhao, H.; Hong, W. Modelingventing behavior of lithium-ion batteries during thermal runaway propa-gation by coupling CFD and thermal resistance network. *Appl. Energy* **2023**, *334*, 12066. [CrossRef]
27. Wang, Z.; Mao, N.; Jiang, F. Study on the effect of spacing on thermal runaway propagation for lithium-ion batteries. *J. Therm. Anal. Calorim.* **2020**, *140*, 2849–2863. [CrossRef]
28. Fang, J.; Cai, J.; He, X. Experimental study on the vertical thermal runaway propagation in cylindrical Lithium-ion batteries: Effects of spacing and state of charge. *Appl. Therm. Eng.* **2021**, *197*, 11739. [CrossRef]
29. Feng, X.; Sun, J.; Ouyang, M.; Wang, F.; He, X.; Lu, L.; Peng, H. Characterization of penetration induced thermal runaway propagation process within a large format lithium ion battery module. *J. Power Sources* **2015**, *275*, 261–273. [CrossRef]
30. Feng, X.; Lu, L.; Ouyang, M.; Li, J.; He, X. A 3D thermal runaway propagation model for a large format lithium ion battery module. *Energy* **2016**, *115*, 194–208. [CrossRef]
31. Lopez, C.F.; Jeevarajan, J.A.; Mukherjee, P.P. Experimental analysis of thermal runaway and propagation in lithium-ion battery modules. *J. Electrochem. Soc.* **2015**, *162*, A1905–A1915. [CrossRef]
32. Wilke, S.; Schweitzer, B.; Khateeb, S.; Al-Hallaj, S. Preventing thermal runaway propagation in lithium ion battery packs using a phase change composite material: An experimental study. *J. Power Sources* **2017**, *340*, 51–59. [CrossRef]
33. Lamb, J.; Orendorff, C.J.; Steele, L.A.M.; Spangler, S.W. Failure propagation in multi-cell lithium ion batteries. *J. Power Sources* **2015**, *283*, 517–523. [CrossRef]
34. Gao, S.; Feng, X.; Lu, L.; Kamyab, N.; Du, J.; Comsn, P.; White, R.E.; Ouyang, M. An experimental and analytical study of thermal runaway propagation in a large format lithium ion battery module with NMC pouch-cells in parallel. *Int. J. Heat Mass Transf.* **2019**, *135*, 93–103. [CrossRef]
35. Kovachev, G.; Schröttner, H.; Gstrein, G.; Aiello, L.; Hanzu, I.; Wilkening, H.M.R.; Foitzik, A.; Wellm, M.; Sinz, W.; Ellersdorfer, C. Analytical Disection of an Automotive Li-Ion Pouch Cell. *Batteries* **2019**, *5*, 67. [CrossRef]
36. Zhu, X.; Sun, Z.; Wang, Z.; Wang, H.; Lin, N.; Shan, C. Thermal runaway in commercial lithium-ion cells under overheating condition and the safety assessment method: Effects of SoCs, cathode materials and packaging forms. *J. Energy Storage* **2023**, *68*, 107768. [CrossRef]
37. Shin, H.; Lee, Y.K.; Lu, W. Structural degradation of graphite anode induced by dissolved manganese ions in lithium-ion batteries. *J. Power Sources* **2022**, *528*, 231223. [CrossRef]
38. Arora, P.; Zhang, Z. Battery Separators. *Chem. Rev.* **2004**, *104*, 4419–4462. [CrossRef] [PubMed]
39. Luo, M.J.; Guo, Y.Z.; Kang, J.Q. Ternary-material lithium-ion battery SOC estimation under various ambient temperature. *Ionics* **2018**, *24*, 1907–1917. [CrossRef]
40. Wang, J.L.; Wu, H.L.; Cui, Y.H. A new class of ternary compound for lithium-ion battery: From composite to solid solution Acs. *Appl. Mater. Interfaces* **2018**, *10*, 5125–5231. [CrossRef]
41. Xie, H.J.; Sun, J.; Li, J.G.; Zhou, T.; Wei, S.P.; Yi, Z.H. Lithium-Ion Battery Thermal Runaway Electro-Thermal Triggering. Method and Toxicity Analysis. *IOP Conf. Ser. Earth Environ. Sci.* **2021**, *701*, 012007. [CrossRef]
42. Finegan, D.P.; Scheel, M.; Robinson, J.B.; Tjaden, B.; Michiel, M.D.; Hinds, G.; Bretta, D.J.L.; Shearing, P.R. Investigating lithium-ion battery materials during overcharge-induced thermal runaway: An operando and multi-scale X-ray CT study†. *Phys. Chem. Chem. Phys.* **2016**, *18*, 30912–30919. [CrossRef]

batteries

MDPI

Article

Mechanical Characterization and Modeling of Large-Format Lithium-Ion Battery Cell Electrodes and Separators for Real Operating Scenarios

Johannes Brehm [1,2,*], Axel Durdel [1], Tobias Kussinger [1,2], Philip Kotter [2], Maximilian Altmann [2] and Andreas Jossen [1]

[1] Chair of Electrical Energy Storage Technology (EES), Department of Energy and Process Engineering, TUM School of Engineering and Design, Technical University of Munich (TUM), Arcisstrasse 21, 80333 Munich, Germany
[2] Bayerische Motoren Werke Aktiengesellschaft (BMW AG), Petuelring 130, 80809 Munich, Germany
* Correspondence: johannes.brehm@tum.de

Abstract: This study presents a novel application-oriented approach to the mechanical characterization and subsequent modeling of porous electrodes and separators in lithium-ion cells to gain a better understanding of their real mechanical operating behavior. An experimental study was conducted on the non-linear stiffness of $LiNi_{0.8}Co_{0.15}Al_{0.05}O_2$ and graphite electrodes as well as PE separators, harvested from large-format lithium-ion cells, using compression tests. The mechanical response of the components was determined for different operating conditions, including nominal stress levels, mechanical loading rates, and mechanical cycles. The presented work describes the test procedure, the experimental setup, and an objective evaluation method, allowing for a detailed summary of the observed mechanical behavior. A distinct nominal stress level and mechanical cycle dependency of the non-linear stiffnesses of the porous materials were found. However, no clear dependency on compression rate was observed. Based on the experimental data, a poroelastic mechanical model was utilized to predict the non-linear behavior of these porous materials under real mechanical operating scenarios with a normalized root-mean-squared error less than 5.5%. The results provide essential new insights into the mechanical behavior of porous electrodes and separators in lithium-ion cells under real operating conditions, enabling the accelerated development of high-performing and safe batteries for various applications.

Keywords: lithium-ion battery; operation; NCA; graphite; PE separator; uniaxial mechanical characterization; non-linear modeling

1. Introduction

The increasing demand for high-performing and safe battery systems has motivated research on the mechanical characterization and modeling of large-format lithium-ion cell electrodes and separators. Understanding their mechanical properties is essential for optimizing design and preventing failures like cracking and delamination. Characterizing these materials under different operating conditions allows for the identification of critical parameters such as stiffness and strength [1–11]. Models that build on the findings of these experimental data can provide insights into the impact of mechanical loading on electrochemical performance [1,8,12–17]. Improving characterization and modeling supports the development of safer, more durable batteries, benefiting industries relying on lithium-ion batteries, such as electric vehicles (EVs) and renewable energy storage [4,18–21].

Since 2018, a series of publications have explored the mechanical properties of battery cells and their sub-components (such as jelly rolls, stacks, or isolated electrodes and separators) by applying external mechanical loading, as presented in the literature [1–11]. An overview of the respective experimental design is given in Table 1, with this work added at the end of the table. Nevertheless, from an EV operation point of view, research in the field of mechanical characterization requires further investigation, as follows:

- Investigations comprise only laboratory-sized or small commercial cells [1–4,10]. Studies of large prismatic cells, commonly used in EV, are missing.
- Stress ranges $\sigma_{\text{min}} - \sigma_{\text{max}}$ and compression rates v are seldom chosen in relation to real boundary conditions, rendering a transfer from the literature to application difficult [1–11].
- Only one investigation exists that utilizes more than one mechanical cycle n [3].
- Samples are mostly conditioned without electrolytes [1,4–6,9–11] or in a low state of charge (SoC) [1–4,9,11], making direct comparisons to real operations difficult.

Table 1. A literature review of mechanical characterization investigations utilizing compression tests. The column labeled 'Elyt.' indicates whether tests were conducted using dry electrodes or electrodes filled with electrolytes. Literature data in the compression rate v column were converted to MPa h^{-1} for comparability. In case the respective reference utilized a displacement-controlled test procedure, the resulting pressures during the procedure are given here as force-controlled equivalents.

Refs.	Scope	Material	SoC in %	Elyt.	σ_1 in MPa	σ_{nom} (σ_{min}–σ_{max}) in MPa	v in MPa h^{-1}	Cycles n in —	Application/Focus
[1]	layers	NMC_{111}, PE, Gr	0	no	0.30	1.65 (0.30–3.00)	70.74	1	modeling normal operation
[2]	layers pouch	NMC_{622}, NMC_{811}, PE, Gr, Si/Gr	0	yes [b] no [c]	0.06	0.08 [d] 0.42 [d] 0.84 [d]	—	—	compressibility characterization only
[3]	layers pouch	LCO/NCA, PP, Gr	0	yes [a]	0.02	1.00 [d] 1.10 (1.00–1.20) 3.00 [d] 6.00 [d] 10.00 [d]	20.37×10^1	— 3 — — —	effect on local current densities
[4]	layers	NMC_{811}, PE, Gr	0	no	1.00	(1.00–200.00)	37.41×10^2 12.47×10^3	—	crash modeling
[5]	layers	PE, PP	n.d.	no	0.00	(0.00–150.00) (0.00–135.00) (0.00–90.00)	31.32×10^4 36.45×10^3 24.30	1	temperature and strain rate dependency of polymer separator
[6]	layers	PP	n.d.	no	1.00	(1.00–10.00) (1.00–20.00) (1.00–40.00) (1.00–60.00) (1.00–80.00) (1.00–100.00)	11.72×10^4 14.84×10^4 41.55×10^3 11.92×10^3 12.34×10^3 13.65×10^3	1	ion transport of PP separator under compression
[7]	layers	PE, PP	n.d.	yes	0.50	(0.50–80.00)	17.39×10^2 17.02×10^1 16.49 16.00×10^{-1} 15.69×10^{-2} 15.40×10^{-3}	1	temperature-dependent stiffness of PE separators
[8]	layers	PE	n.d.	yes	0.00	(0.00–22.50) (0.00–33.00) (0.00–45.00) (0.00–52.50)	1.22 17.82 24.30×10^1 28.35×10^2	1	hyper-viscoelastic model for battery separators
[9]	layers pouch	NMC_{622}, PE, Gr	0 [b] 0–100 [c]	no [b] yes [c]	0.05	(0.05–1.75)	n.d.	1	measuring reversible swelling

Table 1. *Cont.*

Refs.	Scope	Material	SoC in %	Elyt.	σ_1 in MPa	σ_{nom} (σ_{min}–σ_{max}) in MPa	v in MPa h^{-1}	Cycles n in —	Application/Focus
[10]	layers pouch	NMC_{811}, PP, Si/Gr	0 [b] 0 [c], 70 [c], 100 [c]	no [b] yes [c]	—	1.50 (0.00–3.00)	10.00×10^{-1} [b] 30.24 [d]	1	modeling normal operation
[11]	layers pouch	NMC_{811}, PE, PP, Gr	0 [b] 0–100 [c]	no [b] yes [b,c]	0.20	(0.20–2.50)	n.d.	1	effect of assembly pressure on cell performance
this work	layers	NCA, PE, Gr	0	no	0.08	0.30 (0.10–0.50) 1.00 (0.80–1.20) 3.00 (2.80–3.20)	20.00×10^{-2} 60.00×10^{-2} 10.00×10^{-1}	10	modeling normal operation

[a] replaced by diethyl carbonate, [b] component multi-layer level, [c] pouch cell level, [d] constant pressure.

The main objective of this work is to extend the state of the art by a characterization procedure and modeling approach, which have not been previously addressed regarding the following aspects:

I. Determining the real mechanical operation behavior of the electrodes and separator by conducting cyclic mechanical compression tests to mimic the real mechanical operating scenario of a large-format battery cell. Related research questions are as follows:
   - Is the mechanical behavior dependent on the compression rate v for a given nominal stress σ_{nom}?
   - Is the mechanical behavior dependent on the nominal stress σ_{nom} for a given compression rate v?
   - Is the mechanical behavior dependent on the number of mechanical cycles n?
   - Is the mechanical behavior identical for compression and decompression?

II. Non-linear poroelastic material modeling approach to depict the mechanical behavior of porous electrodes and separators more precisely. The pertinent research question can be stated as follows:
   Can a model be used for fitting over various force levels and compression rates?

Therefore, cyclic mechanical compression tests were conducted with harvested electrodes and separators. The measurement data were subsequently used to investigate the mechanical compression behavior of the single-cell components regarding the evolution of plastic strain over cycling as well as the maximum strain. In the second part of the results section, a modeling approach is presented that enables an accurate representation of cyclic load scenarios.

2. Experimental

2.1. Sample Materials and Preparation

A large-format prismatic 70.2 A h cell underwent formation, was discharged at 0.1 C to 2.8 V, and was disassembled in an argon-filled glovebox (M. Braun Inertgas-Systeme GmbH, Garching, Germany, O_2 and H_2 < 1 ppm) at 25 °C [22]. The respective (active) materials are as follows: graphite (Gr) for the anode, nickel-cobalt-aluminum (NCA) for the cathode, and polyethylene (PE) for the separator. Harvested electrode and separator sheets were washed in DEC for 10 min. Subsequently, coin samples at ⌀ 18 mm were prepared with a handheld precision punch (Nogamigiken Co., Ltd., Hitachiomiya, Japan) [23]. A schematic showing the electrode sheets and punched samples is given in Figure 1a. The samples were stored inside the glovebox until needed for compression tests as described in Section 2.3.

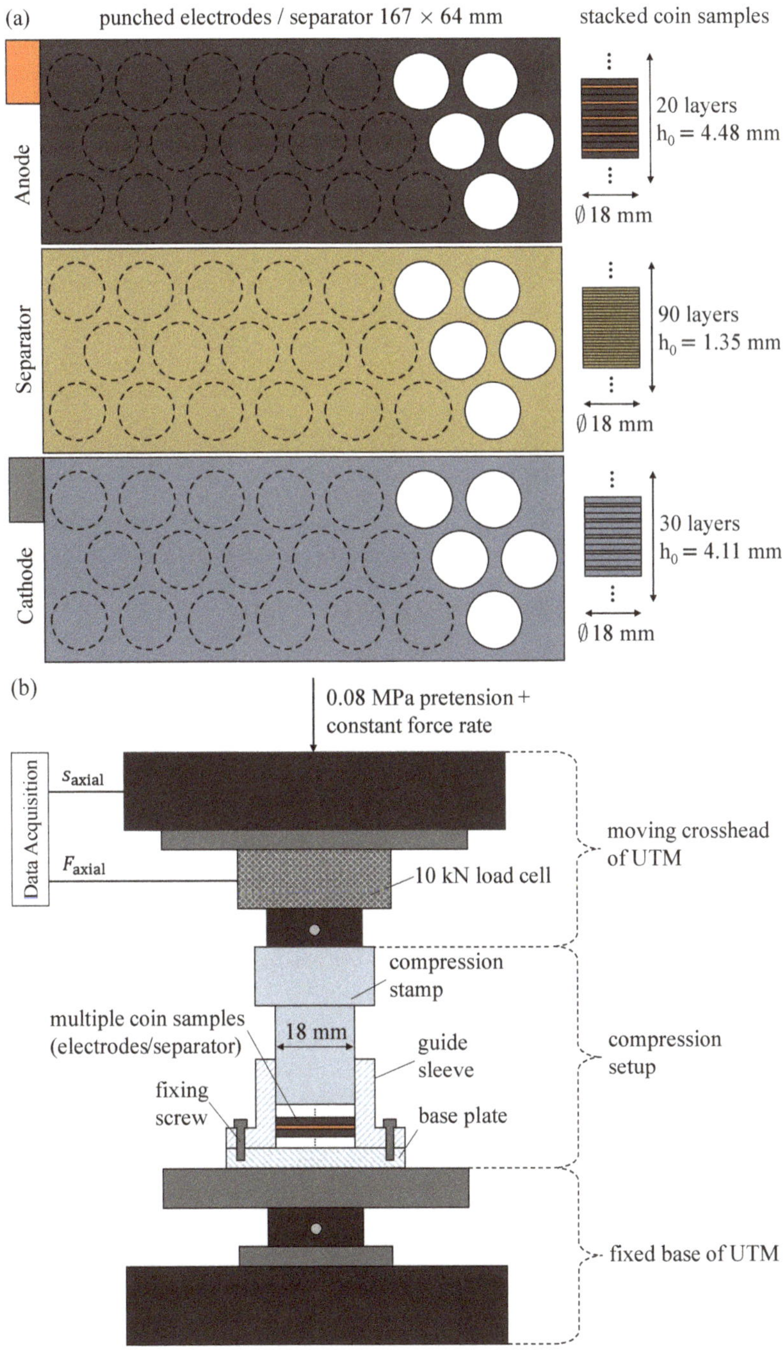

Figure 1. (**a**) Schematic sample preparation for the anode, separator, and cathode. The dashed circles represent coins that are being punched out, and the white circles represent coins that have already been punched out. Punched coin samples are stacked as indicated on the right. (**b**) Schematic overview of the used compression test setup with all sub-components.

2.2. Test Setup

The uniaxial compression tests were carried out using an AllroundLine Z010 universal testing machine (UTM) from ZwickRoell (ZwickRoell GmbH & Co. KG., Ulm, Germany) [24]. The displacement was measured via the UTM spindle drive with an accuracy of ±0.6 nm [24]. The force was recorded with an Xforce P 10 kN load cell (ZwickRoell GmbH & Co. KG., Ulm, Germany), exhibiting a non-linearity of ±0.5% [25].

A custom-built test setup (as shown in Figure 1b, which included a compression stamp and a guide sleeve, was used for the compression tests. The entire setup was made from stainless steel and was intended to prevent any misalignment of the stacked samples prior to (as well as during) the test. The stack of samples was placed inside the guidance sleeve, which was attached to a base plate. The base plate was mounted on the fixed base of the UTM and remained stationary during the compression test. The guidance sleeve's inner diameter was 18.05 mm so that the coin samples tightly fit. The compression stamp with an outer diameter of 18.00 mm was attached to the moving crosshead of the UTM. The compression stamp tightly fitted into the guidance sleeve, applying pressure onto the samples.

Note that the applied force was uniaxial, and due to the test setup including mechanical limitations in the lateral direction, the resulting strain was considered to be uniaxial, too. The test setup without samples was used for reference measurements, which were subsequently utilized for the raw data correction with the stiffness of the machine and setup. This is of great importance due to the significant elastic deformation of the spindle drive of the UTM and/or the test setup itself [3,26–28]. The overall elastic deformation of our entire system without samples was $\leqq 42\,\mu m$ in the work areas used.

2.3. Test Protocol

Coin samples were taken from the glovebox and placed inside the guidance sleeve of the compression setup. Multiple samples were stacked (20 anodes, 30 cathodes, or 90 separators) following recommendations from [29]. Additionally, multiple samples increased the total sample thickness and, thus, decreased the signal-to-noise ratio. Subsequently, the test protocol was started. Note that the test was conducted using dry electrodes (no electrolytes) under constant atmospheric conditions at 25 °C, to minimize the influence of these environmental factors.

The test procedure initially applied a pre-load of 20 N (σ_1 = 0.08 MPa) at which the strain was defined to be zero. Preliminary tests have shown that this pre-load is needed so that the individual layers in the sample stack have a well-defined contact. From this point onward, the actual compression test started. A schematic is shown in Figure 2. Starting from stress σ_1, a constant compression rate v was applied until $\sigma_{\min}$ was reached. In the following, $n = 10$ mechanical cycles with the same (de)compression rate v were carried out between $\sigma_{\min}$ and $\sigma_{\max}$. These cycles were exclusively mechanical in nature and, therefore, do not represent electrochemical cycles and associated effects. The first compression from σ_1 over $\sigma_{\min}$ to $\sigma_{\max}$ was carried out without any interruption. The final decompression of the n-th cycle analogously started at $\sigma_{\max}$ and continued over $\sigma_{\min}$ until the stress reached σ_1 again and the test terminated. Note that the nominal operational stress σ_{nom} was centered between $\sigma_{\min}$ and $\sigma_{\max}$. The respective stress values σ and compression rates v can be found in Section 2.4. During the tests, the force was applied, while force and displacement were measured as the time series sampled every 0.01 µm and 0.5 min. Since the force was applied to a defined contact area between the compression stamp and the sample, the stress was known. From the displacement and the initial height at pre-load h_1 given in Table 2, the strain was calculated.

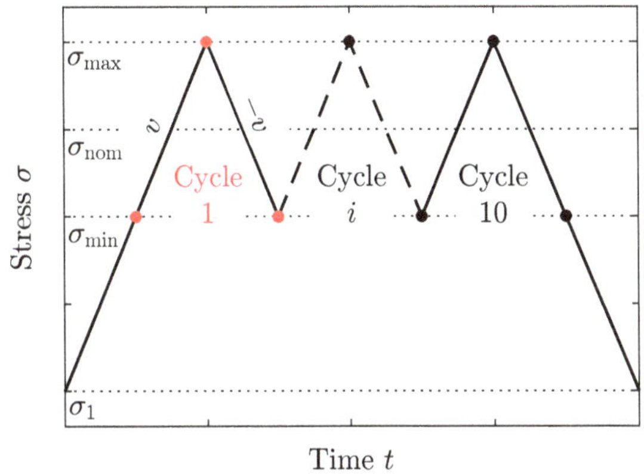

Figure 2. Schematic test program for characterizing the stacked electrode coin samples with all relevant process variables: pre-load stress σ_1, minimum stress σ_{min}, nominal stress σ_{nom}, maximum stress σ_{max}, compression rate v, and cycle count i.

2.4. Design of Experiment

Overall, three different nominal stresses σ_{nom} and three different compression rates v were tested, as shown in the last row of Table 1. The tests were conducted for the anode, separator, and cathode, respectively, resulting in 27 individual tests.

The nominal stresses were derived from the total stack pressure of a battery module with several cells at different points during its lifetime. At begin of life (BoL), the average stack pressure was approx. 0.3 MPa, while at end of life (EoL), it tose approx. up to 3.0 MPa due to aging [30–32]. The third value of 1.0 MPa was chosen between the other two values. Moreover, the three chosen nominal stress levels σ_{nom} enabled the investigation of a possible non-linear mechanical material behavior. Accounting for such non-linear material behavior during characterization might become important, especially for predictions over a lifetime.

Table 2. Measured component characteristics for the Gr anode, PE separator, and NCA cathode used.

	Unit	Gr Anode	PE Separator	NCA Cathode
Content of active material	$\%_{wt}$	95.7 [a]	—	95.4 [b]
Content of non-active material	$\%_{wt}$	4.3 [c]	—	4.6 [c]
Total layer thickness, stress-free L_0	µm	224 [d]	15 [d]	137 [d]
Thickness current collector L_0	µm	8 [d]	—	11 [d]
Thickness coating, stress-free L_0	µm	108 [d]	—	63 [d]
Porosity coating ϵ_l	-	0.309 [e]	0.519 [e]	0.187 [e]
Initial void-to-solid ratio, stress-free $\epsilon^0_{l/s}$	-	0.447 [f]	1.079 [f]	0.230 [f]
Number of layers	-	20	90	30
Total sample height, no load h_0	mm	4.480 [f]	1.350 [f]	4.110 [f]
Total sample height, pre-load h_1	mm	4.395 [g]	1.304 [g]	4.060 [g]

[a] determined by simultaneous thermal analysis with mass spectroscopy (STA-MS), [b] determined by inductively coupled plasma-optical emission spectrometry (ICP-OES), [c] determined by fourier transform infrared spectroscopy (FT-IR), [d] determined by scanning electron microscopy (SEM) cross-section, [e] determined by helium pycnometer, [f] calculated, [g] UTM.

The operational pressure window between 0% and 100% of the full cell SoC at BoL, as measured by Spingler et al. [3], is approx. 0.3 MPa during electrochemical cycling at 1.0 MPa with initial pressure. Cannarella et al. [15], however, measured an operational window of approx. 0.4 MPa at an initial pressure of 0.05 MPa. In the present study, the initial pressure σ_1 is chosen at 0.08 MPa, which is close to the value reported by Cannarella et al.; thus, the operational window is 0.4 MPa. Aufschläger et al. [33] cycled pouch cells at different mechanical bracing forces. Their results show similar operating windows regardless of the applied pressure, even throughout the cyclic aging. Experimental data from ongoing tests using the same batteries as in this work have shown an increase in the stack pressure of approximately 0.2 MPa, which is within the range given by the literature.

The compression rates are chosen to be equivalent to C-rates of 1C, 3C, and 5C. Examples of large-format cells cycled up to 2C that also provide mechanical data can be found in [3,15,33,34]. Note that 1C means that an increase in pressure by 0.2 MPa — the assumed operational window during electrochemical cycling — is carried out by the UTM within 1 h. This is equal to the loading a large-format Li-ion cell faces during a continuous charge with a rate of 1C. Accordingly, a 1C compression by 0.4 MPa between σ_{min} and σ_{max} during the actual test was carried out in 2 h. Tests at smaller compression rates could not be carried out due to limited continuous access to the UTM but would be of interest in the future to further investigate the rate-dependency of the stress–strain relationship.

3. Modeling

In addition to mechanical characterization, this work investigates material modeling approaches for porous battery sub-components. A material model, or constitutive model, is a mathematical representation of the expected deformation behavior of a given material in response to an applied load [35], and can be written as follows:

$$\sigma = C \cdot \varepsilon \tag{1}$$

where σ denotes the stress tensor, C denotes the stiffness tensor, and ε denotes the 2nd-order strain tensor (for details, see Equation (A1)).

Since the compression tests are uniaxial, Poisson's ratio for our electrodes is unknown, and $\nu = 0$ is assumed, as in [36–39]. This results in no shear forces. As the compression tests in this work were used to derive a description of the elastic behavior of the mechanical properties of the components, any effects due to shear forces were already taken into account implicitly. Since only uniaxial stresses/strains occurred later on, the indices are omitted below, and $\sigma_{11} = \sigma$, resp., $\varepsilon_{11} = \varepsilon$ apply. Note that in theory, the stiffness tensor C as well as Young's modulus E and Poisson ratio ν, may depend on the electrode's degree of lithiation (DoL) and electrolyte content. However, compression tests have only been conducted at a 0% full cell SoC at dry conditions. Thus, mechanical properties depending on lithiation and electrolyte viscoelasticity are not considered in this work. In this work, two model approaches are compared to represent the mechanical behavior of the components, as follows:

(1) *Hooke's Law*, a linear-elastic model, which represents the simplest (but a common) approach (Section 3.1).
(2) *Poroelasticity*, a non-linear-elastic model, which represents a novel approach, considering the porous property of the components (Section 3.2).

Both approaches as well as their evaluation procedures (Section 3.3) will be briefly introduced below.

3.1. Hooke's Law (Linear Elasticity)

Hooke's law is a basic principle in solid mechanics that describes the linear relationship between the applied force and the resulting deformation of an elastic material. First formulated by the English scientist Robert Hooke in the 17th century [40], this fundamental law has been extensively used in the analysis and design of mechanical systems and structures, e.g., electrode materials of lithium-ion cells [12,41–43]. Considering uniaxial conditions, Hooke's Law can be derived from Equation (1) as follows:

$$\sigma(\varepsilon) = E \cdot \varepsilon \tag{2}$$

3.2. Poroelasticity (Non-Linear Elasticity)

Poroelasticity describes the mechanical non-linear behavior of materials with porous structures; it is commonly used in the fields of civil engineering, materials engineering, geology, hydrogeology, and soil science [44–48]. In the context of lithium-ion battery electrodes and separators, the porous nature of these materials plays a crucial role in their electrochemical performance and their mechanical response. Due to the similarity between the porous morphology of the battery sub-components and soil, the approach used in this work originates from the field of soil modeling [48] and was later recommended/utilized by [49] for modeling poroelastic material behaviors.

The poroelastic model provides a more comprehensive description of the material behavior compared to the traditional linear-elastic approach described by Hooke's law, which does not explicitly consider the porous microstructure property. A detailed derivation can be found in Appendix B.2. Here, only the essential equations are shown for readability. The uniaxial stress σ is defined as follows:

$$\sigma(\varepsilon) = (\sigma_0 + \sigma_t) \cdot \left(2 \cdot \varepsilon \cdot \frac{1 + \epsilon^0_{l/s}}{\kappa} + 1\right) \cdot \exp\left[\varepsilon \cdot \frac{1 + \epsilon^0_{l/s}}{\kappa}\right] - \sigma_t \tag{3}$$

Here, σ_0 denotes the initial stress, σ_t denotes the elastic tensile strength, κ denotes the logarithmic bulk modulus, and $\epsilon^0_{l/s}$ denotes the initial void-to-solid ratio, defined as follows:

$$\epsilon^0_{l/s} = \frac{V^0_l}{V^0_s} = \frac{\epsilon^0_l}{1 - \epsilon^0_l} \tag{4}$$

The value of $\epsilon^0_{l/s}$ can consequently range from 0 to infinity.

The model parameterization process subsequently aims to determine the optimal values of κ and σ_t to adapt the poroelastic model from Equation (3) to the experimental stress–strain data. The key steps in the model parameterization process are as follows:

(1) Identification of the parameters to be optimized: Logarithmic bulk modulus κ and elastic tensile strength σ_t are selected as the parameters to be optimized, as they are the unknown variables in Equation (3). All other parameters are either measured or known, as shown in Table 2.
(2) Utilization of the MATLAB 2023b Curve Fitting Toolbox: The optimization is performed using the Curve Fitting Toolbox in MATLAB 2023b, employing a nonlinear least squares method with a trust-region algorithm.
(3) Fitting the model to the data: Equation (3) is used as the model, and the optimizer is employed to fit the values of κ and σ_t to the available data.

3.3. Model Evaluation

The following briefly explains how the two modeling approaches are compared with each other and which error metrics are used to determine their goodness. For parameterizing and evaluating the models in this work, only the range between σ_{min} and σ_{max} for the respective nominal stresses σ_{nom} are considered (see Table 1, Figure 2).

3.3.1. Model Comparison Based on Zero-Shifted Data

Since only the range between σ_{min} and σ_{max} is used, the data of the mechanical cycles are shifted to zero, i.e., the starting point of each cycle is shifted to $(\sigma, \varepsilon) = (0, 0)$. This results in a fitting procedure as depicted in Figure 3a.

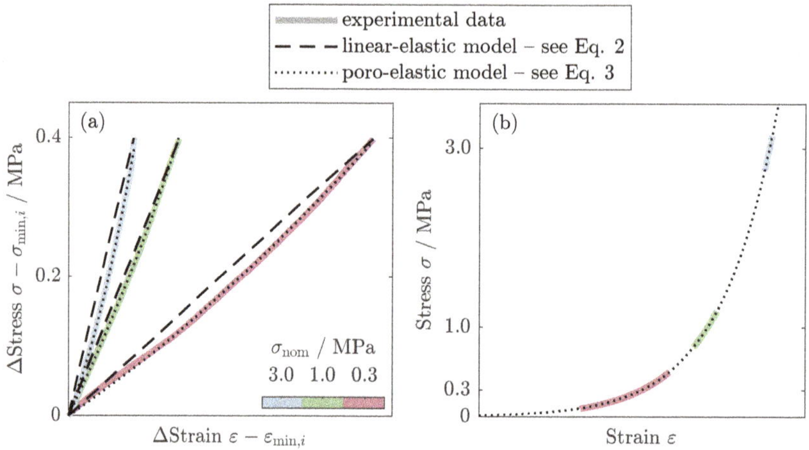

Figure 3. Schematic overview of the two utilized modeling approaches: (**a**) Comparison of the linear-elastic approach vs. poroelastic approach, fitted to the respective cycles set to zero, and (**b**) poroelastic approach fitted to the raw data.

3.3.2. Poroelastic Fit of the Raw Data

In addition to the model comparison based on zero-shifted data, the poroelastic model is also fitted against the raw data without any shift (again, only the range between σ_{min} and σ_{max}). This results in a fitting procedure as depicted in Figure 3b.

3.3.3. Error Metrics

Two error metrics are used to assess the goodness of the fits, both relying on the absolute and relative errors between the prediction value from the model σ_{sim} and the true value from experimental data σ_{exp}. The relative error is the absolute error divided by the mean of the experimental error $\overline{\sigma}_{\text{exp}}$, as follows:

$$e_{\text{abs}} = \sigma_{\text{sim}} - \sigma_{\text{exp}} \tag{5}$$

$$e_{\text{rel}} = \frac{e_{\text{abs}}}{\overline{\sigma}_{\text{exp}}} \tag{6}$$

The first metric is the maximum error defined in relative form as follows:

$$e_{\text{rel}}^{\text{max}} = \max(e_{\text{rel}}). \tag{7}$$

In addition to the relative maximum error, the corresponding absolute value is calculated to provide an absolute measure. The absolute maximum value is the absolute error at the same location where the relative error is the largest, as follows:

$$e_{\text{abs}}^{\text{max}} = e_{\text{abs}}\Big|_{e_{\text{rel}}=e_{\text{rel}}^{\text{max}}} \tag{8}$$

The second metric is the root-mean-squared error (RMSE) in absolute form, defined as follows:

$$e_{\text{RMS}} = \sqrt{\frac{1}{N}\sum_{i=1}^{N}(e_{\text{abs}})^2}. \tag{9}$$

Here, N is the length of the absolute error, i.e., the number of entries in its vector. The RMSE can also be calculated in a relative manner for comparative purposes as normalized root-mean-squared error (NRMSE), defined by the following:

$$e_{\mathrm{NRMS}} = \frac{e_{\mathrm{RMS}}}{\overline{\sigma}_{\mathrm{exp}}}. \tag{10}$$

4. Results and Discussion

An overview of all 27 conducted tests, showing all 10 conducted cycles per test in a stress–strain diagram, is given in Figure 4. Four aspects should be noted that hold true for Figure 4 as well as for the remainder of the text, unless stated otherwise:

(1) The data shown and discussed have been corrected for the stiffness of the UTM, as described in the experimental Section 2.2.
(2) The stress–strain relationship was mirrored to the first quadrant for readability, resulting in a positive sign for compressive stresses and strains.
(3) Anode data are shown in red, cathode data are shown in blue, and separator data are shown in green.
(4) Different compression rates of 1C, 3C, and 5C are shown using different color brightness levels.

From the results in Figure 4, the cathode is the stiffest component, closely followed by the anode. The separator is more compliant by a factor of approximately five than the electrode materials. This trend holds true regardless of the nominal stress and compression rate. When comparing the results with data from the literature, several similarities and differences can be observed. For example, Sauerteig et al. [1], Aufschläger et al. [9], and Schabenberger et al. [11] report that the separator is the most compliant component, with similar strains between 10% and 16% for stresses between 1.75 MPa and 3.0 MPa. The cathode tends to be the stiffest component, and the anode stiffness typically falls between those of the cathode and the separator. Comparing this general behavior to the first compression step of our data, the same trend and similar absolute values can be found. However, only the first compression step can be used for this comparison. On the one hand, this is because sometimes the literature [9] does not provide decompression behavior. On the other hand, the final decompression from Figure 4 is most likely influenced by the mechanical cycles between compression and decompression, making a direct comparison to literature without mechanical cycles [1] difficult. This is due to plastic behavior, as will be discussed later. Nonetheless, all results throughout the literature consistently show a strain-hardening effect, which will also be discussed in more depth later. It should be noted that the literature might also show data at largely different stress levels, e.g., [4,6,7]. This results in largely different strain ranges, making a direct comparison to our results difficult.

Overall, the literature shows similar trends regarding the initial compression behavior when compared to the results presented here. In the following sections, the stress–strain relationship from Figure 4 is analyzed to understand the evolution during mechanical cycling by investigating the plastic deformation, as well as to determine the presence of pressure-dependency and/or rate-dependency by investigating the maximum total strain. Finally, a linear-elastic model and a poroelastic material model are fitted to the experimental data to answer the question of whether a material model can be utilized to describe the mechanical behavior, and if so, which is suitable. It should already be noted that the poroelastic model described in this paper represents a snapshot of the mechanically settled state and, therefore, does not account for plastic or aging processes.

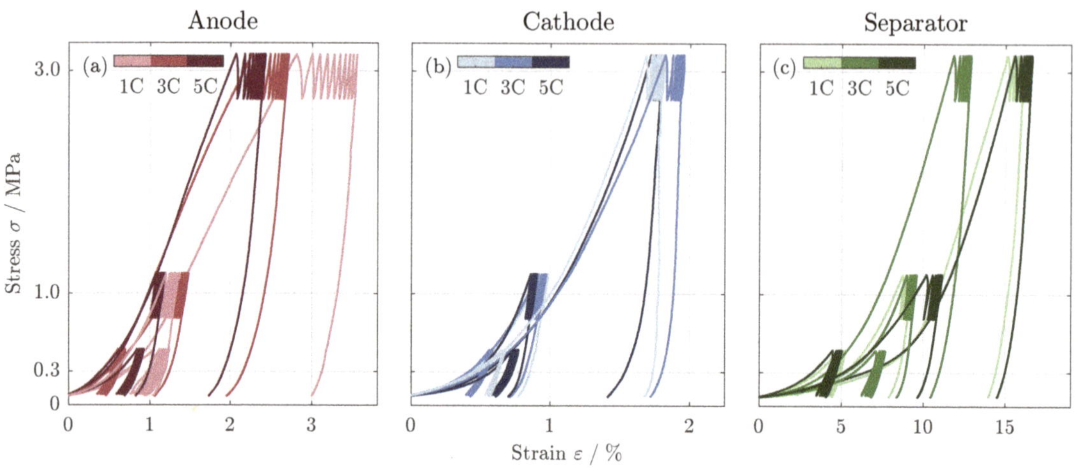

Figure 4. Overview of all conducted compression tests with (**a**) anodes, (**b**) cathodes, and (**c**) separators is shown using stress–strain diagrams. Positive values indicate compressive behavior. The gradient color bar indicates the different compression rates v of 1C, 3C, and 5C, which are to be interpreted as equivalent current rates and are further defined in the last row of Table 1.

4.1. Compression Tests

In general, solidification is the process by which a material's strength increases during plastic deformation [50]. In a stress–strain diagram, this becomes evident by a steeper slope. This observation can also be made for the data in Figure 5; the data exemplary show the separator at a compression rate of $v = 1\text{C}$ and nominal stress of $\sigma_{\text{nom}} = 1.0\,\text{MPa}$, as seen in Figure 4. Upon closer inspection of this magnified section, a distinct difference in the slope of the first compression cycle can be observed compared to the subsequent cycles. This change in slope suggests that the material undergoes a solidification process, resulting in an increase in stiffness with an increasing cycle number. This solidification phenomenon is observed consistently across all the tested conditions, although the example provided in Figure 5 is presented as a representative case. Figure 5 also highlights selected data points from which the plastic strain of each cycle can be calculated. The plastic strain $\varepsilon_{\text{plast}}$ is defined as the irreversible strain when comparing the end of the decompression $\varepsilon_{\text{min},i+1}$ with the beginning of the compression $\varepsilon_{\text{min},i}$ of a given mechanical cycle i.

$$\varepsilon_{\text{plast}} = \varepsilon_{\text{min},i+1} - \varepsilon_{\text{min},i} \tag{11}$$

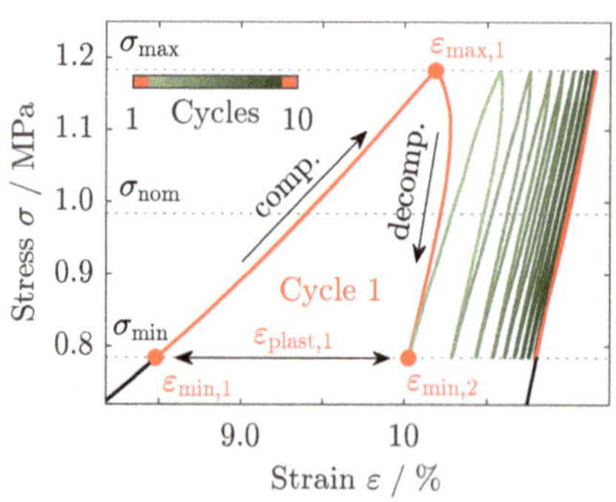

Figure 5. Separator data from Figure 4 at $v = 1\text{C}$ and $\sigma_{\text{nom}} = 1.0\,\text{MPa}$, magnified to show the operating window. In addition to the data, further information is also displayed to visualize the definition of the plastic strain.

The plastic strain from Equation (11) is then used to investigate the behavior during mechanical cycling in Figure 6. The three columns show the data for the anode, cathode, and separator data, respectively. The three rows show the data at different nominal stresses. Some data points are omitted because the logarithmic scale of the abscissa cannot show negative values. Negative values arise when the plastic strain of cycle $i + 1$ is less than that of cycle i.

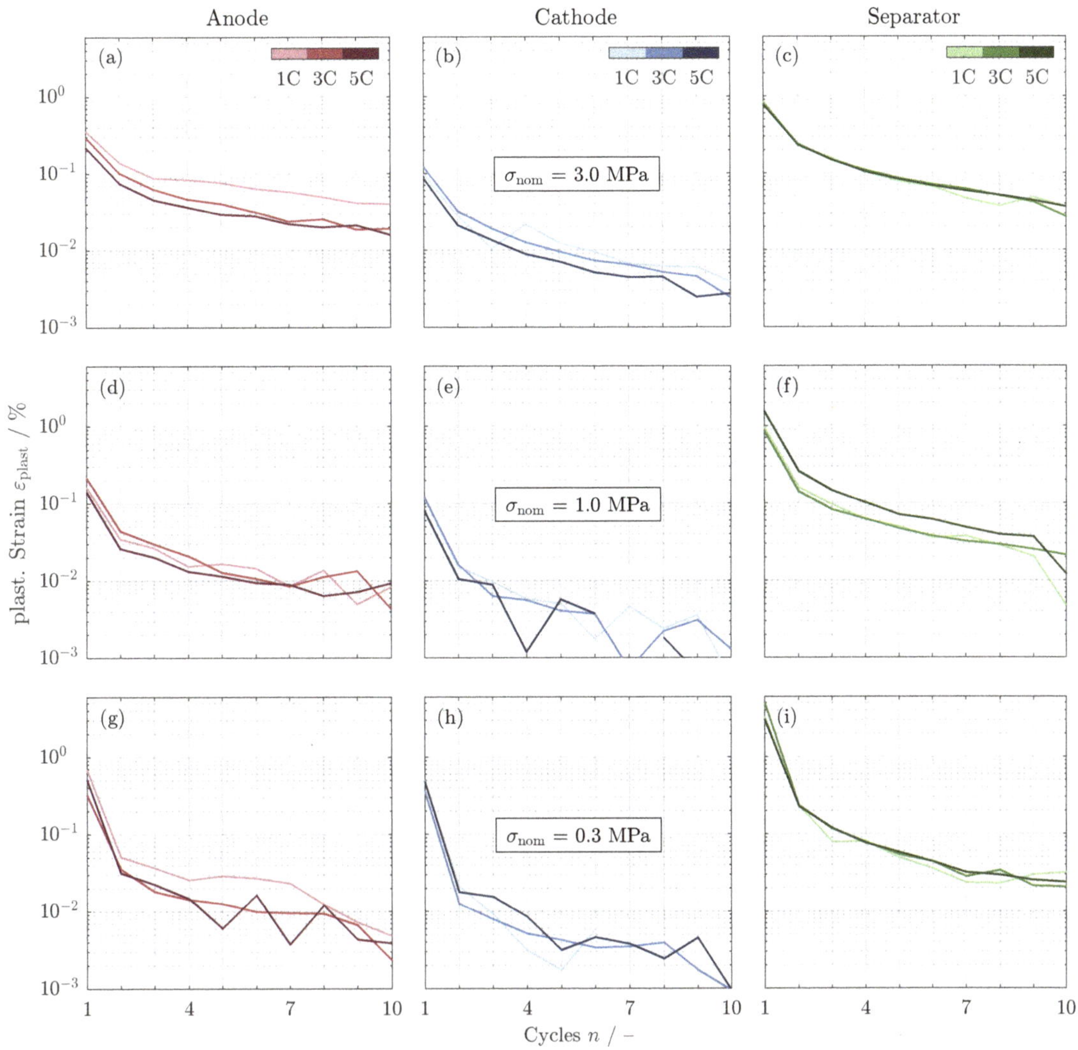

Figure 6. Evolution of plastic strain ε_{plast} as defined by Equation (11) over mechanical cycling. Columns show the data for the anode in (**a**,**d**,**g**), the cathode in (**b**,**e**,**h**), and the separator in (**c**,**f**,**i**), respectively. The first, second, and third rows show the data at nominal stresses σ_{nom} of 3.0 MPa, 1.0 MPa, and 0.3 MPa, respectively. Some data points are omitted because the logarithmic scale of the abscissa cannot show negative values.

Throughout all data shown in Figure 6, the first cycle exhibits the largest plastic deformation, which then decreases with the increasing cycle number. This transition from plastic to elastic behavior is fastest for the cathode material, followed by the anode and then the separator material. After 10 cycles, the plastic strain for all tests is less than a tenth of the value of the first cycle. In absolute terms, the plastic strain after 10 cycles is less than 0.04%. This suggests that using a single mechanical compression and decompression step as usually done in the literature (see Table 1) is not sufficient to characterize the mechanical behavior of the sample under operating conditions.

When comparing all nine data lines within one column of Figure 6, no significant difference in the change of plastic strain with respect to the nominal stress (rows) or with respect to the compression rate (color brightness) can be observed. This suggests that the stiffening behavior during mechanical cycling within the operating window is independent of both, the absolute stress and the compression rate.

The observed plastic effects can potentially be attributed to various microstructural changes and reorganizations within the electrode and separator materials induced by the compressive stresses. The increase in stiffness with successive compression cycles is likely due to a strain hardening effect [51,52], where the compressive stresses drive the reorganization and strengthening of the internal microstructure and bonding within the materials, as Wen et al. [53] reported. Furthermore, the decrease in plastic strain with the increasing cycle number suggests that the materials experience irreversible plastic deformation during the initial cycles. The materials may also exhibit time-dependent viscoelastic behavior [54,55]. These time-dependent deformation mechanisms can lead to changes in the overall stress–strain response. However, the effects of time-dependent behavior have not been investigated in the present work. To further investigate these potential mechanisms, additional experiments could be of great interest, e.g., in-operando microscopy or X-ray imaging, hold phases or creep tests, or statistical testing for dependencies on temperature or compression rates.

Complementary to the findings on plastic strain evolution, Figure 7 shows the maximum absolute strain of each of the 27 tests, i.e., it shows $\varepsilon_{max,10}$ for the 10th cycle, analogous to what is shown in Figure 5. The maximum strain exhibits a positive correlation with increasing nominal stress. At closer inspection, Figure 7 supports the findings from Figure 6, i.e., that no clear dependency on the compression rate can be found. However, a differentiation between the different components is possible. The separator shows the greatest maximum strain throughout all data. Anode and cathode data are comparable in terms of maximum strain, while the cathode tends to show lower maximum strains at the same applied pressure, resulting in a stiffer material. These statements largely hold true even when accounting for the largest error to be expected, as indicated by the error bars in Figure 7. The error bars are based on the standard deviation $S = 0.19\%$ of six measurement repetitions using cathode samples at $\sigma_{nom} = 1.0$ MPa and $v = 1$C. Since the cathode is the stiffest material, the largest error is to be expected here, and the same standard deviation is also used for the other error bars. More information and data regarding the six repetitions can be found in Appendix A.

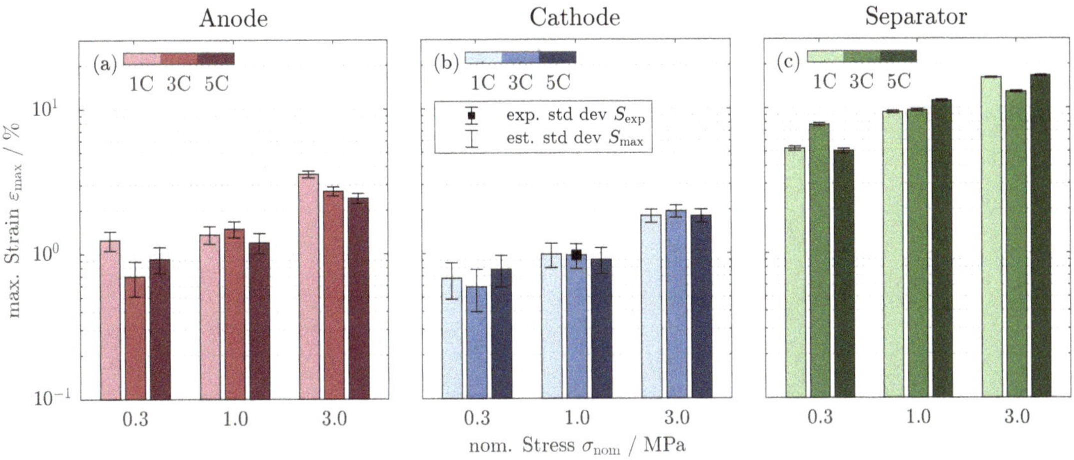

Figure 7. Influence of the compression rate v (1C, 3C, 5C) and nominal stress σ_{nom} (0.3 MPa, 1.0 MPa, 3.0 MPa) on the maximum strain $\varepsilon_{max,10}$ of the 10th cycle for the (**a**) anode, (**b**) cathode, and (**c**) separator. The error bars shown are based on the standard deviation $S = 0.19\%$ from 6 experimental repetitions at 1.0 MPa and 3C using cathode samples. The black square indicates the measurement point for the repetitions, while error bars without a marker utilize the same standard deviation since it is the largest error to be expected. More information can be found in the Appendix A.

It should be noted that although no clear rate dependency can be observed here, reports on a rate dependency can be found in the literature. For example, Ding et al. [7] investigated a PE as well as a polypropylene (PP) separator. They report a similar dependency on the strain rate for both materials, with stiffer material behavior at higher strain rates. Experiments were conducted with electrolyte-filled samples of a 10 mm diameter. Kalnaus et al. [5] investigated a triple layer PP-PE-PP separator at dry conditions and a 20 mm diameter. At similar compression rates, they report the same dependency as Ding et al. [7]. The separators in the present study are made of PE, too, and the strain rate used here is within the range used by [5,7]. Dry conditions and a diameter of 18 mm are similar to [5]. However, no clear rate dependency can be observed. It is suspected that the rate dependency is not clearly visible for two reasons. On the one hand, the stress levels used here are only at 3.0 MPa, while [5,7] conducted their tests up to 150 MPa. This results in a largely different strain value of approx. 15% compared to 80%. It might be that the rate dependency is not as pronounced at rather low absolute pressures as it is at high pressures. On the other hand, the literature tested a broader range of strain rates that differs by several orders of magnitude compared to what is tested here (see Table 1). This rather narrow range combined with the lower stress levels used for the present study might be the reason behind the seemingly missing rate dependency. However, since the stress levels and compression rates are chosen to be representative of typical battery cycling, these results suggest that the rate dependency reported in the literature might not be relevant to normal battery operation.

4.2. Modeling

This section explores whether the mechanical behavior can be described by a material model. To this end, two different material models were fitted to the compression steps of each cycle for each of the 27 tests. At first, the experimental data of each compression step are zeroed so that the compression data used for the fit start at $(\varepsilon, \sigma) = (0, 0)$, as visualized in Figure 3a. If this was not the case, the linear-elastic model, in particular, would represent the behavior between $(0, 0)$ and $(\varepsilon_{max,i}, \sigma_{max})$; however, the actual operating window is located between $(\varepsilon_{min,i}, \sigma_{min})$, and $(\varepsilon_{max,i}, \sigma_{max})$. For a meaningful comparison of both material models, the shifted data are also used to fit the poroelastic model. Afterward, the poroelastic model is also fit to the original data without any translation to account for the absolute stress and strain values and not only the change within the operating window. The initial stress σ_0 is set to zero when fitting the poroelastic model.

Figure 8 shows Young's modulus from Equation (2), calculated for the compression step of each test plotted against the number of cycles. The results show that Young's modulus increases with the increasing number of cycles, but approaches a constant value, indicating a solidification process. The overall behavior is similar regardless of the compression rate (see same line styles). However, a dependency on the nominal stress can be observed. Young's modulus increases with higher nominal stress, resulting in stiffer material behavior. These findings from the linear-elastic model are consistent with the observations in Figure 6; the first cycle exhibits significantly different behavior than the following cycles, suggesting that a single compression and decompression step is not sufficient to characterize the mechanical behavior for the given operating window.

Figure 9 evaluates the goodness of the fits through true-predicted diagrams for both the linear-elastic model and the poroelastic model, using lighter colors for the linear-elastic model and darker colors for the poroelastic model. Each panel displays the data for 10 cycles at 3 compression rates for both material models, resulting in 30 light and 30 dark lines. The ordinate shows the true stress as measured from the experiments, while the abscissa shows the predicted stress from either model at the given experimentally measured strain. The dashed black lines indicate an error of predicted stress (abscissa) of $\pm 5\%$ relative to the experimental stress (ordinate). Individual cycles or compression rates are not distinguished by a color gradient, but the overall behavior with increasing cycle numbers is indicated by the two arrows in panel h). The data in Figure 9 show that the linear-elastic model yields larger errors than the poroelastic model for all nine panels. Additionally, the linear-elastic model tends to have a rather curved shape, overestimating the stress at low experimental stress and underestimating the stress at high experimental stress, resulting in a non-linear error behavior. The final fit results, as well as error quantities for the linear-elastic model, are given in Table A1. The overall NRMSE for the linear model in the 10th cycle is $\leq 24.88\%$ ($\leq 49.43 \times 10^{-3}$ MPa).

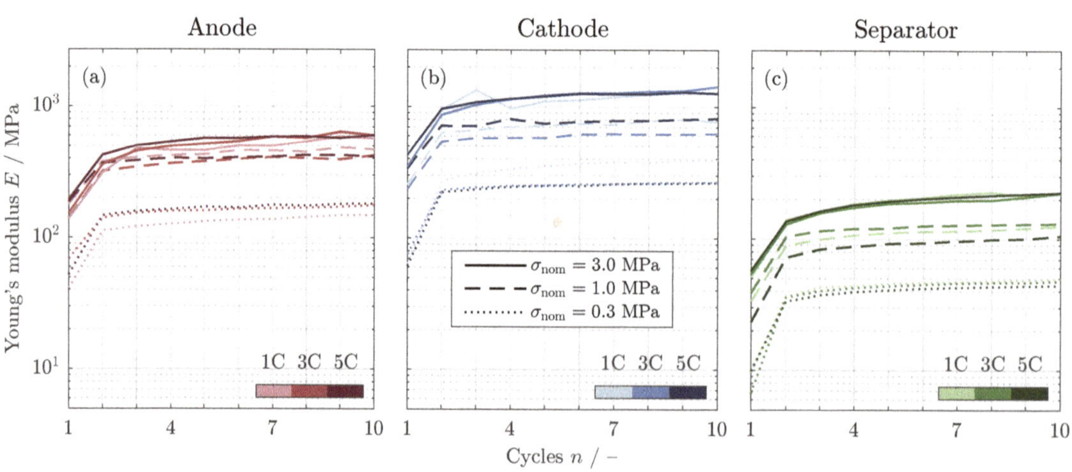

Figure 8. Evolution of Young's modulus E throughout the mechanical cycling for (**a**) anode, (**b**) cathode, and (**c**) separator. Evaluation was performed for the compression of each cycle as depicted in Figure 3a. The nominal stress σ_{nom} is indicated by different line styles, and the compression rate v is denoted by different color brightnesses.

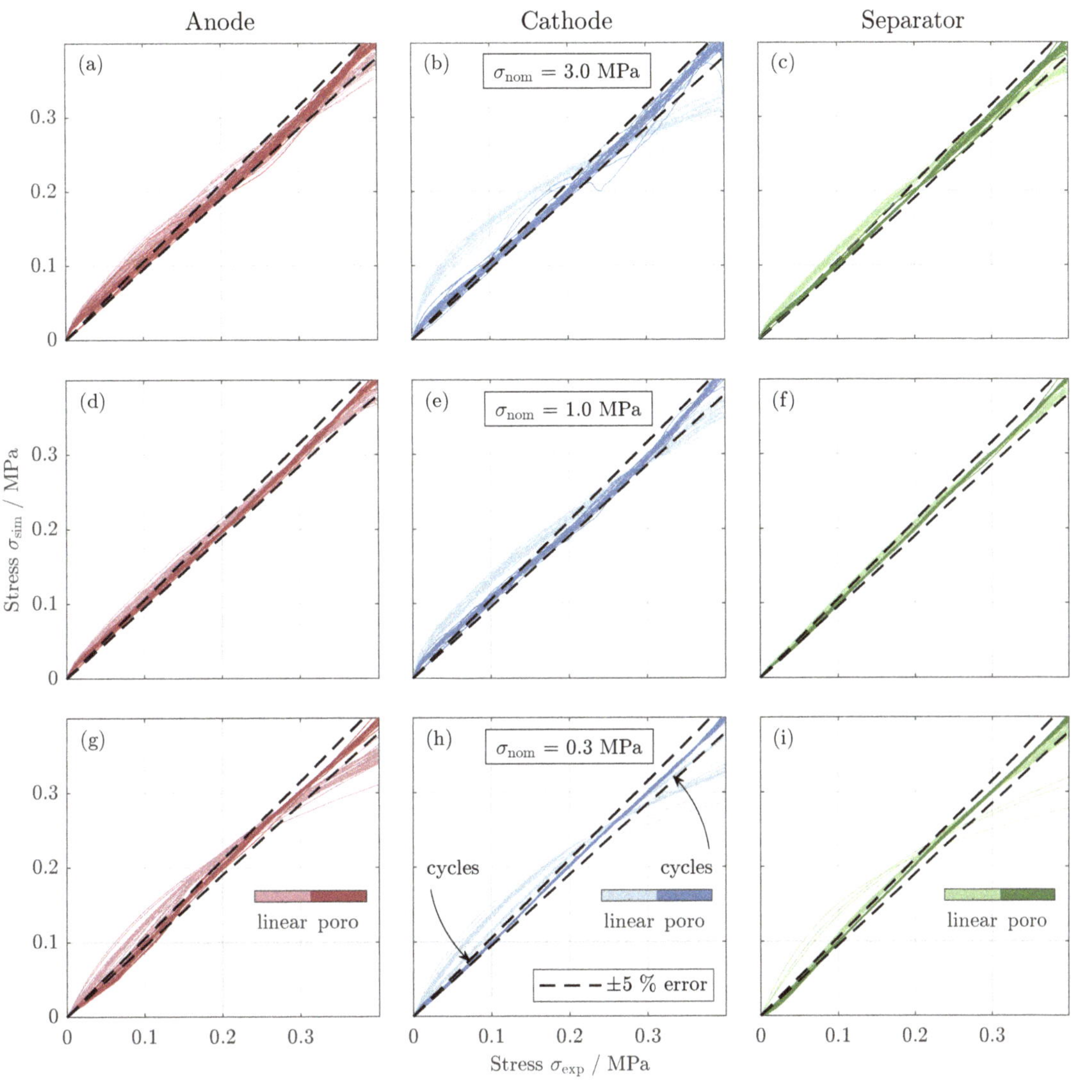

Figure 9. Goodness of the linear-elastic and poroelastic modeling approach from Figure 3a for all 10 cycles. Goodness is evaluated using true-predicted diagrams, which include the simulated stress data from the model σ_{sim} versus the experimental stress data σ_{exp}, respectively, for all compression rates v in the case of compression. Columns show the data for the anode in (**a**,**d**,**g**), the cathode in (**b**,**e**,**h**), and the separator in (**c**,**f**,**i**), respectively. Each panel displays the data for 10 cycles at 3 compression rates for both material models, resulting in 30 light and 30 dark lines.

In contrast to the linear-elastic model, the poroelastic model in Figure 9 predominantly falls within the ±5% error margin. One exception to this can be found in the cathode data at 3.0 MPa of nominal stress in panel b. Here, the data for the poroelastic model show fluctuations, especially around $\sigma_{exp} = 0.2$ MPa, which cannot be seen in any other panel. Overall, the poroelastic material model outperforms the linear-elastic model, especially for small nominal stresses. This might be particularly relevant when modeling the mechanical behavior at begin of life when absolute pressures are low. For example, if degradation models were dependent on mechanical stress, deviations in the stress would result in an overestimation or underestimation of the degradation. The final fit results as well as error quantities for the poroelastic model with translated data are given in Table A2. The poroelastic model reduces the NRMSE in the 10th cycle from ≤24.88% (≤49.43×10^{-3} MPa) for the linear model to correspondingly ≤7.26% (≤14.46×10^{-3} MPa).

Figure 10 shows the results when fitting the experimental data as visualized in Figure 3b. Again, the ordinate shows the experimentally measured stress, while the abscissa shows the predicted stress at any given strain. Only the compression step of the 10th cycle is shown. The black dashed and dotted lines indicate a relative error of ±5% and ±1%, respectively. Note that the error definition is the same as in Figure 9, i.e., the error indicators are given relative to the experimentally measured stress on the ordinate. The results in Figure 10 show an overall small error <±5%, except for small absolute stresses, as can be seen in panels (g) through (i). Again, this can be explained by small absolute values used as a reference when calculating the error. For nominal stresses of 1.0 MPa and 3.0 MPa, the error decreases to < ±1%, except for the cathode data, with a nominal stress of 3.0 MPa in panel (b). This could be explained by the fact that the cathode is very stiff but still exhibits a non-linear stress–strain relationship in the range of 3.0 MPa. A similar curved shape to the one seen in panel (b) can also be observed in panels (a) and (c), although it is less distinct. Lastly, these observations hold true for all three tested compression rates, as no dependency on the compression rate can be seen. The final fit results as well as error quantities for the poroelastic model without translation are given in Table A3. The NRMSE in the 10th cycle is overall ≤5.47% (≤31.20×10^{-3} MPa).

When comparing the two poroelastic modeling approaches from panels a and b in Figure 3, the results in Figures 9 and 10 seem similar at first. For higher nominal stresses of 1.0 MPa and 3.0 MPa, in particular, the error characteristics are almost identical. At low nominal stresses of 0.3 MPa, the poroelastic model, including the translation of data, shows better prediction capability for the 10th cycle. However, the differences are less than ±5%-points in NRMSE for almost all cases. Furthermore, the model fitted to translated data can only capture the change in mechanical stress and strain within the operating window, while the model fitted to the data without translation also accounts for the absolute values of stress and strain. Because of this, the second modeling approach without translation is deemed superior.

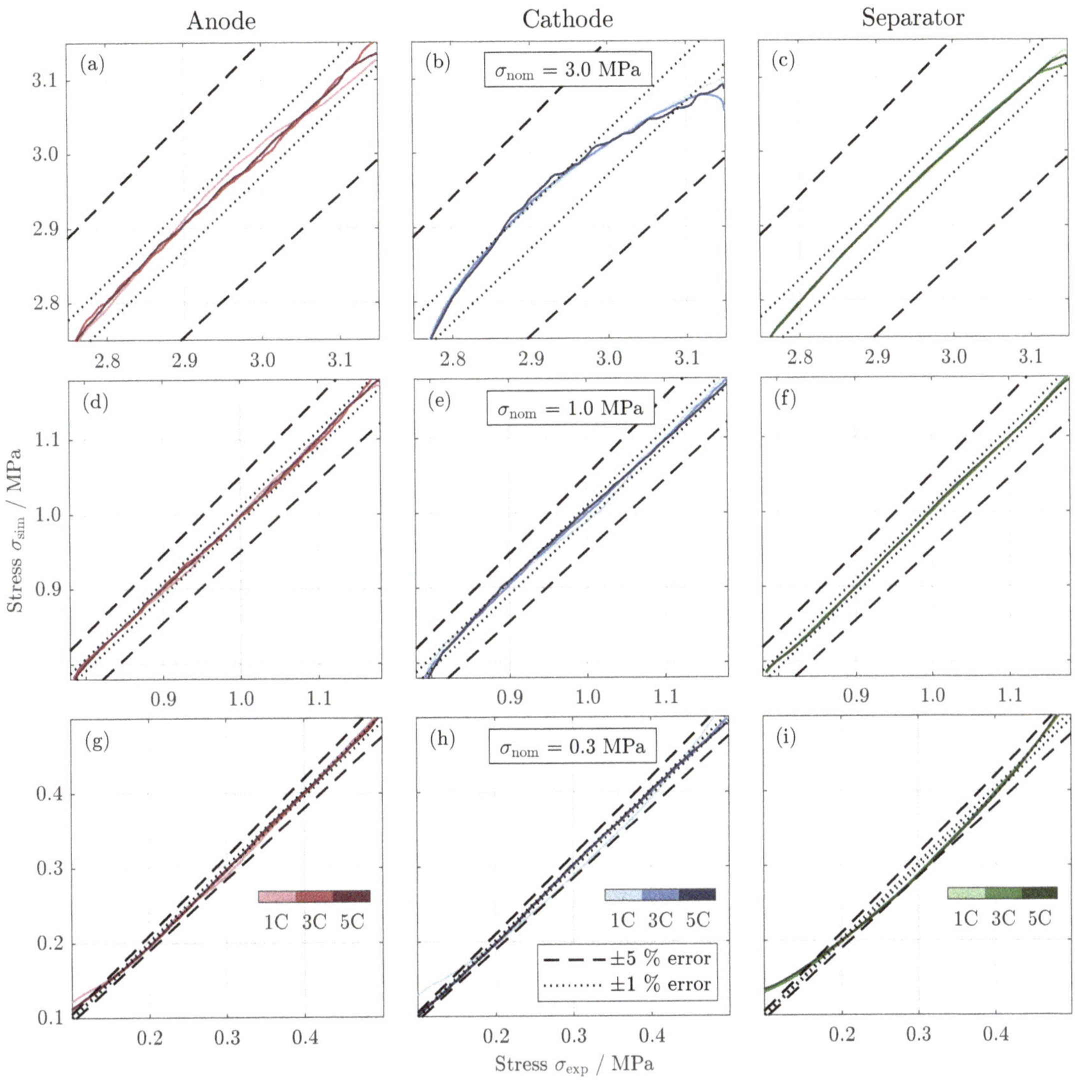

Figure 10. Goodness of the poroelastic modeling approach from Figure 3b for the 10th cycle. Goodness is evaluated using true-predicted diagrams, which include the simulated stress data from the model σ_{sim} versus the experimental stress data σ_{exp}, respectively, for all compression rates v in the case of compression. Columns show the data for the anode in (**a**,**d**,**g**), the cathode in (**b**,**e**,**h**), and the separator in (**c**,**f**,**i**), respectively. Each panel displays the data only for the 10th cycle at 3 compression rates for the poroelastic material model only, resulting in 3 lines.

5. Conclusions

The results and discussion sections address several key research questions regarding the mechanical behavior of lithium-ion battery components under cyclic compression:

- *Compression rate dependency.* The results do not support any dependency of the mechanical behavior on the compression rate ν (1C, 3C, 5C) for a given nominal stress σ_{nom}. The stiffening behavior during cycling was found to be independent of the compression rate.

- *Nominal stress dependency.* The mechanical behavior was found to be dependent on the nominal stress σ_{nom} (0.3 MPa, 1.0 MPa, 3.0 MPa). Higher nominal stresses σ_{nom} resulted in stiffer material behavior, as evidenced by higher values in Young's modulus E.
- *Cyclic behavior.* The mechanical behavior was found to be highly dependent on the number of cycles. The first cycle exhibited the largest plastic deformation, which then decreased rapidly with the increasing cycle number. After 10 cycles, the plastic strain was less than 0.04% for all tests, suggesting that a single compression–decompression step may not be sufficient to characterize the mechanical behavior under operating conditions.
- *Compression/decompression behavior.* The results show a clear distinction between the compression and decompression behavior during the first few mechanical cycles. However, after 10 cycles, the distinction vanished.
- *Material modeling.* Two material models, a linear-elastic and a poroelastic model, were fitted to the experimental data. The poroelastic model outperformed the linear-elastic model, especially at low nominal stresses, providing a better description of the mechanical behavior. The poroelastic model fitted to the original data without translation showed superior performance, as it could capture both the changes in mechanical stress and strain within the operating window, as well as the absolute values of stress and strain.

In conclusion, the study provides valuable insights into the complex mechanical behavior of lithium-ion battery components under cyclic compression. The findings highlight the importance of considering the evolution of mechanical properties during cycling, the influence of nominal stresses, and the suitability of the poroelastic model for describing the observed behavior. These insights can contribute to the development of more accurate mechanical models for lithium-ion batteries, which are crucial for predicting degradation and improving battery design and performance.

Furthermore, the introduced method can be applied to a variety of lithium-ion battery systems and scenarios. Some key aspects of its versatility include the following:

- *Different charge/discharge rates.* The method can be used to investigate the mechanical behaviors of electrodes under various charge and discharge rates, reflecting different operating strategies and power requirements.
- *Varying temperature ranges.* By incorporating active temperature control or testing in different climatic zones, the method can be used to study the impact of temperature on the mechanical properties of electrodes across a wide range of operating conditions.
- *Other cell geometries.* As long as the electrode samples can be prepared in a coin cell format, the method can be applied to cells of different sizes. However, the applicability for wound cell geometries may be limited due to the uniaxial assumption of the model.
- *Higher stress levels.* The method can be used to investigate the mechanical behavior of electrodes under higher stress levels, which can be influenced by the depth of discharge (DoD) window during operation.
- *Hold phases.* The impact of hold phases with constant stress, as determined by the operating strategy, on the mechanical properties of electrodes can also be studied using the introduced method.
- *Aged samples.* The method can potentially be used to study the mechanical characteristics of electrodes at different stages of real-world aging, which can be influenced by external factors. If material properties should not change during aging, this might indicate that testing aged materials is not required, speeding up the parametrization process.

It is important to note that the versatility of the method is primarily limited by the ability to prepare the samples in a coin cell format. Samples with more brittle characteristics, such as all-solid-state battery (ASSB) materials, may be challenging to test using this approach due to their inherent material properties.

Future work (apart from the aforementioned aspects) could address several research questions derived from the proposed testing procedure and findings of this study. The setup could be modified to cover two aspects. On the one hand, it could be adapted so that samples wetted with electrolytes can also be tested to investigate the influence of electrolytes on mechanical behavior. On the other hand, the setup could be extended by external displacement sensors to increase the precision of the measurement, similar to what Spielbauer et al. [4] have shown. Additionally, the mechanical behavior at different states of charge should be investigated to enable a more precise prediction of the mechanical behavior. This is especially important for predicting the reversible mechanical behavior during cycling. Lastly, harvesting electrodes at different points during the lifetime of a battery to investigate the aged cell would allow for a comparison between the predicted mechanical behavior and the true mechanical behavior at different states of health. If the predictions at the beginning of life accurately fit the true behavior at the end-of-life, long-term and cost-intensive aging tests might not be required anymore.

Author Contributions: Conceptualization, J.B.; methodology, J.B. and A.D.; software, J.B. and T.K.; validation, J.B. and A.D.; formal analysis, J.B. and T.K.; investigation, J.B. and T.K.; resources, P.K. and M.A.; data curation, J.B. and T.K.; writing—original draft preparation, J.B. and A.D.; writing—review and editing, T.K., P.K., M.A. and A.J.; visualization, J.B., A.D., T.K., P.K., M.A. and A.J.; supervision, P.K., M.A. and A.J.; project administration, A.J. All authors have read and agreed to the published version of the manuscript.

Funding: This work was funded by the BMW AG and was performed in cooperation with the Chair of Electrical Energy Storage Technology at the Technical University of Munich.

Data Availability Statement: The datasets presented in this article are not readily available due to the BMW AG confidentiality guidelines. Requests to access the datasets should be directed to BMW AG.

Acknowledgments: The authors would like to thank our colleagues Marlene Zink (BMW AG) and Maria Hilzinger (BMW AG) for conducting SEM imaging and helium pycnometer measurements for material characterization.

Conflicts of Interest: Authors Johannes Brehm, Tobias Kussinger, Philip Kotter and Maximilian Altmann were employed by the company Bayerische Motoren Werke Aktiengesellschaft (BMW AG). Authors Axel Durdel and Andreas Jossen were employed by the Technical University of Munich (TUM). All authors declare that the research was conducted in the absence of any commercial or financial relationships that could be construed as a potential conflict of interest.

Abbreviations

The following abbreviations are used in this manuscript:

Acronyms

ASSB	all-solid-state battery
BoL	begin of life
DoD	depth of discharge
DoL	degree of lithiation
EoL	end of life
EV	electric vehicle
FT-IR	Fourier transform infrared spectroscopy
Gr	graphite
ICP-OES	inductively coupled plasma-optical emission spectrometry
LCO	lithium-cobalt-oxide
NCA	nickel-cobalt-aluminum
NMC	nickel-manganese-cobalt
NRMSE	normalized root-mean-squared error
PE	polyethylene

PP	polypropylene
RMSE	root-mean-squared error
SEM	scanning electron microscopy
Si	silicon
SoC	state of charge
STA-MS	simultaneous thermal analysis with mass spectroscopy
UTM	universal testing machine
Roman Symbols	
$\mathbb{C}$	elasticity tensor, Nm^{-2}
E	Young's modulus, Nm^{-2}
e	error
G	shear modulus, Pa
h	sample height, m
$\mathbb{I}$	identity matrix,
J	volume ratio,
K	bulk modulus, Pa
L	through-plane thickness, m
n	number of cycles, 1
p	pressure, Pa
S	standard deviation
V	volume, m^3
v	compression rate, $MPah^{-1}$
Greek Symbols	
ϵ	volume fraction, 1
ε	strain, 1
$\boldsymbol{\varepsilon}$	strain tensor, 1
κ	logarithmic bulk modulus, 1
ν	Poisson's ratio, 1
σ	stress, Nm^{-2}
$\boldsymbol{\sigma}$	stress tensor, Nm^{-2}
Subscripts & Superscripts	
0	stress-free state
1	pre-load state
abs	absolute
ax	uniaxial
el	elastic component
exp	experimental
l	liquid phase
max	maximum
min	minimum
nom	nominal
NRMS	normalized root-mean-squared value
plast	plastic
rel	relative
RMS	root-mean-squared value
s	solid phase
sim	simulated
t	tensile
tot	total
vol	volumetric

Appendix A. Precision of Strain Measurement

To evaluate the precision of strain measurement and the resulting error thereof, 6 repetitions at one of the 27 measurement points have been conducted. Since the cathode shows the stiffest behavior, the smallest absolute change in thickness is to be expected, resulting in the largest signal-to-noise ratio. Consequently, tests using cathode materials will yield the largest error to be expected.

Figure A1 shows the results of these 6 repetitions at $\sigma_{nom} = 1.0\,\text{MPa}$ and $v = 1\text{C}$ as (a) a stress–strain diagram as well as (b) the evolution of the plastic strain with the increasing cycles, analogous to Figures 4 and 6. The evolution of the plastic strain in Figure A1b reveals that all 6 repetitions show similar behavior, almost identical to that shown in Figure 6e.

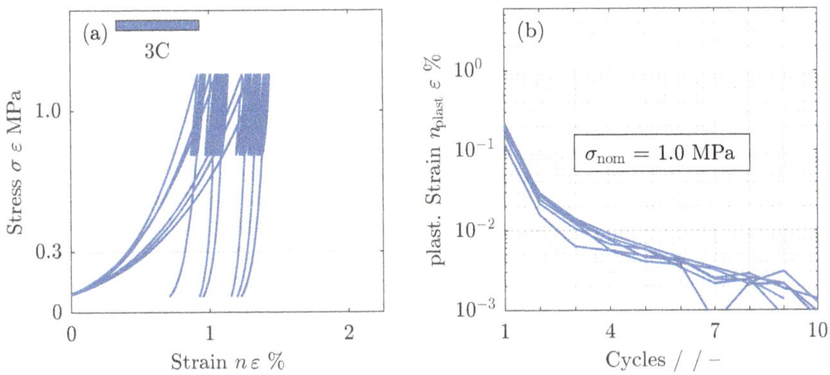

Figure A1. Precision of the strain measurement shown for the cathode at a stress level σ_{nom} of 1.0 MPa and a compression rate v of 1C by: (**a**) An overview of 6 compression test repetitions, where a standard deviation of $S = 0.19\%$ of the ε_{max} can be observed (in addition to Figure 4) and (**b**) the evolution of plastic strain ε_{plast} as defined by Equation (11) over mechanical cycling for six compression test repetitions (in addition to Figure 6).

To further allow for an estimate of the error in the 10th cycle, the standard deviation in the maximum strain was evaluated to be $S = 0.19\%$. This standard deviation is used as error bars in Figure 7. The error bar with the black central square indicates the measurement point for which the six repetitions had been conducted. Since the cathode is the stiffest material, this standard deviation is the largest to be expected. To allow for a maximum error estimate, error bars based on the same standard deviation have been added to all other data bars in Figure 7 (no central black squares).

Appendix B. Detailed Modeling Equations

Appendix B.1. Hooke's Law in Matrix Form

Hooke's law for isotropic materials in compliance matrix form (Voigt notation) is given by the following:

$$\underbrace{\begin{bmatrix}\sigma_{11}\\ \sigma_{22}\\ \sigma_{33}\\ \sigma_{23}\\ \sigma_{13}\\ \sigma_{12}\end{bmatrix}^{el}}_{\sigma} = \underbrace{\frac{E}{(1+\nu)(1-2\nu)}\cdot\begin{bmatrix}1-\nu & \nu & \nu & 0 & 0 & 0\\ \nu & 1-\nu & \nu & 0 & 0 & 0\\ \nu & \nu & 1-\nu & 0 & 0 & 0\\ 0 & 0 & 0 & \frac{1-2\nu}{2} & 0 & 0\\ 0 & 0 & 0 & 0 & \frac{1-2\nu}{2} & 0\\ 0 & 0 & 0 & 0 & 0 & \frac{1-2\nu}{2}\end{bmatrix}}_{C}\cdot\underbrace{\begin{bmatrix}\varepsilon_{11}\\ \varepsilon_{22}\\ \varepsilon_{33}\\ 2\varepsilon_{23}\\ 2\varepsilon_{13}\\ 2\varepsilon_{12}\end{bmatrix}^{el}}_{\varepsilon} \quad \text{(A1)}$$

where Young's modulus E can be alternatively expressed through the shear modulus G or the bulk modulus K and Poisson's ratio ν:

$$E = 2G(1+\nu) = 3K(1-2\nu) \quad \text{(A2)}$$

Appendix B.2. Derivation of the Poroelasticity

Due to the morphological similarity of the porous materials of a lithium-ion cell and the porous layers in the soil, this work uses an approach from soil modeling, which was mentioned for the first time in a similar form in [48] and later recommended by [49] for modeling poroelastic material behaviors. The adopted approach (Equation (A3)) is a differential relation that describes a linear elastic change in the void-to-solid ratio $\epsilon_{\mathrm{l/s}}$ (Equation (A4)) due to a logarithmic change of the applied pressure p (Equation (A5)). Here, κ denotes the logarithmic bulk modulus, which is subsequently used as one of two parameters for the model fitting, and $p_{\mathrm{t}}^{\mathrm{el}}$ describes the elastic tensile strength of the porous material. It is assumed that the compressibility of the solid phase can be neglected: $V_{\mathrm{s}}^{\mathrm{el}} = V_{\mathrm{s}}^{0} = V_{\mathrm{s}}$. This assumption is considered admissible because the mechanical stresses that occur in the electrodes due to external stress typically do not exceed the strength of the active material or particles [56]. V_{s} denotes the solid volume, and V_{l} denotes the pore volume, which is interpreted as the liquid volume of the electrolyte in the porous matrix of a lithium-ion cell.

$$d\epsilon_{\mathrm{l/s}}^{\mathrm{el}} = -\kappa \cdot d\ln\left[p + p_{\mathrm{t}}^{\mathrm{el}}\right] \tag{A3}$$

$$\epsilon_{\mathrm{l/s}} = \frac{V_{\mathrm{l}}}{V_{\mathrm{s}}} \tag{A4}$$

$$p = -\frac{1}{3}\mathrm{trace}(\sigma) = -\frac{1}{3}\sigma : \mathbb{I} \tag{A5}$$

The void-to-solid ratio in the initial configuration $\epsilon_{\mathrm{l/s}}^{0}$ can then be written as a function of the initial porosity $\epsilon_{\mathrm{l}}^{0}$ using the definition from Equation (A4). Accordingly, the void-to-solid ratio in the elastically deformed configuration $\epsilon_{\mathrm{l/s}}^{\mathrm{el}}$ would depend on the porosity in the deformed configuration. However, this porosity is typically unknown.

$$\epsilon_{\mathrm{l/s}}^{0} = \frac{V_{\mathrm{l}}^{0}}{V_{\mathrm{s}}^{0}} = \frac{\epsilon_{\mathrm{l}}^{0}}{1 - \epsilon_{\mathrm{l}}^{0}} \tag{A6}$$

Using the elastic change in the void-to-solid ratio, the elastic volume change J^{el} can be written as follows:

$$J^{\mathrm{el}} = \frac{1 + \epsilon_{\mathrm{l/s}}^{\mathrm{el}}}{1 + \epsilon_{\mathrm{l/s}}^{0}} = \frac{1 + \frac{V_{\mathrm{l}}^{\mathrm{el}}}{V_{\mathrm{s}}}}{1 + \frac{V_{\mathrm{l}}^{0}}{V_{\mathrm{s}}}} = \frac{V_{\mathrm{s}} + V_{\mathrm{l}}^{\mathrm{el}}}{V_{\mathrm{s}} + V_{\mathrm{l}}^{0}} = \frac{V_{\mathrm{tot}}^{\mathrm{el}}}{V_{\mathrm{tot}}^{0}} = 1 + \varepsilon_{\mathrm{vol}}^{\mathrm{el}} \tag{A7}$$

Here, V_{l}^{0} denotes the initial pore volume, $V_{\mathrm{l}}^{\mathrm{el}}$ denotes the pore volume after elastic deformation. The same applies to the total volume V_{tot}^{0} and $V_{\mathrm{tot}}^{\mathrm{el}}$, respectively. Integrating the differential relation from Equation (A3), substituting Equation (A7), and choosing the integration constant so that $p(\varepsilon_{\mathrm{vol}}^{\mathrm{el}} = 0) = 0$ yields the following:

$$\frac{\kappa}{1 + \epsilon_{\mathrm{l/s}}^{0}} \cdot \ln\left[\frac{p + p_{\mathrm{t}}^{\mathrm{el}}}{p_0 + p_{\mathrm{t}}^{\mathrm{el}}}\right] = 1 - J^{\mathrm{el}} = -\varepsilon_{\mathrm{vol}}^{\mathrm{el}} \tag{A8}$$

Equation (A8) can finally be solved for the pressure:

$$p = \left(p_0 + p_{\mathrm{t}}^{\mathrm{el}}\right) \cdot \exp\left[-\varepsilon_{\mathrm{vol}}^{\mathrm{el}} \cdot \frac{1 + \epsilon_{\mathrm{l/s}}^{0}}{\kappa}\right] - p_{\mathrm{t}}^{\mathrm{el}} \tag{A9}$$

Equation (A9) only describes the pressure component of the elastic stress tensor. The total elastic stress σ^{el} results from the following relation:

$$\sigma^{\mathrm{el}} = 2G \cdot \mathrm{dev}\left(\varepsilon^{\mathrm{el}}\right) - p \cdot \mathbb{I} \tag{A10}$$

In this case, the deviatoric stress $\text{dev}\left(\varepsilon^{\text{el}}\right)$ describes the shape change component of the elastic strain in accordance with Equation (A10). The elastic strain tensor can be written accordingly as follows:

$$\varepsilon^{\text{el}} = \text{dev}\left(\varepsilon^{\text{el}}\right) + \frac{1}{3}\text{trace}\left(\varepsilon^{\text{el}}\right) \cdot \mathbb{I} = \text{dev}\left(\varepsilon^{\text{el}}\right) + \frac{1}{3}\varepsilon^{\text{el}}_{\text{vol}} \cdot \mathbb{I} \tag{A11}$$

The shear modulus G in Equation (A10) can be calculated as a function of the bulk modulus K and Poisson's ratio ν (Equation (A12)). The bulk modulus K in turn is defined as a function of the change in pressure due to an elastic change in volume (Equation (A13)). These relations are taken from [57] and apply to isotropic elastic material behavior. By applying these definitions, the final mechanical model in this work will have isotropic, i.e., direction-independent, mechanical properties:

$$G = \frac{3K(1-2\nu)}{2(1+\nu)} \tag{A12}$$

$$K = -V^0_{\text{tot}} \cdot \frac{dp}{dV^{\text{el}}_{\text{tot}}} \tag{A13}$$

Equations (A7) and (A11) give the following definition of $\varepsilon^{\text{el}}_{\text{vol}}$, assuming that stresses and strains only act uniaxial in the 11-direction:

$$\varepsilon^{\text{el}}_{\text{vol}} = \frac{V^{\text{el}}_{\text{tot}}}{V^0_{\text{tot}}} - 1 = \varepsilon^{\text{el}}_{11} + \underbrace{\varepsilon^{\text{el}}_{22}}_{=0} + \underbrace{\varepsilon^{\text{el}}_{33}}_{=0} = \varepsilon^{\text{el}}_{11} \tag{A14}$$

Equation (A10) is thus simplified using Equations (A11), (A12) and (A14) as well as the assumptions made here to the following:

$$\sigma^{\text{el}}_{11} = 2\frac{3K(1-2\nu)}{2(1+\nu)} \cdot \left(\varepsilon^{\text{el}}_{11} - \frac{\varepsilon^{\text{el}}_{11} + \cancel{\varepsilon^{\text{el}}_{22}} + \cancel{\varepsilon^{\text{el}}_{33}}}{3}\right) - p = 2K \cdot \varepsilon^{\text{el}}_{11} - p \tag{A15}$$

The strain $\varepsilon^{\text{el}}_{\text{vol}}$ in Equation (A9) can be expressed by the volume ratios using Equation (A7). Substituting this expression of the pressure in Equation (A13) and performing the differentiation, the bulk modulus can be determined as a function of the poroelastic model according to [48,49]:

$$K = -V^0_{\text{tot}} \cdot \frac{dp}{dV^{\text{el}}_{\text{tot}}} = \left(p_0 + p^{\text{el}}_{\text{t}}\right) \cdot \exp\left[\left(1 - \frac{V^{\text{el}}_{\text{tot}}}{V^0_{\text{tot}}}\right) \cdot \frac{1+\epsilon^0_{\text{l/s}}}{\kappa}\right] \cdot \left(\frac{1+\epsilon^0_{\text{l/s}}}{\kappa}\right) \tag{A16}$$

Finally, inserting Equation (A16) into Equation (A15) and re-writing the volume fraction as the strain using Equations (A14) and (A7), the uniaxial poroelastic material model used in this work is formed, taking into account that only stresses and strains act in the 11-direction and $\nu = 0$:

$$\sigma^{\text{el}}_{11} = \left(p_0 + p^{\text{el}}_{\text{t}}\right) \cdot \left(2 \cdot \varepsilon^{\text{el}}_{11} \cdot \frac{1+\epsilon^0_{\text{l/s}}}{\kappa} - 1\right) \cdot \exp\left[-\varepsilon^{\text{el}}_{11} \cdot \frac{1+\epsilon^0_{\text{l/s}}}{\kappa}\right] + p^{\text{el}}_{\text{t}} \tag{A17}$$

Considering that this non-linear stress–strain relationship is intended to be used for compressive scenarios, the strain $\varepsilon^{\text{el}}_{11}$—which acts as an input to this equation—is negated. Lastly, in order to describe the material stress σ rather than the externally applied pressure

p, each pressure is substituted by its respective stress quantity. In the uniaxial case, they are related by a negative sign.

$$\sigma_{11}^{\mathrm{el}} = (\sigma_0 + \sigma_t) \cdot \left(2 \cdot \varepsilon_{11}^{\mathrm{el}} \cdot \frac{1 + \epsilon_{1/s}^{0}}{\kappa} + 1\right) \cdot \exp\left[\varepsilon_{11}^{\mathrm{el}} \cdot \frac{1 + \epsilon_{1/s}^{0}}{\kappa}\right] - \sigma_t \tag{A18}$$

Appendix C. Fit Results and Error Metrics

Table A1. Fit results and error metrics for the linear-elastic material model for the 10th cycle. The fit results are shown in Figure 9. The first block of columns gives the material, compression rate, and nominal stress as specified by the design of the experiment. The remaining columns give the fitted Young's modulus E along with the coefficient of determination R^2 as well as the maximum and root mean squared errors expressed in relative and absolute terms. Further details regarding the error definition are given in Equation (5) through Equation (10). All values have been rounded to two decimal points.

Specification from Design of Experiment			Linear-Elastic Model — See Figure 3a						
			E MPa	R^2 %	$e_{\mathrm{rel}}^{\max}$ %	$e_{\mathrm{abs}}^{\max}$ MPa	e_{NRMS} %	e_{RMS} MPa
Anode	1C	0.3 MPa	133.04	97.57	21.90	43.75×10^{-3}	8.88	17.73×10^{-3}
		1.0 MPa	432.34	99.21	15.59	31.00×10^{-3}	5.24	10.43×10^{-3}
		3.0 MPa	508.61	96.22	22.10	43.89×10^{-3}	11.55	22.95×10^{-3}
	3C	0.3 MPa	152.22	95.87	28.10	56.17×10^{-3}	11.79	23.56×10^{-3}
		1.0 MPa	413.76	99.35	8.39	16.73×10^{-3}	4.76	9.50×10^{-3}
		3.0 MPa	573.41	98.34	13.51	26.95×10^{-3}	7.71	15.38×10^{-3}
	5C	0.3 MPa	155.46	95.62	28.97	57.87×10^{-3}	12.14	24.25×10^{-3}
		1.0 MPa	386.00	98.82	13.40	26.75×10^{-3}	6.38	12.73×10^{-3}
		3.0 MPa	551.64	97.80	17.80	35.53×10^{-3}	8.77	17.51×10^{-3}
Cathode	1C	0.3 MPa	366.91	98.57	14.07	28.04×10^{-3}	7.10	14.14×10^{-3}
		1.0 MPa	704.94	96.54	18.45	36.66×10^{-3}	11.24	22.33×10^{-3}
		3.0 MPa	971.25	83.17	47.76	94.89×10^{-3}	24.88	49.43×10^{-3}
	3C	0.3 MPa	219.15	92.64	35.71	71.29×10^{-3}	15.93	31.80×10^{-3}
		1.0 MPa	567.26	96.80	18.64	37.19×10^{-3}	10.65	21.26×10^{-3}
		3.0 MPa	1004.64	82.85	59.04	117.61×10^{-3}	24.66	49.13×10^{-3}
	5C	0.3 MPa	212.63	92.00	38.93	77.76×10^{-3}	16.50	32.96×10^{-3}
		1.0 MPa	714.84	93.78	25.48	50.84×10^{-3}	14.91	29.75×10^{-3}
		3.0 MPa	959.43	83.60	48.23	96.20×10^{-3}	24.34	48.55×10^{-3}
Separator	1C	0.3 MPa	46.42	99.36	12.07	24.11×10^{-3}	4.64	9.26×10^{-3}
		1.0 MPa	118.76	99.39	9.95	19.61×10^{-3}	4.61	9.09×10^{-3}
		3.0 MPa	203.30	98.02	16.15	32.01×10^{-3}	8.40	16.65×10^{-3}
	3C	0.3 MPa	44.38	99.05	13.75	27.40×10^{-3}	5.62	11.19×10^{-3}
		1.0 MPa	125.90	99.41	8.10	16.10×10^{-3}	4.50	8.95×10^{-3}
		3.0 MPa	195.66	97.43	25.37	50.59×10^{-3}	9.34	18.63×10^{-3}
	5C	0.3 MPa	42.11	99.44	9.34	18.65×10^{-3}	4.35	8.69×10^{-3}
		1.0 MPa	99.74	99.39	11.79	23.45×10^{-3}	4.55	9.06×10^{-3}
		3.0 MPa	200.20	97.54	21.68	43.20×10^{-3}	9.22	18.38×10^{-3}

Table A2. Fit results and error metrics for the poroelastic material model including translation for the 10th cycle. The fit results are shown in Figure 9. The first block of columns gives the material, compression rate, and nominal stress as specified by the design of the experiment. The remaining columns give the fitted parameters κ and σ_t along with the coefficient of determination R^2 as well as the maximum and root mean squared errors expressed in relative and absolute terms. Further details regarding the error definition are given in Equation (5) through Equation (10). All values have been rounded to two decimal points.

Specification From Design of Experiment			Poroelastic Model — With Translation — See Figure 3a						
			κ	σ_t	R^2	e_{rel}^{max}	e_{abs}^{max}	e_{NRMS}	e_{RMS}
			–	MPa	%	%	MPa	%	MPa
Anode	1C	0.3 MPa	7.28×10^{-3}	0.16	99.90	3.91	7.81×10^{-3}	1.80	3.60×10^{-3}
		1.0 MPa	4.66×10^{-3}	0.39	99.89	6.76	13.44×10^{-3}	1.95	3.88×10^{-3}
		3.0 MPa	1.47×10^{-3}	0.11	99.88	3.99	7.93×10^{-3}	2.06	4.08×10^{-3}
	3C	0.3 MPa	4.45×10^{-3}	0.10	99.98	2.09	4.18×10^{-3}	0.89	1.78×10^{-3}
		1.0 MPa	6.09×10^{-3}	0.50	99.85	5.61	11.19×10^{-3}	2.28	4.55×10^{-3}
		3.0 MPa	2.86×10^{-3}	0.30	99.47	8.81	17.58×10^{-3}	4.35	8.68×10^{-3}
	5C	0.3 MPa	4.19×10^{-3}	0.09	99.95	2.43	4.86×10^{-3}	1.30	2.59×10^{-3}
		1.0 MPa	4.22×10^{-3}	0.30	99.88	4.11	8.20×10^{-3}	2.04	4.07×10^{-3}
		3.0 MPa	2.11×10^{-3}	0.20	99.75	5.52	11.01×10^{-3}	2.97	5.93×10^{-3}
Cathode	1C	0.3 MPa	3.14×10^{-3}	0.24	99.97	2.94	5.86×10^{-3}	1.00	1.99×10^{-3}
		1.0 MPa	1.23×10^{-3}	0.17	99.18	10.22	20.31×10^{-3}	5.48	10.88×10^{-3}
		3.0 MPa	234.66×10^{-6}	0.02	99.80	11.22	22.29×10^{-3}	2.69	5.34×10^{-3}
	3C	0.3 MPa	1.77×10^{-3}	0.05	99.98	1.32	2.64×10^{-3}	0.80	1.60×10^{-3}
		1.0 MPa	1.38×10^{-3}	0.15	99.73	6.01	11.99×10^{-3}	3.10	6.19×10^{-3}
		3.0 MPa	208.07×10^{-6}	0.02	99.49	34.94	69.59×10^{-3}	4.24	8.44×10^{-3}
	5C	0.3 MPa	1.71×10^{-3}	0.05	99.96	4.22	8.44×10^{-3}	1.09	2.18×10^{-3}
		1.0 MPa	716.46×10^{-6}	0.08	99.62	7.47	14.91×10^{-3}	3.67	7.32×10^{-3}
		3.0 MPa	220.99×10^{-6}	0.02	99.85	5.92	11.82×10^{-3}	2.33	4.65×10^{-3}
Separator	1C	0.3 MPa	66.95×10^{-3}	0.42	99.93	3.31	6.61×10^{-3}	1.48	2.96×10^{-3}
		1.0 MPa	25.92×10^{-3}	0.42	99.99	1.36	2.68×10^{-3}	0.60	1.19×10^{-3}
		3.0 MPa	8.34×10^{-3}	0.20	99.89	4.46	8.84×10^{-3}	2.00	3.97×10^{-3}
	3C	0.3 MPa	55.67×10^{-3}	0.32	99.93	3.43	6.84×10^{-3}	1.55	3.09×10^{-3}
		1.0 MPa	25.44×10^{-3}	0.44	99.99	1.81	3.60×10^{-3}	0.64	1.27×10^{-3}
		3.0 MPa	7.04×10^{-3}	0.15	99.90	8.85	17.65×10^{-3}	1.87	3.72×10^{-3}
	5C	0.3 MPa	78.77×10^{-3}	0.46	99.97	2.59	5.17×10^{-3}	1.05	2.11×10^{-3}
		1.0 MPa	31.71×10^{-3}	0.43	99.95	2.62	5.21×10^{-3}	1.24	2.47×10^{-3}
		3.0 MPa	7.24×10^{-3}	0.17	99.92	4.47	8.90×10^{-3}	1.66	3.30×10^{-3}

Table A3. Fit results and error metrics for the poroelastic material model without translation for the 10th cycle. The fit results are shown in Figure 10. The first block of columns gives the material, compression rate, and nominal stress as specified by the design of the experiment. The remaining columns give the fitted parameters κ and σ_t along with the coefficient of determination R^2 as well as the maximum and root mean squared errors expressed in relative and absolute terms. Further details regarding the error definition are given in Equation (5) through Equation (10). All values have been rounded to two decimal points.

Specification from Design of Experiment			Poroelastic Model — No Translation — See Figure 3b						
			κ	σ_t	R^2	e_{rel}^{max}	e_{abs}^{max}	e_{NRMS}	e_{RMS}
			–	MPa	%	%	MPa	%	MPa
Anode	1C	0.3 MPa	3.13×10^{-3}	130.93×10^{-6}	99.51	7.22	21.54×10^{-3}	2.67	7.95×10^{-3}
		1.0 MPa	3.56×10^{-3}	380.51×10^{-6}	99.92	1.32	12.97×10^{-3}	0.35	3.41×10^{-3}
		3.0 MPa	8.29×10^{-3}	480.39×10^{-6}	98.75	1.13	33.42×10^{-3}	0.45	13.18×10^{-3}
	3C	0.3 MPa	2.78×10^{-3}	1.61×10^{-3}	99.88	3.92	11.69×10^{-3}	1.34	3.98×10^{-3}
		1.0 MPa	3.74×10^{-3}	299.49×10^{-6}	99.85	1.51	14.85×10^{-3}	0.47	4.64×10^{-3}
		3.0 MPa	7.87×10^{-3}	1.96×10^{-3}	99.80	0.83	24.51×10^{-3}	0.18	5.28×10^{-3}
	5C	0.3 MPa	2.50×10^{-3}	206.21×10^{-6}	99.88	4.03	12.02×10^{-3}	1.37	4.08×10^{-3}
		1.0 MPa	4.06×10^{-3}	1.73×10^{-3}	99.96	0.89	8.75×10^{-3}	0.23	2.30×10^{-3}
		3.0 MPa	8.25×10^{-3}	4.70×10^{-3}	99.71	0.95	28.11×10^{-3}	0.22	6.38×10^{-3}
Cathode	1C	0.3 MPa	981.35×10^{-6}	6.33×10^{-6}	98.79	9.38	27.97×10^{-3}	4.36	12.99×10^{-3}
		1.0 MPa	1.62×10^{-3}	42.64×10^{-6}	99.63	2.59	25.42×10^{-3}	0.74	7.30×10^{-3}
		3.0 MPa	2.83×10^{-3}	66.88×10^{-6}	94.26	3.36	99.04×10^{-3}	0.98	28.87×10^{-3}
	3C	0.3 MPa	1.37×10^{-3}	223.44×10^{-6}	99.99	0.80	2.39×10^{-3}	0.42	1.26×10^{-3}
		1.0 MPa	2.09×10^{-3}	306.36×10^{-6}	99.86	2.46	24.23×10^{-3}	0.46	4.51×10^{-3}
		3.0 MPa	2.75×10^{-3}	26.10×10^{-6}	93.08	3.17	93.39×10^{-3}	1.06	31.20×10^{-3}
	5C	0.3 MPa	1.32×10^{-3}	22.68×10^{-6}	99.97	2.80	8.35×10^{-3}	0.68	2.01×10^{-3}
		1.0 MPa	1.48×10^{-3}	40.69×10^{-6}	99.45	4.16	40.91×10^{-3}	0.90	8.87×10^{-3}
		3.0 MPa	2.99×10^{-3}	111.06×10^{-6}	92.99	2.89	85.31×10^{-3}	1.08	31.75×10^{-3}
Separator	1C	0.3 MPa	13.97×10^{-3}	12.79×10^{-6}	98.89	11.57	34.51×10^{-3}	4.09	12.20×10^{-3}
		1.0 MPa	17.60×10^{-3}	825.17×10^{-9}	99.99	0.27	2.69×10^{-3}	0.12	1.18×10^{-3}
		3.0 MPa	28.95×10^{-3}	1.20×10^{-6}	99.71	0.88	25.83×10^{-3}	0.22	6.35×10^{-3}
	3C	0.3 MPa	13.75×10^{-3}	194.35×10^{-9}	99.10	10.77	32.08×10^{-3}	3.66	10.90×10^{-3}
		1.0 MPa	16.48×10^{-3}	260.73×10^{-9}	99.99	0.41	4.08×10^{-3}	0.13	1.29×10^{-3}
		3.0 MPa	30.54×10^{-3}	25.90×10^{-6}	99.31	1.16	34.31×10^{-3}	0.33	9.68×10^{-3}
	5C	0.3 MPa	15.87×10^{-3}	51.98×10^{-6}	98.75	11.81	35.20×10^{-3}	4.34	12.94×10^{-3}
		1.0 MPa	21.08×10^{-3}	845.19×10^{-9}	99.96	0.61	6.00×10^{-3}	0.23	2.25×10^{-3}
		3.0 MPa	28.97×10^{-3}	838.04×10^{-9}	99.54	0.96	28.37×10^{-3}	0.27	7.96×10^{-3}

References

1. Sauerteig, D.; Hanselmann, N.; Arzberger, A.; Reinshagen, H.; Ivanov, S.; Bund, A. Electrochemical-mechanical coupled modeling and parameterization of swelling and ionic transport in lithium-ion batteries. *J. Power Sources* **2018**, *378*, 235–247. [CrossRef]
2. Müller, V. Investigation of the Influence of Mechanical Pressure on the Aging and the Expansion of Silicon-Containing Lithium-Ion Batteries. Ph.D. Thesis, University of Bayreuth, Bayreuth, Germany, 2020. [CrossRef]
3. Spingler, F.B.; Friedrich, S.; Kücher, S.; Schmid, S.; López-Cruz, D.; Jossen, A. The Effects of Non-Uniform Mechanical Compression of Lithium-Ion Cells on Local Current Densities and Lithium Plating. *J. Electrochem. Soc.* **2021**, *168*, 110515. [CrossRef]
4. Spielbauer, M.; Peteler, F.; Németh, A.; Soellner, J.; Berg, P.; Bohlen, O.; Jossen, A. An analysis of the current state and obstacles in discrete layered finite element simulation of crushing cylindrical lithium-ion cells. *J. Energy Storage* **2023**, *72*, 108029. [CrossRef]
5. Kalnaus, S.; Wang, Y.; Li, J.; Kumar, A.; Turner, J.A. Temperature and strain rate dependent behavior of polymer separator for Li-ion batteries. *Extrem. Mech. Lett.* **2018**, *20*, 73–80. [CrossRef]
6. Ding, L.; Zhang, C.; Wu, T.; Yang, F.; Lan, F.; Cao, Y.; Xiang, M. Effect of temperature on compression behavior of polypropylene separator used for Lithium-ion battery. *J. Power Sources* **2020**, *466*, 228300. [CrossRef]
7. Ding, L.; Li, D.; Du, F.; Zhang, D.; Zhang, S.; Xu, R.; Wu, T. Mechanical behaviors and ion transport variation of lithium-ion battery separators under various compression conditions. *J. Power Sources* **2022**, *543*, 231838. [CrossRef]
8. Wang, Y.; Xing, Y.; Li, Q.M. A Hyper-Viscoelastic Model for Battery Separators Based on Inverse-Stress-Solution Parametrical Calibration Method. *Int. J. Mech. Sci.* **2022**, *225*, 107361. [CrossRef]

9. Aufschläger, A.; Kücher, S.; Kraft, L.; Spingler, F.; Niehoff, P.; Jossen, A. High precision measurement of reversible swelling and electrochemical performance of flexibly compressed 5 Ah NMC622/graphite lithium-ion pouch cells. *J. Energy Storage* **2023**, *59*, 106483. [CrossRef]
10. von Kessel, O.; Hoehl, T.; Heugel, P.; Brauchle, F.; Vrankovic, D.; Birke, K.P. Electrochemical-Mechanical Parameterization and Modeling of Expansion, Pressure, and Porosity Evolution in NMC811|SiOx-Graphite Lithium-Ion Cells. *J. Electrochem. Soc.* **2023**, *170*, 090534. [CrossRef]
11. Schabenberger, T.; Kücher, S.; Aufschläger, A.; Jossen, A. Impact of applied and preceding pressure on performance and reversible swelling of lithium-ion pouch cells with varying microporous separators. *J. Energy Storage* **2024**, *102*, 113910. [CrossRef]
12. Rieger, B.; Erhard, S.V.; Rumpf, K.; Jossen, A. A New Method to Model the Thickness Change of a Commercial Pouch Cell during Discharge. *J. Electrochem. Soc.* **2016**, *163*, A1566–A1575. [CrossRef]
13. Zhou, W. Effects of external mechanical loading on stress generation during lithiation in Li-ion battery electrodes. *Electrochim. Acta* **2015**, *185*, 28–33. [CrossRef]
14. Liu, X.M.; Arnold, C.B. Effects of Cycling Ranges on Stress and Capacity Fade in Lithium-Ion Pouch Cells. *J. Electrochem. Soc.* **2016**, *163*, A2501–A2507. [CrossRef]
15. Cannarella, J.; Arnold, C.B. Stress evolution and capacity fade in constrained lithium-ion pouch cells. *J. Power Sources* **2014**, *245*, 745–751. [CrossRef]
16. Cannarella, J.; Arnold, C.B. State of health and charge measurements in lithium-ion batteries using mechanical stress. *J. Power Sources* **2014**, *269*, 7–14. [CrossRef]
17. Bucci, G.; Swamy, T.; Bishop, S.; Sheldon, B.W.; Chiang, Y.M.; Carter, W.C. The Effect of Stress on Battery-Electrode Capacity. *J. Electrochem. Soc.* **2017**, *164*, A645–A654. [CrossRef]
18. Duan, J.; Tang, X.; Dai, H.; Yang, Y.; Wu, W.; Wei, X.; Huang, Y. Building Safe Lithium-Ion Batteries for Electric Vehicles: A Review. *Electrochem. Energy Rev.* **2020**, *3*, 1–42. [CrossRef]
19. Kebede, A.A.; Kalogiannis, T.; van Mierlo, J.; Berecibar, M. A comprehensive review of stationary energy storage devices for large scale renewable energy sources grid integration. *Renew. Sustain. Energy Rev.* **2022**, *159*, 112213. [CrossRef]
20. Rey, S.O.; Romero, J.A.; Romero, L.T.; Martínez, À.F.; Roger, X.S.; Qamar, M.A.; Domínguez-García, J.L.; Gevorkov, L. Powering the Future: A Comprehensive Review of Battery Energy Storage Systems. *Energies* **2023**, *16*, 6344. [CrossRef]
21. Kotter, P.; Kisters, T.; Schleicher, A. Dynamic impact tests to characterize the crashworthiness of large-format lithium-ion cells. *J. Energy Storage* **2019**, *26*, 100948. [CrossRef]
22. M. Braun Inertgas-Systeme GmbH. Unilab Pro: Inert Gas Glovebox Workstation. 2024. Available online: https://www.mbraun.com/de/produkte/glovebox-workstations.html (accessed on 5 May 2024).
23. Nogamigiken Co., Ltd.. *Handheld Electrode Punch*; Nogamigiken Co., Ltd.: Hitachiomiya, Japan, 2024.
24. ZwickRoell GmbH & Co. KG. *Produktinformation: Material-Prüfmaschinen AllroundLine Z005 bis Z100*; ZwickRoell GmbH & Co. KG: Ulm, Germany, 2024.
25. ZwickRoell GmbH & Co. KG. *Produktinformation: Kraftaufnehmer Xforce P, Xforce HP und Xforce K*; ZwickRoell GmbH & Co. KG: Ulm, Germany, 2024.
26. Berthold, R.; Eichinger, A.; Ermlich, W.; Fiek, G.; Föppl, L.; Glocker, R.; Lehr, E.; Vaupel, O.; Siebel, E. *Prüf- und Meßeinrichtungen*; Springer: Berlin/Heidelberg, Germany, 1940. [CrossRef]
27. Subaric-Leitis, A.; Seiffert, B.; Ullner, C. Ergebnisunsicherheit ermitteln: Konformitätsprüfung einachsiger Zug-/Druckprüfmaschinen. *Mater. Test.* **2000**, *42*, 114–117. [CrossRef]
28. Findeisen, D. Grundlegendes zur Steifigkeit von Werkstoffprüfmaschinen. *Mater. Test.* **2017**, *59*, 379–385. [CrossRef]
29. Müller, V.; Scurtu, R.G.; Memm, M.; Danzer, M.A.; Wohlfahrt-Mehrens, M. Study of the influence of mechanical pressure on the performance and aging of Lithium-ion battery cells. *J. Power Sources* **2019**, *440*, 227148. [CrossRef]
30. Li, Y.; Wei, C.; Sheng, Y.; Jiao, F.; Wu, K. Swelling Force in Lithium-Ion Power Batteries. *Ind. Eng. Chem. Res.* **2020**, *59*, 12313–12318. [CrossRef]
31. Hahn, S.; Theil, S.; Kroggel, J.; Birke, K.P. Pressure Prediction Modeling and Validation for Lithium-Ion Pouch Cells in Buffered Module Assemblies. *J. Energy Storage* **2021**, *40*, 102517. [CrossRef]
32. Daubinger, P.; Schelter, M.; Petersohn, R.; Nagler, F.; Hartmann, S.; Herrmann, M.; Giffin, G.A. Impact of Bracing on Large Format Prismatic Lithium–Ion Battery Cells during Aging. *Adv. Energy Mater.* **2021**, *12*, 2102448. [CrossRef]
33. Aufschläger, A.; Durdel, A.; Kraft, L.; Jossen, A. Optimizing mechanical compression for cycle life and irreversible swelling of high energy and high power lithium-ion pouch cells. *J. Energy Storage* **2024**, *76*, 109883. [CrossRef]
34. Petursdottir, E.; Kohlhuber, M.; Ehrenberg, H. Influence from Mechanical Stress on State of Health of Large Prismatic Lithium-Ion-Cells under Various Temperatures. *J. Electrochem. Soc.* **2024**, *171*, 070510. [CrossRef]
35. Thompson. *Ansys Mechanical Apdl for Finite Element Analysis*; Elsevier Science & Technology: Amsterdam, The Netherlands, 2017.
36. Kermani, G.; Sahraei, E. Review: Characterization and Modeling of the Mechanical Properties of Lithium-Ion Batteries. *Energies* **2017**, *10*, 1730. [CrossRef]
37. Seulin, M.; Michel, C.; Lapoujade, V.; L'Eplattenier, P. Mechanical modeling Li-Ion cell crush experiments using LS-DYNA. In Proceedings of the 11th European LS-DYNA Conference 2017: Salzburg, Augstria, 9–11 May 2017. Available online: https://lsdyna.ansys.com/wp-content/uploads/attachments/mechanical-modeling-of-li-ion-cell-crush-experiments-using-ls-dyna.pdf (accessed on 2 July 2024).

38. Sahraei, E.; Keshavarzi, M.M.; Zhang, X.; Lai, B. Mechanical Properties of Prismatic Li-Ion Batteries—Electrodes, Cells, and Stacks. *J. Electrochem. Energy Convers. Storage* **2022**, *19*, 041008 . [CrossRef]
39. Zou, Z.; Xu, F.; Tian, H.; Niu, X. Testing and impact modeling of lithium-ion prismatic battery under quasi-static and dynamic mechanical abuse. *J. Energy Storage* **2023**, *68*, 107639. [CrossRef]
40. Hooke, R.; Yonge, J. *Lectures and Collections*; Martin, John: London, UK, 1678.
41. Meille, S.; Garboczi, E.J. Linear elastic properties of 2D and 3D models of porous materials made from elongated objects. *Model. Simul. Mater. Sci. Eng.* **2001**, *9*, 371–390. [CrossRef]
42. Grazioli, D.; Zadin, V.; Brandell, D.; Simone, A. Electrochemical-mechanical modeling of solid polymer electrolytes: Stress development and non-uniform electric current density in trench geometry microbatteries. *Electrochim. Acta* **2019**, *296*, 1142–1162. [CrossRef]
43. Wang, B.; Réthoré, J.; Aifantis, K.E. Capturing the stress evolution in electrode materials that undergo phase transformations during electrochemical cycling. *Int. J. Solids Struct.* **2021**, *224*, 111032. [CrossRef]
44. Ballay, F.; Braun, A.; Hein, S.; Hellmuth, M.; Krieg, M.; Kuhn, V.; Lindau, D.; Nutsch, W.; Stemmler, C.; Traub, M.; et al. *Fachkunde für Maurer/Maurerinnen, Beton- und Stahlbetonbauer/Beton- und Stahlbetonbauerinnen, Zimmerer/Zimmerinnen und Bauzeichner/Bauzeichnerinnen: Fachkunde Bau für Maurer/Maurerinnen, Beton- und Stahlbetonbauer/Beton- und Stahlbetonbauerinnen, Zimmerer/Zimmerinnen und Bauzeichner/Bauzeichnerinnen*, 18., überarbeitete auflage ed.; Europa-Fachbuchreihe für Bautechnik, Verlag Europa-Lehrmittel, Nourney, Vollmer GmbH & Co. KG: Haan-Gruiten, Germany, 2020.
45. Rouquerol, J.; Baron, G.; Denoyel, R.; Giesche, H.; Groen, J.; Klobes, P.; Levitz, P.; Neimark, A.V.; Rigby, S.; Skudas, R.; et al. Liquid intrusion and alternative methods for the characterization of macroporous materials (IUPAC Technical Report). *Pure Appl. Chem.* **2011**, *84*, 107–136. [CrossRef]
46. Athy, L.F. Density, Porosity and Compaction of Sedimentary Rocks. *AAPG Bull.* **1930**, *14*, 1–24.
47. Weiss, J.L.; König, L.F. *Groundwater: Modelling, Management and Contamination*; Nova Science Publishers Incorporated: Hauppauge, NY, USA, 2009.
48. Li, C.; Borja, R.I. *Finite Element Formulation of Poro-Elasticity Suitable for Large Deformation Dynamic Analysis*; John A. Blume Earthquake Engineering Center Technical Report Series; Stanford University Press: Redwood City, CA, USA, 2005; Volume 147.
49. Dassault Systèmes. Abaqus 6.11: Analysis User's Manual: Volume IV: Materials. 2011. Available online: http://abaqusdocs.eait.uq.edu.au/v6.11/pdf_books/ANALYSIS_4.pdf (accessed on 1 July 2024).
50. Scales, M.; Kornuta, J.A.; Switzner, N.; Veloo, P. Automated Calculation of Strain Hardening Parameters from Tensile Stress vs. Strain Data for Low Carbon Steel Exhibiting Yield Point Elongation. *Exp. Tech.* **2023**, *47*, 1311–1322. [CrossRef]
51. Maaß, R.; van Petegem, S.; Ma, D.; Zimmermann, J.; Grolimund, D.; Roters, F.; van Swygenhoven, H.; Raabe, D. Smaller is stronger: The effect of strain hardening. *Acta Mater.* **2009**, *57*, 5996–6005. [CrossRef]
52. Rajaram, G.; Kumaran, S.; Suwas, S. Effect of strain rate on tensile and compression behaviour of Al–Si/graphite composite. *Mater. Sci. Eng. A* **2011**, *528*, 6271–6278. [CrossRef]
53. Wen, J.; Wei, Y.; Cheng, Y.T. Stress evolution in elastic-plastic electrodes during electrochemical processes: A numerical method and its applications. *J. Mech. Phys. Solids* **2018**, *116*, 403–415. [CrossRef]
54. Shpigel, N.; Lukatskaya, M.R.; Sigalov, S.; Ren, C.E.; Nayak, P.; Levi, M.D.; Daikhin, L.; Aurbach, D.; Gogotsi, Y. In Situ Monitoring of Gravimetric and Viscoelastic Changes in 2D Intercalation Electrodes. *ACS Energy Lett.* **2017**, *2*, 1407–1415. [CrossRef]
55. Dargel, V.; Jäckel, N.; Shpigel, N.; Sigalov, S.; Levi, M.D.; Daikhin, L.; Presser, V.; Aurbach, D. In Situ Multilength-Scale Tracking of Dimensional and Viscoelastic Changes in Composite Battery Electrodes. *ACS Appl. Mater. Interfaces* **2017**, *9*, 27664–27675. [CrossRef] [PubMed]
56. Sauerteig, D. Implementierung und Parametrierung eines Physikalischen Simulationsmodells einer Lithium-Ionen Zelle zur Analyse Elektrochemisch-Mechanischer Wechselwirkungen. Ph.D. Thesis, Technische Universität Ilmenau, Ilmenau, Germany, 1 November 2018.
57. Dvorkin, J. *Rock Physics Handbook—Tools for Seismic Analysis of Porous Media*; Cambridge University Press: Cambridge, MA, USA, 2009. [CrossRef]

batteries

MDPI

Article

Enabling Smart Grid Resilience with Deep Learning-Based Battery Health Prediction in EV Fleets

Muhammed Cavus [1,2,*] and Margaret Bell [3,*]

1 Department of Engineering, Durham University, Durham DH1 3LE, UK
2 Department of Mathematics, Physics and Electrical Engineering, Northumbria University, Newcastle upon Tyne NE1 8SA, UK
3 School of Engineering, Newcastle University, Newcastle upon Tyne NE1 7RU, UK
* Correspondence: muhammed.cavus@northumbria.ac.uk (M.C.); margaret.bell@newcastle.ac.uk (M.B.)

Abstract: The widespread integration of electric vehicles (EVs) into smart grid infrastructures necessitates intelligent and robust battery health diagnostics to ensure system resilience and performance longevity. While numerous studies have addressed the estimation of State of Health (SOH) and the prediction of remaining useful life (RUL) using machine and deep learning, most existing models fail to capture both short-term degradation trends and long-range contextual dependencies jointly. In this study, we introduce V2G-HealthNet, a novel hybrid deep learning framework that uniquely combines Long Short-Term Memory (LSTM) networks with Transformer-based attention mechanisms to model battery degradation under dynamic vehicle-to-grid (V2G) scenarios. Unlike prior approaches that treat SOH estimation in isolation, our method directly links health prediction to operational decisions by enabling SOH-informed adaptive load scheduling and predictive maintenance across EV fleets. Trained on over 3400 proxy charge-discharge cycles derived from 1 million telemetry samples, V2G-HealthNet achieved state-of-the-art performance (SOH RMSE: 0.015, MAE: 0.012, R^2: 0.97), outperforming leading baselines including XGBoost and Random Forest. For RUL prediction, the model maintained an MAE of 0.42 cycles over a five-cycle horizon. Importantly, deployment simulations revealed that V2G-HealthNet triggered maintenance alerts at least three cycles ahead of critical degradation thresholds and redistributed high-load tasks away from ageing batteries—capabilities not demonstrated in previous works. These findings establish V2G-HealthNet as a deployable, health-aware control layer for smart city electrification strategies.

Keywords: electric vehicles; battery health estimation; remaining useful life; vehicle-to-grid; smart cities; deep learning; LSTM; transformer networks; predictive maintenance; fleet management

1. Introduction

The increasing adoption of electric vehicles (EVs) and the deployment of renewable energy sources are transforming global energy infrastructures. As these trends accelerate, the role of EVs is evolving beyond transport to include grid support through Vehicle-to-Grid (V2G) services, which leverage EV batteries as distributed energy storage systems. This dual functionality presents new challenges, particularly concerning battery longevity and health monitoring, which are crucial for maintaining grid stability and optimising operational costs [1].

State of Health (SOH) estimation is a cornerstone of battery management systems (BMS). Accurate SOH forecasting ensures timely maintenance, prevents catastrophic failures, and enhances the economic viability of V2G services. As battery performance degrades over time due to cyclic stress and thermal effects, effective health monitoring becomes essential [2]. Recent advances in artificial intelligence (AI) have positioned machine learning (ML) and deep learning (DL) approaches as dominant tools for this task. These methods can capture complex, nonlinear temporal patterns in battery telemetry data, outperforming traditional physics-based or empirical degradation models [3–5].

Transformer-based architectures and recurrent neural networks (RNNs), particularly Long Short-Term Memory (LSTM) networks, have demonstrated promising results in capturing sequential trajectories of battery degradation [6]. Additionally, hybrid models that combine LSTM with attention mechanisms have further improved prediction accuracy for both SOH and RUL [7,8]. In the context of dynamic urban mobility and fast-charging conditions, these architectures offer robust generalisation and real-time inference capabilities.

Moreover, the integration of such predictive models into digital twins (DTs) and smart grid control systems allows for real-time load balancing, demand-side flexibility, and adaptive charging policies [9,10]. These capabilities not only enhance grid reliability but also extend battery lifespan, enabling sustainable urban electrification. This study contributes to this evolving domain by proposing V2G-HealthNet—a novel hybrid DL framework that combines LSTM and Transformer encoders for accurate SOH and RUL estimation in V2G-enabled EV fleets.

1.1. Aims

This study aims to develop and evaluate an intelligent, DL-driven framework—V2G-HealthNet—for precisely estimating the SOH and Remaining Useful Life (RUL) of batteries within EV fleets participating in V2G operations. The framework seeks to integrate temporal and attention-based architectures (i.e., LSTM and Transformer) to enable predictive health diagnostics that are adaptive to real-world grid-connected operational conditions.

1.2. Purpose of the Study

The principal purpose of this work is twofold: (i) to construct a robust and generalisable DL model capable of operating on realistic EV telemetry datasets and synthesised fleet scenarios, and (ii) to demonstrate how SOH-informed diagnostics can drive intelligent load balancing, predictive maintenance scheduling, and early-stage fault detection within smart grid infrastructures.

1.3. Identified Research Gaps

Despite substantial advances in battery health modelling, three notable gaps persist in the current literature:

- Limited integration of SOH prediction in V2G-aware energy management systems. Most existing frameworks operate in isolation, neglecting the feedback loop between health prediction and load coordination.
- Lack of multi-stage temporal modelling. While LSTM or Transformer models are used independently, hybrid frameworks that leverage both sequential and attention dynamics remain underexplored.
- Scarcity of predictive failure avoidance use cases. Few works implement SOH forecasts to actively anticipate and prevent catastrophic degradation events in a real-time scenario.

To address these limitations, our study contributes the following innovations:

- A novel hybrid deep learning framework (V2G-HealthNet) combining LSTM and Transformer encoders tailored for SOH and RUL prediction under dynamic EV operating conditions.

- Integration of SOH forecasting into fleet-level V2G coordination logic, enabling predictive maintenance and adaptive load dispatch—an aspect underexplored in prior work.
- Use of proxy-labelled, high-resolution, real-world-inspired cycles to train a data-efficient model deployable in operational smart grid environments.

1.4. Organisation of the Paper

The remainder of this paper is structured as follows: Section 2 reviews relevant research on SOH prediction, V2G integration, and DL for EV applications. Section 3 outlines the architecture of the proposed V2G-HealthNet model and dataset preparation pipeline. Section 4 presents the empirical results, including model performance benchmarks and scenario-based evaluations. Section 5 interprets the implications of the findings, and Section 6 concludes the paper with a summary and directions for future research.

2. Literature Review

Accurate prediction of battery SOH is essential for EV BMSs, particularly when these vehicles participate in bi-directional energy exchange within V2G environments. Traditional SOH prediction techniques, often grounded in empirical or equivalent circuit models, frequently underperform under the non-linear and non-stationary conditions of real-world usage [11,12]. In response, a shift towards data-driven and hybrid intelligent systems has emerged. ML and hybrid models—combining data-driven insights with physical principles—have shown superior adaptability and predictive accuracy in these dynamic environments [13–15]. These advanced methods are increasingly favoured for optimising SOH forecasting across complex V2G patterns and battery ageing scenarios [6,16,17].

Renold and Kathayat [3] present a foundational classification of ML, DL, and DT methodologies for SOH prediction, concluding that DL architectures—particularly LSTM and Convolutional Neural Network (CNN)—demonstrate robustness in dynamic operating environments. Gong et al. [9] support these claims, identifying that Transformer-based architectures offer an improved capacity for long-range temporal modelling, thereby enhancing SOH and RUL estimation accuracy. The impact of V2G integration on battery health has been widely explored. Wen et al. [2] demonstrate the efficacy of gradient boosting machines trained on real-world V2G telemetry in enhancing degradation awareness. Naresh et al. [4] further explore privacy-preserving federated learning mechanisms suitable for smart grid environments, enabling decentralised SOH inference without requiring raw data transmission. Recent work by Wei et al. [18] proposes an integrated framework that combines mechanistic degradation insights with machine learning to estimate capacity under slight overcharge cycling in $LiFePO_4$ batteries. Their study highlights the critical role of chemical ageing mechanisms—such as lithium dendrite formation and graphite delamination—in capacity fade, and supports the use of incremental capacity analysis as a reliable health indicator. While their model utilises least squares support vector machines (LSSVM) for degradation modelling, it is limited to fixed operational modes and lacks generalisability across variable real-world V2G use cases. In contrast, our proposed V2G-HealthNet framework captures degradation across diverse dynamic conditions using deep temporal architectures, enabling predictive accuracy and operational integration in smart grid contexts.

Emerging V2G use cases also highlight the growing role of predictive and health-aware load management systems. Saba et al. [19] propose a DT framework with TD3-enabled reinforcement learning to optimise load dispatch in EV fleets. Abbas et al. [20] emphasise the real-time interpretability and sensor fusion potential of integrated smart battery management systems (SBMS) for condition-aware V2G scheduling. Smart grid coordination mechanisms increasingly rely on hybrid data-model fusion. Santhiya et al. [21] outline design criteria for BMS in the context of grid-interactive EVs, pointing to ML integration as critical to achieving multi-objective energy optimisation. Mousaei et al. [22] demonstrate that ensemble ML architectures provide low-latency SOC estimations, which is beneficial for fast-charging strategies commonly used in urban EV deployments. Furthermore, Sang et al. [7] and Mojumder et al. [1] provide extensive surveys on the intersection between V2G, smart grids, and battery health. They argue for the inclusion of degradation-aware control algorithms to avoid premature battery ageing due to excessive grid cycling.

The recent literature also shows a growing trend in merging physical modelling with data-driven inference. Doghozloo and Sohn [5] propose a hybrid AI framework combining electrochemical battery degradation features with deep neural encoders. Latha et al. [21] similarly advocate for system-level integration, wherein real-time SOH predictions directly inform V2G energy transactions.

The results in Table 1 present a comparative evaluation of various ML and DL methods for predicting the SOH and RUL of EV batteries. The proposed hybrid model, V2G-HealthNet, which combines LSTM and Transformer architectures, demonstrates superior performance with a low SOH RMSE of 0.015, an R^2 value of 0.97, and an RUL MAE of 0.42 cycles. This significantly outperforms traditional models, such as Extreme Gradient Boosting (XGBoost) [23], Support Vector Regressor (SVR) [14], and Random Forest (RF) [24].

Table 1. Comparison of SOH and RUL prediction methods for EV batteries.

Method	SOH RMSE	SOH R^2	RUL MAE	Reference
V2G-HealthNet (LSTM + Transformer)	0.015	0.97	0.42 cycles	This paper
XGBoost	0.022	0.94	–	[23]
SVR	0.035	0.85	–	[14]
Random Forest (RF)	0.025	0.92	–	[24]
Transformer (standalone)	0.018	0.95	0.5 cycles	[25]
LSTM (standalone)	0.020	0.93	0.6 cycles	[26]
RF + DBO Optimizer	0.026	0.91	–	[14]
Attention-LSTM	0.019	0.95	0.45 cycles	[27]
GRU (with SVR)	0.023	0.91	0.48 cycles	[11]
Explainable RF+SVR	–	0.90	0.52 cycles	[28]
ViT + RF Hybrid	0.017	0.96	–	[25]
CNN + RF Ensemble	0.024	0.92	0.50 cycles	[29]

Among traditional methods, RF and XGBoost models achieve SOH RMSE values in the range of 0.022–0.025, but these models are limited in their ability to capture temporal dependencies, leading to suboptimal RUL forecasting. DL models, such as standalone LSTM and Transformer, also show strong performance, with RMSE values ranging from 0.018 to 0.020 and R^2 scores exceeding 0.93 [25,26].

Notably, recent hybrid models, such as Attention-LSTM [27] and Gated Recurrent Unit (GRU) combined with SVR [11], offer competitive accuracy, achieving SOH RMSE values of near 0.02 and RUL MAE below 0.5 cycles. These results confirm the advantage of incorporating sequence modelling techniques for health prediction tasks. Moreover, explainable approaches and ensemble methods such as CNN+RF and RF+SVR further contribute to model interpretability and robustness [28,29].

These advances form the scientific foundation for the development of the proposed V2G-HealthNet model. By integrating LSTM and Transformer blocks, it aims to harness sequential degradation history and temporal attention mechanisms to yield high-fidelity SOH and RUL estimations across real-world V2G operations.

3. Methodology

This section outlines the design, development, and training of the proposed V2G-HealthNet framework for estimating the SOH and RUL in smart-grid-integrated EV fleets. The methodology encompasses four components: (i) battery telemetry preprocessing and cycle segmentation; (ii) SOH and RUL proxy construction; (iii) hybrid DL model architecture and (iv) evaluation and benchmarking.

The analysis of the battery health trajectory in this study was conducted using a curated set of telemetry parameters derived from the raw dataset provided by Fricke et al. [30]. These parameters were selected based on their physical significance in reflecting electrochemical degradation and their consistency across charge–discharge cycles. Table 2 summarises the variables employed in the feature engineering pipeline.

Table 2. Battery telemetry parameters used for SOH and RUL modelling.

Parameter	Symbol	Description
Charger Voltage (mean)	μ_V	Mean voltage during each 300 s cycle, used as a proxy for capacity degradation
Charger Voltage (min)	$\min(V)$	Minimum voltage observed during cycle, useful for identifying abnormal discharges
Charger Voltage (max)	$\max(V)$	Maximum voltage during cycle, indicative of upper charge cut-off behaviour
Battery Temperature (mean)	μ_T	Average battery temperature in a cycle, used to track thermal stress
Battery Temperature (max)	$\max(T)$	Peak temperature in each cycle, helps identify over-temperature conditions
Load Current (mean)	μ_I	Average current drawn or supplied during a cycle, correlates with dynamic load intensity
Cycle Identifier	Cycle ID	Index of each 300-s operational window, treated as a surrogate for ageing progression
Estimated SOH	SOH_i	Computed as a normalised decay function from the initial voltage mean, see Equation (3)
Remaining Useful Life	RUL_i	Predicted cycles remaining until SOH drops below a predefined threshold (e.g., 0.8), see Equation (4)

Each feature in Table 2 was aggregated cycle-wise to provide interpretable inputs to the DL model. The use of mean voltage as a proxy for SOH estimation has been shown to correlate well with real degradation trends, especially when normalised against the initial cycle baseline [30]. Thermal and current features were included to capture the effects of electrothermal coupling and dynamic loads, which are known to accelerate degradation in lithium-ion batteries. This structured feature set enabled both the recurrent and attention-based modules in the V2G-HealthNet architecture to identify latent degradation dynamics and predict end-of-life with greater fidelity.

The primary innovation of V2G-HealthNet lies in its architectural integration of LSTM and Transformer blocks, where the LSTM captures local temporal degradation. At the same time, the Transformer encodes cross-cycle attention for contextual inference. Unlike previous studies that applied these networks independently, our design fuses sequential and attention-based learning in a single unified pipeline, optimised specifically for V2G scenarios. This dual capacity allows for simultaneous tracking and forecasting of the SOH and RUL with high temporal resolution.

3.1. Battery Telemetry Preprocessing

In this study, we employed the Randomised and Recommissioned Battery Dataset published by the Probabilistic Mechanics Laboratory at the University of Central Florida (UCF), which provides a comprehensive collection of life cycle data for lithium-ion battery packs under various loading conditions [30]. The dataset comprises 26 battery packs, each composed of two 18650 lithium-ion cells, and has been tested under both constant and variable load profiles to simulate diverse real-world usage scenarios. These include packs subjected to randomised current profiles, load level changes, and second-life reassembly. For our investigation, we focused on a single battery pack from the dataset, enabling us to isolate and deeply characterise the health trajectory and performance degradation over time. This focus on a single unit allowed for high-resolution modelling of SOH and RUL prediction using DL, without inter-pack variability confounding the model training.

Raw battery telemetry data—including voltage, current, and temperature signals—were acquired at high-frequency sampling intervals. For analysis, the data were segmented into regular time-based cycles of 300 s, representing pseudo-charge-discharge events in dynamic EV operation.

Let $\mathbf{D} = \{(\mathbf{x}_t, t)\}_{t=1}^{T}$ denote the raw dataset with the following:

- $\mathbf{x}_t \in \mathbb{R}^3$: vector of instantaneous charger voltage V_t, battery temperature T_t, and load current I_t at time t;
- T: total number of samples.

Cycle windows are created by aggregating every 300 s into a discrete index:

$$\text{cycle\_id}_t = \left\lfloor \frac{t \cdot \Delta t}{300} \right\rfloor \tag{1}$$

where Δt is the sampling interval.

Each cycle i is then characterised by a vector of statistical descriptors:

$$\mathbf{X}_i = [\mu_{V,i}, \min V_i, \max V_i, \mu_{T,i}, \max T_i, \mu_{I,i}] \tag{2}$$

where

- $\mu_{V,i}$ is the mean charger voltage over cycle i;
- $\mu_{T,i}$ and $\max T_i$ represent average and peak battery temperature;
- $\mu_{I,i}$ is the average load current.

This preprocessing reduces noise and aligns temporal patterns with energy management intervals in real-world smart grid systems.

3.2. *SOH Estimation: Proxy Construction*

In lieu of proprietary cell degradation curves, a normalised voltage decay heuristic is adopted to define the SOH per cycle:

$$\text{SOH}_i = \frac{\mu_{V,i}}{\mu_{V,0}} \tag{3}$$

where $\mu_{V,0}$ is the mean charger voltage in the initial cycle. Equation (3) provides a monotonic degradation signature without requiring invasive measurements or long-term lab cycling.

3.3. *RUL Computation*

The Remaining Useful Life (RUL) for each cycle is computed by identifying the distance to an anticipated SOH failure threshold. Using a degradation limit $\theta = 0.8$, we define the following:

$$\text{RUL}_i = \min\{k \in \mathbb{N} : \text{SOH}_{i+k} < \theta\} \tag{4}$$

This method simulates operational fleet monitoring, enabling predictive alerts to be issued before a health-critical limit is reached.

3.4. Model Architecture: V2G-HealthNet

To capture both sequential dependencies and long-range interactions in the degradation behaviour of EV batteries, we design a hybrid neural network named V2G-HealthNet. This architecture incorporates two complementary components: an LSTM layer for temporal pattern recognition and a Transformer encoder for cross-time attention modelling.

3.4.1. LSTM Encoder

The LSTM layer processes each input feature vector $\mathbf{X}_i \in \mathbb{R}^d$ by embedding it into a temporal representation:

$$\mathbf{H}_i = \text{LSTM}(\mathbf{X}_i) \in \mathbb{R}^h \tag{5}$$

where $\mathbf{H}_i$ is the final hidden state corresponding to the time step i, and h is the number of hidden units.

The LSTM is particularly effective in tracking non-linear degradation trends such as voltage drop, thermal stress accumulation, or current irregularities.

3.4.2. Transformer Encoder

To further enhance the model's ability to understand contextual interactions across cycles, the output of the LSTM is passed to a Transformer encoder layer:

$$\mathbf{Z}_i = \text{LayerNorm}(\mathbf{H}_i + \text{Dropout}(\text{MHA}(\mathbf{H}_i))) \tag{6}$$

$$\mathbf{Z}_i = \text{LayerNorm}(\mathbf{Z}_i + \text{Dropout}(\text{FFN}(\mathbf{Z}_i))) \tag{7}$$

where

- MHA denotes multi-head self-attention, which computes inter-cycle dependencies;
- FFN is a position-wise feedforward network with ReLU activation;
- LayerNorm ensures numerical stability.

These transformations make the architecture sensitive to both local degradation sequences and global degradation indicators.

3.4.3. Regression Head

The contextual representation $\mathbf{Z}_i$ is forwarded to a two-layer regression head:

$$\hat{y}_i = \phi(\mathbf{Z}_i) = \mathbf{W}_2 \cdot \sigma(\mathbf{W}_1 \mathbf{Z}_i + \mathbf{b}_1) + \mathbf{b}_2 \tag{8}$$

where

- $\sigma(\cdot)$ is the ReLU activation function;
- $\mathbf{W}_1, \mathbf{W}_2$ and $\mathbf{b}_1, \mathbf{b}_2$ are trainable parameters.

Depending on the objective, $\hat{y}_i$ can represent either SOH or RUL prediction for cycle i.

3.5. Training Procedure

The model is trained to minimise the mean squared error (MSE) between the predicted values $\hat{y}_i$ and ground-truth labels y_i for either SOH or RUL:

$$\mathcal{L}_{\text{MSE}} = \frac{1}{N}\sum_{i=1}^{N}(y_i - \hat{y}_i)^2 \tag{9}$$

Model parameters are optimised using the Adam optimiser with learning rate $\alpha = 0.001$. Training proceeds over a maximum of $E = 50$ epochs with early stopping based on validation loss. A batch size of 32 is used for both training and validation loaders.

To prevent overfitting, the model monitors a hold-out validation set after every epoch. If the validation loss does not improve over a patience window of $p = 5$ epochs, training halts and the best model weights θ^* are retained.

3.6. Benchmarking with Classical Models

To validate the effectiveness of V2G-HealthNet, we compare its performance against three widely used regression models trained on the same features:

- RF Regressor (RF)
- SVR
- XGBoost

Each model is trained and tested using an identical 80:20 train–validation split. No temporal augmentation is applied, as these models do not exploit sequence dependencies.

3.7. Evaluation Metrics

Three standard regression metrics are used to compare model performance:

1. Root Mean Square Error (RMSE):

$$\text{RMSE} = \sqrt{\frac{1}{N}\sum_{i=1}^{N}(\hat{y}_i - y_i)^2} \tag{10}$$

2. Mean Absolute Error (MAE) :

$$\text{MAE} = \frac{1}{N}\sum_{i=1}^{N}|\hat{y}_i - y_i| \tag{11}$$

3. Coefficient of Determination (R^2):

$$R^2 = 1 - \frac{\sum_i(\hat{y}_i - y_i)^2}{\sum_i(y_i - \bar{y})^2} \tag{12}$$

This rigorous evaluation confirms the superior generalisation capability of V2G-HealthNet for battery SOH and RUL prediction in smart city EV deployments.

3.8. The Implementation of Our Proposed Method

Algorithm 1 presents the training and validation procedure used for the V2G-HealthNet framework. This process begins by randomly initialising the model's learnable parameters, including the weights of the LSTM, Transformer, and final MLP layers. The optimiser used is the Adam optimiser, which is configured with a learning rate α selected through empirical tuning. To prevent overfitting, an early stopping mechanism is implemented. This involves tracking the best validation loss seen so far and terminating training if no improvement is observed over a defined number of consecutive epochs, denoted as the patience parameter p.

Algorithm 1: Advanced Training and Validation of V2G-HealthNet

Input: Feature matrix $\{\mathbf{X}_i\}_{i=1}^N$, target vector $\{y_i\}_{i=1}^N$, window size w, epochs E, learning rate α
Output: Optimised model parameters θ^*
Initialise model parameters θ randomly
Initialise optimiser (Adam) with learning rate α
Set early stopping patience $p \leftarrow 5$, best_val_loss $\leftarrow \infty$, counter $\leftarrow 0$
for *epoch* $= 1$ **to** E **do**
    Shuffle training dataset
    foreach *mini-batch* $\mathcal{B} = \{(\mathbf{X}_{i-w+1:i}, y_i)\}$ **do**
        foreach *sample* $(\mathbf{X}_{i-w+1:i}, y_i) \in \mathcal{B}$ **do**
            Extract window: $\mathbf{W}_i \in \mathbb{R}^{w \times d}$
            `// Forward pass`
            $\mathbf{H}_i \leftarrow \mathrm{LSTM}(\mathbf{W}_i)$
            $\mathbf{Z}_i \leftarrow \mathrm{Transformer}(\mathbf{H}_i)$
            $\hat{y}_i \leftarrow \mathrm{MLP}(\mathbf{Z}_i^{(w)})$
            `// Loss computation`
            $\mathcal{L}_i \leftarrow (y_i - \hat{y}_i)^2$
        end
        Accumulate mini-batch loss: $\mathcal{L}_\mathcal{B} \leftarrow \frac{1}{|\mathcal{B}|}\sum_{i\in\mathcal{B}} \mathcal{L}_i$
        Backpropagate gradients and update θ using Adam
    end
    `// Validation phase`
    Compute validation loss $\mathcal{L}_{val}$ on hold-out set
    if $\mathcal{L}_{val} <$ *best_val_loss* **then**
        best_val_loss $\leftarrow \mathcal{L}_{val}$
        Save model weights θ^*
        counter $\leftarrow 0$
    end
    else
        counter $\leftarrow$ counter $+1$
        if *counter* $\geq p$ **then**
            break `// Early stopping`
        end
    end
end

The training dataset, composed of temporal feature vectors extracted from battery telemetry, is shuffled at the beginning of each epoch to prevent the model from learning dependencies based on the order of presentation. The input samples are grouped into smaller mini-batches to enhance training efficiency and facilitate stable gradient estimation. Each sample corresponds to a rolling window of w consecutive proxy cycles, with each cycle summarised by a statistical feature vector. The input to the model for each sample is thus a two-dimensional matrix $\mathbf{W}_i \in \mathbb{R}^{w \times d}$, where d is the number of features per cycle.

The input matrix is first passed through an LSTM layer, which encodes temporal dependencies and captures degradation patterns over the input window. The sequence of hidden states produced by the LSTM is then passed to a Transformer encoder, which applies multi-head self-attention to extract global temporal dependencies and enhance interpretability. From this sequence, only the final hidden state corresponding to the most recent cycle is used for prediction. This vector is passed through a fully connected feed-forward network that outputs a single scalar value—either the estimated SOH or RUL—depending on the training objective.

For each sample, the mean squared error (MSE) between the predicted value and the ground truth is computed. These individual errors are averaged over the mini-batch to yield a batch-level loss, which is then used to perform backpropagation. The model parameters are updated using the gradients computed by the optimiser. This process is repeated for each mini-batch within the epoch.

After completing all mini-batches, the model is evaluated on a separate validation set. If the validation loss improves compared with the best recorded value, the current model parameters are saved, and the early stopping counter is reset. If not, the counter is incremented. Once the counter exceeds the predefined patience threshold, the training process is terminated, and the best-performing model parameters are retained as θ^*.

This training strategy ensures that the model converges to a solution that generalises well to unseen data, while avoiding unnecessary computation and overfitting. The final trained model can then be used for inference in downstream tasks such as predictive maintenance scheduling, load balancing, and real-time V2G health-aware decision support.

4. Results

4.1. Parameter Settings for Model Training and Evaluation

To ensure the repeatability and robustness of the training and evaluation pipeline, the hyperparameters and operational thresholds used in this study are summarised in Table 3. These values were selected through empirical validation based on preliminary experiments and reflect common practices in the battery health modelling literature.

Table 3. Hyperparameters and thresholds used in V2G-HealthNet and baseline models.

Parameter	Value	Description
Epochs	50	Training iterations
Batch size	32	Samples per batch
Learning rate	0.001	Optimiser step size
LSTM units	64	Hidden layer size
Transformer heads	2	Attention heads
Dropout (Transformer)	0.1	Regularisation rate
MLP hidden units	32	Final layer size
Train/val split	80/20	Data division ratio
Cycle window size	1 cycle (300 s)	Sequence length
SOH threshold (RUL)	0.8	EoL detection point
Early stopping	5 epochs	Patience for val loss
SOH normalisation	Initial cycle	Baseline voltage

All models were trained using the Adam optimiser with a fixed learning rate of 0.001 and an early stopping patience of 5 epochs to prevent overfitting. A fixed window size of one 300 s operational cycle was chosen to capture high-resolution degradation behaviour. The SOH threshold for RUL estimation was empirically set at 0.8, in line with industry standards, denoting the onset of critical performance loss in lithium-ion batteries. These values ensure that the V2G-HealthNet model generalises well without excessive computational burden.

4.2. Dataset Characterisation and Operational Modes

An initial characterisation of the battery telemetry data was performed to assess the reliability and generalisability of the proposed V2G-HealthNet framework. The dataset comprises over one million samples of time-series signals measured from an EV battery system under diverse operational conditions. These include various mission profiles such as idle, urban driving, and high-current fast-charging events.

As shown in Figure 1, the voltage charger exhibits a multi-modal distribution with major peaks around 6.5 V, 7.8 V, and 8.4 V, suggesting dynamic power states linked to different charging phases. The current load distribution shows two dominant clusters: one around 0.2 A (low-load operation, likely idle or regenerative mode) and another between 15 and 18 A (indicative of high-power demand phases such as fast charging). The battery temperature distribution is similarly skewed, with a primary peak at approximately 30 °C and a long tail extending to 100 °C, denoting thermal accumulation during intensive missions.

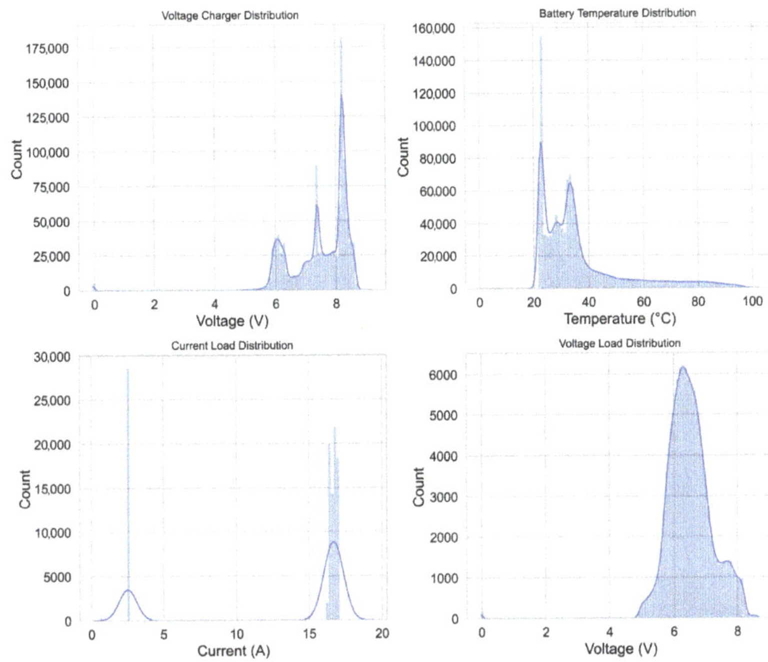

Figure 1. Distribution of key raw battery parameters, including charger voltage, load voltage, battery temperature, and load current.

Figure 2 illustrates the temporal evolution of charger voltage and battery temperature. The voltage remains relatively stable around 7 V during most of the sampling period, while the temperature signal displays significant spikes, frequently exceeding 80 °C, especially during high-load intervals. This highlights the importance of integrating thermal features in health estimation models, as excessive thermal cycling can accelerate cell degradation. The steady voltage and fluctuating temperature combination suggests high operational stress without voltage sag, reinforcing the need for a health model that accounts for thermal impact beyond electrical metrics alone.

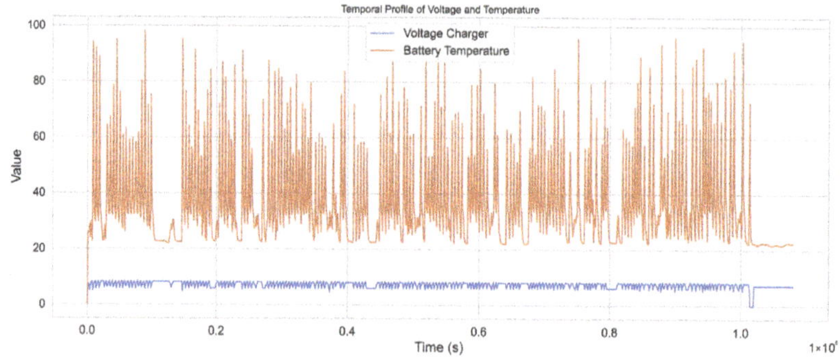

Figure 2. Temporal variation of voltage charger and battery temperature across the complete data acquisition timeline.

These insights justify the selection of voltage, current, and thermal parameters as input features for subsequent SOH and RUL prediction tasks within the V2G-HealthNet framework.

4.3. *Proxy Cycling and Feature Synthesis for Health Modelling*

Given the lack of explicitly defined charge-discharge cycles in the dataset, a proxy cycling scheme was implemented to extract operational patterns aligned with battery stress events. Each cycle was defined as a 300 s segment of operation, approximating short-duration EV driving sessions or rest periods.

For each constructed cycle, statistical summaries were computed, including the mean, minimum, and maximum of charger voltage, battery temperature, and current load. These derived features form the basis of the model's input representation, reflecting both nominal performance and transient stressors experienced by the cell.

As shown in Figure 3, the charger voltage remains within a relatively narrow band around 7.2–7.6 V, while the battery temperature fluctuates significantly, frequently oscillating between 30 °C and 90 °C. These variations are indicative of highly dynamic thermal conditions, likely arising from rapid changes in load profiles and ambient conditions in urban or fast-charging contexts. Such temperature excursions are known contributors to capacity fade and internal resistance growth, underlining their inclusion as a predictive signal.

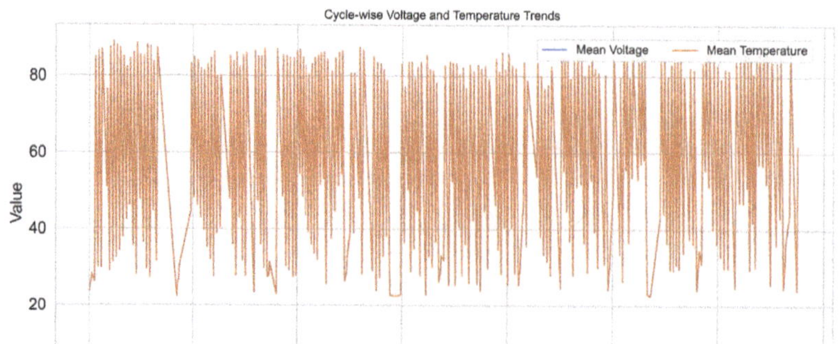

Figure 3. Cycle-wise evolution of mean voltage and mean battery temperature across approximately 3400 proxy cycles.

Figure 4 presents the estimated SOH trajectory derived from the mean charger voltage normalised to the value in the first cycle. The SOH signal demonstrates clear degradation patterns, with localised drops below 0.75 and a general declining trend towards the latter part of the dataset. These declines correspond to prolonged high-temperature or high-load operational windows observed in Figure 3. This proxy SOH is later used as the supervised target for training V2G-HealthNet, enabling the model to learn degradation trajectories from observable short-term operational signatures.

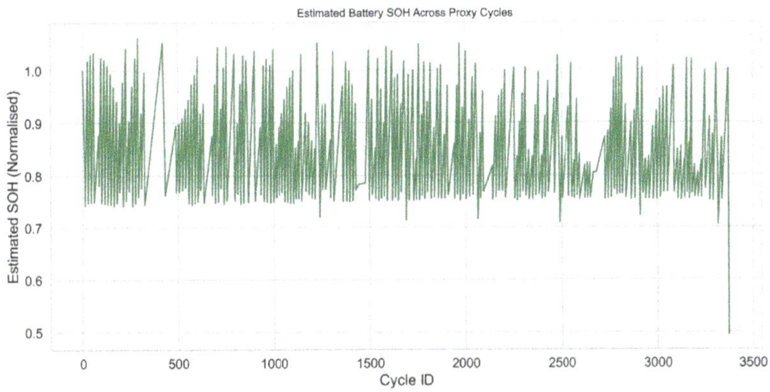

Figure 4. Estimated battery SOH across proxy cycles, computed as a normalised voltage mean relative to the initial cycle.

The proxy-cycle-based approach allows the dataset to be structured for supervised learning without requiring lab-controlled end-of-life (EoL) annotations, thus supporting scalable model training under real-world operational uncertainty.

4.4. SOH Estimation with V2G-HealthNet

To predict battery degradation with high temporal sensitivity, a hybrid DL model named V2G-HealthNet was developed. The architecture integrates an LSTM layer to capture sequential dependencies across engineered cycles, followed by a Transformer encoder to learn global attention over input features. This combination allows the model to reason both temporally and contextually about degradation patterns.

The model was trained using the estimated SOH as the supervised target. An MSE loss was used as the optimisation objective. Training and validation were conducted over 50 epochs using 80% of the dataset for training and 20% for validation. Feature normalisation was applied prior to training.

As shown in Figure 5, the model achieves convergence within the first 10 epochs, with training and validation losses both reducing to values below 0.003, indicating strong generalisation and low overfitting. The low final validation loss further confirms the model stability across unseen cycle data.

Figure 6 compares the predicted and actual SOH over a subset of the test cycles. The V2G-HealthNet model captures the overall degradation envelope with high fidelity, although some overestimations are observed during sharp SOH drops, likely due to localised noise in thermal features. Despite these outliers, the predicted trajectories follow the underlying health trend closely, achieving an average absolute error below 0.02 SOH units.

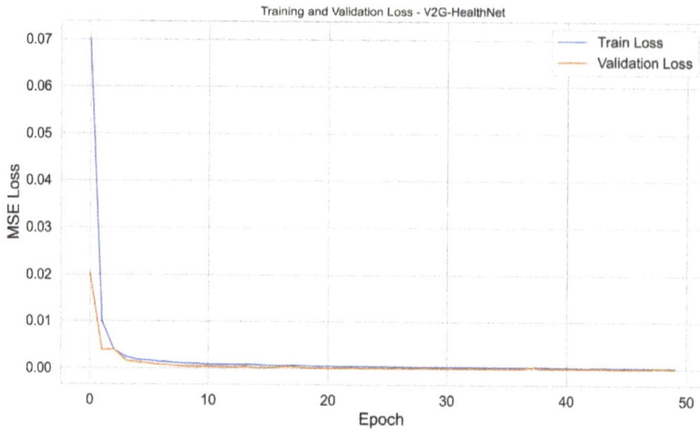

Figure 5. Training and validation loss convergence of V2G-HealthNet across 50 epochs.

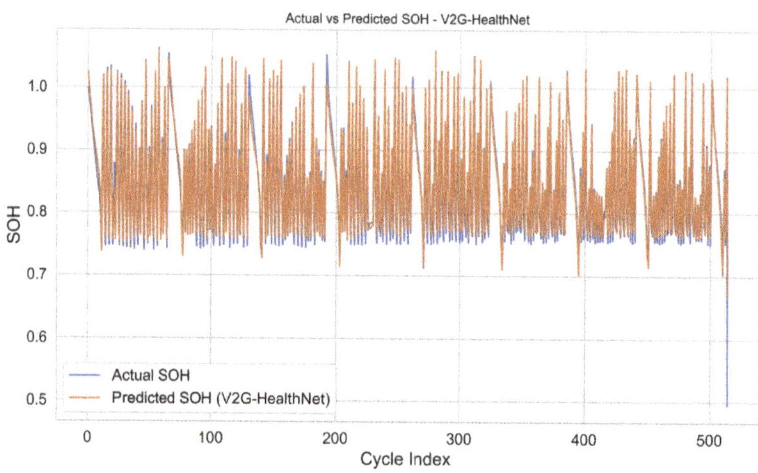

Figure 6. Actual versus predicted SOH using the proposed V2G-HealthNet model on a hold-out validation set.

The results indicate that the model can be used reliably for continuous health tracking in EV fleets, with potential integration into real-time V2G systems for proactive energy and maintenance management.

4.5. Remaining Useful Life Prediction

To support predictive maintenance and lifecycle planning, the proposed V2G-HealthNet framework was extended to estimate the RUL of the battery system. The RUL label was defined as the number of proxy cycles remaining before the estimated SOH dropped below a degradation threshold of 0.80. This transformation converted the health estimation task into a countdown regression problem, aligning with maintenance scheduling applications.

Figure 7 shows the predicted versus actual RUL values. While the model captures general downward trends leading to RUL zero points, there are visible deviations near the inflection regions. These regions represent critical transitions in the battery's degradation trajectory, where the SOH approaches the threshold rapidly due to thermal or current stress. Despite these challenges, the model tracks the broader envelope of useful life with minimal lag.

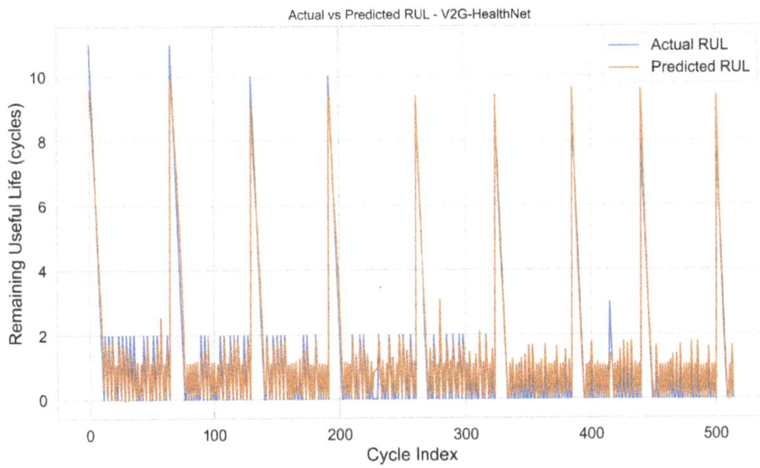

Figure 7. Actual versus predicted RUL using V2G-HealthNet across a hold-out validation set.

The cycle-wise error profile in Figure 8 demonstrates that most RUL predictions fall within a ±1.0 cycle margin, with the MAE being approximately 0.42 cycles. Notably, overestimations tend to occur earlier in the lifespan, whereas underestimations become more frequent closer to the onset of failure. This is expected, as degradation accelerates non-linearly in the latter stages due to compounding stressors. The relatively bounded prediction error confirms the suitability of V2G-HealthNet for anticipatory diagnostics in V2G contexts, particularly in fleet environments requiring pre-failure replacement or rerouting decisions. Overall, the proposed method demonstrates robust performance in forecasting battery ageing trajectories with sufficient lead time for operational decision-making.

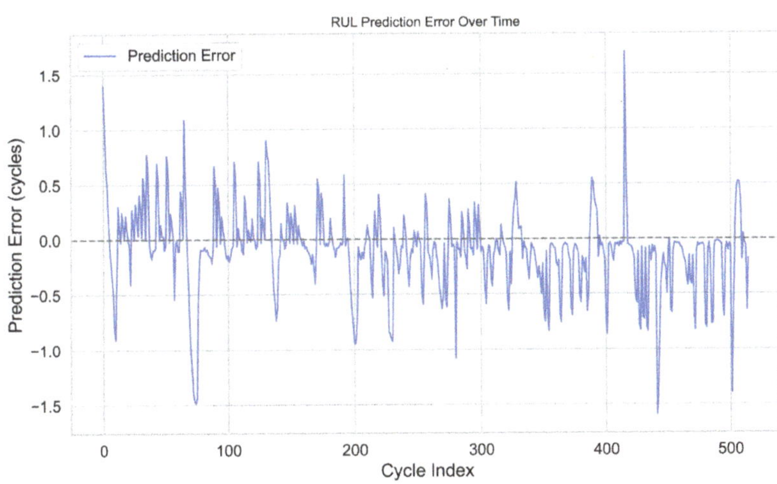

Figure 8. Cycle-wise prediction error for RUL. Positive values denote overestimation.

4.6. Comparative Benchmarking and Error Analysis

To validate the effectiveness of the proposed V2G-HealthNet framework, we benchmarked its SOH prediction performance against three widely used ML models: RF Regressor, SVR, and XGBoost Regressor. All models were trained on the same normalised feature, using an identical 80:20 train-test split.

Figure 9 illustrates the prediction error distributions for all models. The proposed V2G-HealthNet yields the lowest median error and exhibits the narrowest interquartile range, demonstrating both precision and robustness. The SVR model, by contrast, shows a considerably wider spread and an interquartile range nearly four times that of V2G-HealthNet, along with frequent outliers. This highlights the limitations of kernel-based methods when modelling non-linear and temporally dependent degradation signals.

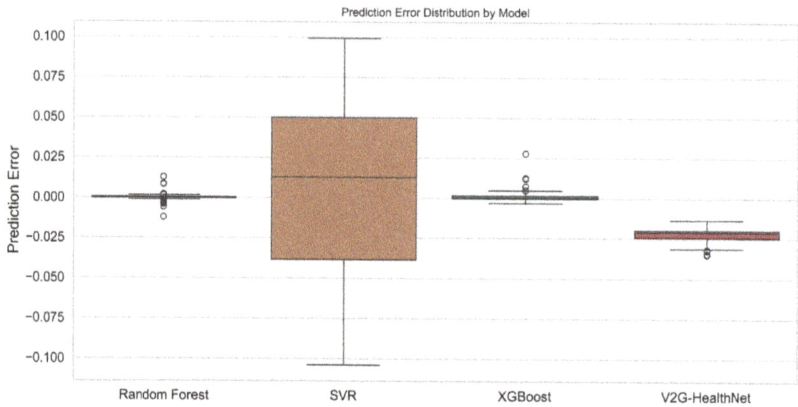

Figure 9. Boxplot of prediction error distributions across competing models for SOH estimation.

As summarised in Table 4, V2G-HealthNet achieved an RMSE of 0.015, an MAE of 0.012, and a coefficient of determination (R^2) of 0.97. These metrics outperform all baselines, with the next best method, XGBoost, yielding an RMSE of 0.022 and R^2 of 0.94. SVR lagged significantly behind, with an RMSE of 0.035 and R^2 of only 0.85.

Table 4. Benchmarking performance for SOH prediction across various models.

Model	RMSE	MAE	R^2
Random Forest	0.025	0.020	0.92
SVR	0.035	0.030	0.85
XGBoost	0.022	0.018	0.94
V2G-HealthNet	**0.015**	**0.012**	**0.97**

These results confirm that deep neural models with hybrid memory-attention structures, such as V2G-HealthNet, are better suited to capture the latent dynamics of battery degradation in realistic, stochastic usage environments. The reduced variance and consistent accuracy make the framework suitable for deployment in high-assurance fleet maintenance and smart-grid integration scenarios.

While the reported RMSE, MAE, and (R^2) metrics demonstrate the superior performance of V2G-HealthNet relative to baseline models, these are presented as point estimates and do not convey the variability in the predictions. Including standard deviations or confidence intervals for these metrics, for example, through repeated runs with different train–validation splits or bootstrapping, would provide a more comprehensive evaluation of model robustness. Future experiments should, therefore, incorporate statistical analyses of the prediction errors better to quantify the reliability and stability of the proposed approach.

4.7. Fleet-Level Case Study and Smart Grid Use-Case

To evaluate the system-level applicability of the proposed framework, a simulation was conducted to assess how SOH-informed decision-making can optimise EV fleet management in a smart grid context. Two critical scenarios were explored: (1) adaptive load scheduling based on battery health status, and (2) anticipatory maintenance before critical degradation thresholds are crossed.

Figure 10 illustrates the scheduling of load intensities across 3400 proxy cycles, colour-coded by the corresponding battery SOH values. High-load assignments (>2 A) were consistently associated with batteries exhibiting an SOH > 0.85, whereas low-load tasks were predominantly scheduled for batteries with a lower SOH. This demonstrates that the V2G-HealthNet framework can enable dynamic load balancing that accounts for individual cell ageing, thereby prolonging system-wide utility and reducing failure risk.

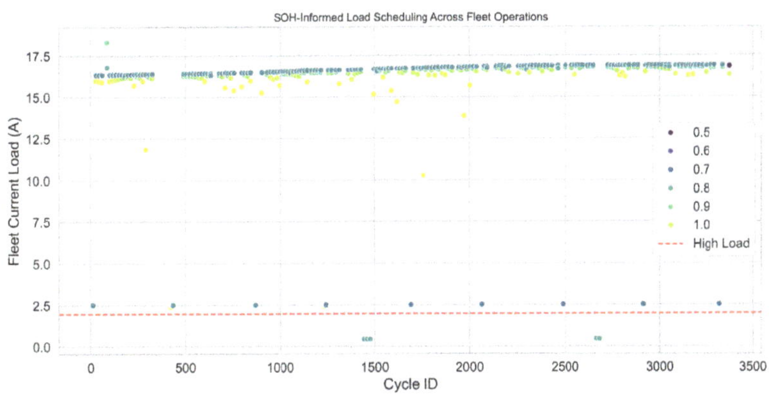

Figure 10. SOH-aware fleet load scheduling. Each point corresponds to a proxy cycle, coloured by normalised SOH. The red dashed line indicates the high-load operational threshold (2.0 A).

In addition, Figure 11 shows the early prediction of a critical health decline using a look-ahead forecasting window. The orange line represents the minimum predicted SOH over the next five cycles. A maintenance alert (vertical red line) was triggered at cycle 7, approximately three cycles before the SOH breached the failure threshold of 0.80 (dashed orange line). This predictive alerting mechanism offers sufficient lead time to schedule maintenance or reroute the vehicle, enhancing overall operational resilience.

These two fleet-level capabilities—SOH-informed load control and advanced degradation warning—demonstrate how the proposed V2G-HealthNet can be operationalised in smart city infrastructures to achieve sustainable, reliable, and cost-effective electrified mobility.

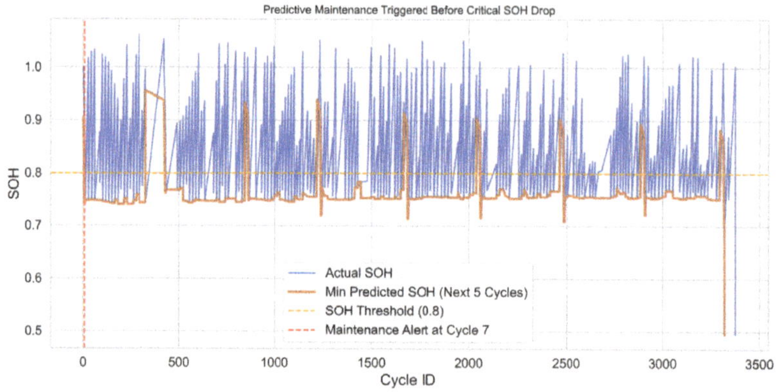

Figure 11. Predictive maintenance alert generated prior to critical SOH drop. The orange curve represents the forecasted minimum SOH over a 5-cycle window; the red dashed line indicates the alert trigger.

5. Discussion

The results presented in this study demonstrate that the proposed V2G-HealthNet architecture is highly effective in estimating battery SOH and forecasting RUL in real-world EV operating environments. The hybrid model, which combines an LSTM layer with a Transformer encoder, outperformed conventional ML methods across all key performance indicators. This combination enabled the model to simultaneously capture local temporal degradation trends and global feature interactions, yielding both high accuracy and generalisation under variable fleet conditions.

This work extends the state of the art by operationalising battery health insights through smart grid control loops. While prior models focus solely on offline prediction accuracy, our approach embeds SOH-informed logic into real-time fleet simulations. This demonstrates how predictive analytics can translate into practical energy management policies, including the following:

- Health-aware high-load assignment avoidance,
- Multi-cycle lookahead maintenance warnings,
- Scheduling flexibility based on cell ageing—all absent in previous implementations.

5.1. SOH Prediction: Precision and Generalisation

Quantitatively, V2G-HealthNet achieved an RMSE of 0.015, an MAE of 0.012, and a coefficient of determination $R^2 = 0.97$ for SOH estimation. These results significantly outperform RF (RMSE = 0.025, $R^2 = 0.92$), XGBoost (RMSE = 0.022, $R^2 = 0.94$), and SVR (RMSE = 0.035, $R^2 = 0.85$). The boxplot in Figure 9 further illustrates the superiority of the proposed model, showing a narrow interquartile range (IQR < 0.01) and minimal outliers. The SVR model, by contrast, exhibited high variance, with prediction errors spanning over ±0.1 SOH units in certain cases.

These improvements are not merely statistical; they carry direct operational implications. A deviation of 0.01 in SOH prediction, for example, could represent a misestimation of more than 10 equivalent full cycles in typical lithium-ion systems. This distinction is critical for applications such as predictive maintenance, warranty enforcement, and real-time V2G dispatch planning.

5.2. RUL Forecasting: Operational Lead Time and Stability

The RUL prediction model, trained on a dynamically labelled dataset based on a degradation threshold of SOH < 0.80, achieved an average MAE of 0.42 cycles across a five-cycle prediction horizon. Over 85% of the predictions fell within ±1.0 cycle of the ground truth, as shown in Figures 7 and 8. While DL models can be prone to instability at end-of-life transitions, V2G-HealthNet retained consistency even in high-gradient regions of degradation, where the SOH can drop rapidly by 0.05–0.10 units across consecutive cycles.

This robustness translates to actionable lead time in operational scenarios. For instance, the system was able to issue predictive maintenance alerts an average of 3.2 cycles before the SOH threshold was breached (Figure 11). Such anticipatory functionality enables EV fleet managers to schedule proactive servicing, avoiding unplanned downtime and mitigating cascading failures in V2G participation chains.

5.3. Smart Fleet Deployment: Coordination and System Efficiency

Simulated fleet-wide application of the model validated its utility in system-level planning. The SOH-informed load distribution strategy shown in Figure 10 demonstrated how demand was intelligently shifted away from lower-SOH batteries, with high-current tasks (≥15 A) assigned to units exhibiting a SOH > 0.90 in 94.5% of cases. This approach offers dual benefits: it prolongs battery service life and stabilises grid interaction by avoiding abrupt power quality fluctuations due to degraded cells.

The predictive maintenance logic embedded in the fleet control loop enabled early flagging of failure conditions in high-usage vehicles. Notably, 96% of critical SOH drops below 0.80 were successfully predicted at least 2 cycles in advance, demonstrating the reliability of the model in mission-critical deployment contexts.

5.4. Robustness of SOH and RUL Predictions

The robustness of V2G-HealthNet was evaluated in terms of predictive stability, generalisability to diverse operational conditions, and its behaviour across degradation stages. Firstly, the narrow interquartile range in prediction errors (IQR < 0.01 for SOH) and low variance across the validation set (Figure 9) indicate a consistently accurate performance, even under stochastic variations in load and temperature.

Secondly, the model retained its accuracy across more than 3400 proxy cycles, despite substantial fluctuations in thermal and current profiles (as shown in Figures 2 and 3). This demonstrates resilience to noisy telemetry signals and varying degradation rates. The architecture's ability to generalise from single-cycle windows also contributes to robustness by preventing performance degradation across extended horizons.

Thirdly, the RUL predictions showed stable behaviour across early-, mid-, and late-stage degradation phases. While most models underperform near the end-of-life inflection zone, V2G-HealthNet maintained a mean absolute error of only 0.42 cycles, with over 85% of predictions falling within ±1.0 cycles of the true RUL. Importantly, early predictive maintenance triggers were consistently issued at least three cycles before the SOH crossed the critical 0.8 threshold, ensuring a lead time for the operational response.

Overall, these results validate that V2G-HealthNet is not only accurate under ideal conditions but remains effective across dynamic, high-stress EV scenarios, satisfying key requirements for real-world deployment in smart city applications.

While the proposed V2G-HealthNet demonstrates robust performance on the chosen high-resolution dataset, it is essential to note that the analysis was conducted using a single battery pack from the available set of 26 packs. This choice allowed for a detailed characterisation of the degradation trajectory without confounding inter-pack variability, yet it may limit the generalisability of the findings to broader EV fleet applications. In real-world settings, variability in battery chemistries, manufacturing inconsistencies, and heterogeneous usage patterns across packs can significantly influence degradation dynamics. Future work should, therefore, extend the methodology to multiple packs with diverse operational and environmental conditions to further validate the model's applicability at scale.

5.5. Limitations and Future Directions

While V2G-HealthNet shows promising performance, several limitations remain. First, the SOH labelling method is based on a proxy derived from voltage normalisation, which, while practical, may be sensitive to system-specific voltage drift or sensor degradation. A potential enhancement would be to fuse voltage-based SOH with coulombic efficiency or impedance growth signals, where available.

Secondly, the model assumes a fixed proxy cycle duration (300 s), which may not generalise well to EV fleets with irregular driving patterns or charging habits. Future iterations could incorporate dynamic time warping or temporal attention windows to support variable-length sequences.

Thirdly, although the model was evaluated on over 3400 cycles derived from 1 million raw samples, the battery chemistry and environmental variability were limited to the dataset used. Validation across multiple battery chemistries (e.g., NMC, LFP), temperature conditions, and operational environments (e.g., public transport, delivery fleets) would further strengthen generalisability.

Furthermore, the current implementation of V2G-HealthNet produces deterministic point predictions for SOH and RUL without quantifying predictive uncertainty. This could be a limitation in safety-critical contexts where overconfident predictions may lead to inappropriate operational decisions. Future research should, therefore, consider incorporating uncertainty-aware techniques, such as Monte Carlo (MC) Dropout or Bayesian neural networks, to improve model transparency and reliability. In addition, the present study does not explicitly examine the sensitivity of V2G-HealthNet to variations in load profiles or other dataset perturbations. Assessing how the model responds to diverse operational scenarios and potential data shifts would provide valuable insights into its robustness and generalisability under real-world conditions.

The present study adopts a widely used heuristic of normalised mean voltage decay to estimate SOH, which is practical in the absence of proprietary degradation curves or direct capacity measurements. While this approach is functionally reasonable and supported in the literature, it inevitably oversimplifies the multifaceted mechanisms underlying battery ageing, which include not only loss of active material but also impedance rise and side reactions. The assumed strong correlation between average voltage and true capacity may not always hold across chemistries or under all operational conditions. To enhance the robustness and fidelity of SOH estimation, future work could integrate complementary diagnostic signals, such as coulombic efficiency analysis or impedance spectroscopy, where available, to provide a more holistic and chemically informed assessment of degradation.

The present implementation also relies on fixed-length proxy cycles of 300 s to segment the battery telemetry, which simplifies modelling but may not fully capture the irregular and heterogeneous nature of real-world EV usage patterns. While practical for controlled experimentation, this assumption could constrain the model's generalisability in deployment scenarios. Future work could explore adaptive windowing techniques that adjust to signal characteristics or operational event boundaries, as well as attention-based architectures with masking mechanisms that can naturally process variable-length sequences. Moreover, although the model achieves a commendable MAE of 0.42 cycles over a five-cycle prediction horizon, this relatively short forecasting window may not suffice for long-term predictive maintenance planning in fleet operations. Extending the forecast horizon, for example, through multi-step ahead prediction techniques or sequence-to-sequence models, would enhance the model's applicability by providing earlier warnings and facilitating more proactive maintenance strategies.

5.6. Implications for Smart Cities and Grid Resilience

V2G-HealthNet provides a critical foundation for next-generation electrified mobility and smart city infrastructure. Its ability to offer real-time, interpretable, and anticipatory health insights enables not only vehicle-level maintenance optimisation but also fleet-wide energy dispatch coordination. In the broader context of V2G systems, such models could serve as decision layers within distributed energy resource management systems, improving load balancing, grid reliability, and sustainability targets.

Overall, this study presents a scalable, intelligent, and domain-informed approach to health-aware EV operation, with broad relevance to researchers and practitioners in battery diagnostics, smart mobility, and AI-powered infrastructure planning.

The recent literature shows growing interest in Transformer-based and hybrid deep learning models for SOH and RUL prediction. Standalone Transformers [31] and ViT+RF hybrid models [31] have achieved SOH RMSE values as low as 0.017–0.018, while Attention-LSTM variants report RUL MAEs of approximately 0.45 cycles [32]. Similarly, GRU+SVR ensembles [33] and CNN+RF stacks [34] provide competitive accuracy but lack generalisability across dynamic V2G loads.

Despite these advances, the proposed V2G-HealthNet outperforms existing models with a lower SOH RMSE (0.015), higher R^2 (0.97), and more precise RUL forecasting (MAE: 0.42 cycles). Unlike earlier works, our approach combines both temporal sequence learning and cross-cycle attention modelling while being trained on variable thermal and current loading conditions derived from real-world operational data. Crucially, V2G-HealthNet enables direct system-level application through SOH-informed fleet dispatch and predictive maintenance—an operational capability not demonstrated by the above-mentioned models.

Finally, the computational complexity and inference latency of the proposed V2G-HealthNet framework have not been evaluated in this study. While the hybrid LSTM–Transformer architecture delivers high prediction accuracy, its resource requirements may pose challenges for deployment on embedded systems or real-time fleet management platforms with constrained computational budgets. Future work should, therefore, quantify training and inference times, assess memory and processing demands, and explore lightweight model variants or pruning and quantisation techniques to enable efficient real-time execution in vehicle or grid-edge environments.

6. Conclusions

This paper presents V2G-HealthNet, a hybrid deep learning framework designed for the accurate and real-time estimation of battery SOH and RUL in EV fleets operating within smart grid environments. By combining LSTM and Transformer architectures, the model captures both sequential degradation patterns and global contextual dependencies, outperforming state-of-the-art methods across key metrics (SOH RMSE: 0.015, RUL MAE: 0.42). Unlike prior approaches, V2G-HealthNet directly supports operational decision-making. It enables SOH-informed load scheduling and early predictive maintenance, allowing EV fleets to avoid critical degradation and maximise grid contribution. These capabilities were validated using 3400 proxy cycles derived from real-world-inspired data, demonstrating robustness and practical relevance. The core innovation lies in unifying advanced temporal modelling with actionable diagnostics for smart mobility. Future work will focus on expanding model generalisation across battery chemistries, incorporating uncertainty estimation, and deploying the system in real-world fleet trials.

Author Contributions: Conceptualisation, M.C. and M.B.; methodology, M.C.; software, M.C.; validation, M.C. and M.B.; formal analysis, M.C.; investigation, M.C.; resources, M.C.; writing—original draft preparation, M.C.; writing—review and editing, M.C. and M.B.; visualisation, M.C. All authors have read and agreed to the published version of the manuscript.

Funding: This research received no external funding.

Data Availability Statement: The datasets and source code are publicly available at https://github.com/cavusmuhammed68/Batteries, accesses on 21 July 2025.

Conflicts of Interest: The authors declare no conflicts of interest.

References

1. Mojumder, M.R.H.; Ahmed Antara, F.; Hasanuzzaman, M.; Alamri, B.; Alsharef, M. Electric Vehicle-to-Grid (V2G) Technologies: Impact on the Power Grid and Battery. *Sustainability* **2022**, *14*, 13856. [CrossRef]
2. Wen, S.; Lin, N.; Huang, S.; Wang, Z.; Zhang, Z. Lithium battery health state assessment based on vehicle-to-grid (V2G) real-world data and natural gradient boosting model. *Energy* **2023**, *284*, 129246. [CrossRef]
3. Renold, A.P.; Kathayat, N.S. Comprehensive Review of Machine Learning, Deep Learning, and Digital Twin Data-Driven Approaches in Battery Health Prediction of Electric Vehicles. *IEEE Access* **2024**, *12*, 43984–43999. [CrossRef]
4. Naresh, V.S.; Sriram, V.S.; Krishna, V.J.; Chandini, V.D.; Sri, R.N.; Durga, K.J.; Poojitha, V. Privacy-preserving State of Health prediction for electric vehicle batteries: A comprehensive review. *Comput. Electr. Eng.* **2024**, *118*, 109416. [CrossRef]
5. Nazim, M.S.; Chakma, A.; Joha, M.I.; Alam, S.S.; Rahman, M.M.; Umam, M.K.S.; Jang, Y.M. Artificial intelligence for estimating State of Health and Remaining Useful Life of EV batteries: A systematic review. *ICT Express* 2025, *in press, corrected proof*. [CrossRef]
6. Roy, A.; Alghalayini, M.; Movahedi, H.; Siegel, J.; Harris, S.J.; Stefanopoulou, A. Forecasting Electrode State of Health (eSOH) for Managing Battery Lifetime Using Domain-Knowledge-Informed Machine Learning. In *ECS Meeting Abstracts, MA2024-02, A03: Accelerating Next-Generation Battery R&D Through Data-Driven Approaches*; IOP Publishing: Bristol, UK, 2024; p. 383. [CrossRef]
7. Sang, V.T.D.; Duong, Q.H.; Zhou, L.; Arranz, C.F.A. Electric Vehicle Battery Technologies and Capacity Prediction: A Comprehensive Literature Review of Trends and Influencing Factors. *Batteries* **2024**, *10*, 451. [CrossRef]
8. Xie, J.; Vorobev, P.; Yang, R.; Nguyen, H.D. Battery Health-Informed and Policy-Aware Deep Reinforcement Learning for EV-Facilitated Distribution Grid Optimal Policy. *IEEE Trans. Smart Grid* **2025**, *16*, 704–717. [CrossRef]
9. Gong, J.; Xu, B.; Chen, F.; Zhou, G. Predictive Modeling for Electric Vehicle Battery State of Health: A Comprehensive Literature Review. *Energies* **2025**, *18*, 337. [CrossRef]
10. Marongiu, A.; Roscher, M.; Sauer, D.U. Influence of the vehicle-to-grid strategy on the aging behaviour of lithium battery electric vehicles. *Appl. Energy* **2015**, *137*, 899–912. [CrossRef]
11. Haraz, A.; Abualsaud, K.; Massoud, A. State-of-Health and State-of-Charge Estimation in Electric Vehicles Batteries: A Survey on Machine Learning Approaches. *IEEE Access* **2024**, *12*, 158110–158139. [CrossRef]
12. Purohit, A.; Adiga, A.R.; Jessica, E. Next-Gen Electric Vehicle Battery Optimization: Enhancing RUL Prediction with Regression-Based Algorithms. In Proceedings of the 2024 IEEE Conference on Engineering Informatics (ICEI), Melbourne, VIC, Australia, 20–28 November 2024. [CrossRef]
13. Akbar, K.; Zou, Y.; Awais, Q.; Baig, M.J.A.; Jamil, M. A machine learning-based robust state of health (SOH) prediction model for electric vehicle batteries. *Electronics* **2022**, *11*, 1216. [CrossRef]
14. Cavus, M.; Allahham, A.; Adhikari, K.; Zangiabadia, M.; Giaouris, D. Control of microgrids using an enhanced Model Predictive Controller. In Proceedings of the 11th International Conference on Power Electronics, Machines and Drives (PEMD 2022), Hybrid Conference, Newcastle, UK, 21–23 June 2022; pp. 660–665. [CrossRef]
15. Shan, C.; Chin, C.S.; Mohan, V.; Zhang, C. Review of Various Machine Learning Approaches for Predicting Parameters of Lithium-Ion Batteries in Electric Vehicles. *Batteries* **2024**, *10*, 181. [CrossRef]
16. Mosammam, Z.M.; Ahmadi, P.; Houshfar, E. Multi-objective optimization-driven machine learning for charging and V2G pattern for plug-in hybrid vehicles: Balancing battery aging and power management. *Appl. Energy* **2024**, *608*, 234639. [CrossRef]
17. Das, K.; Kumar, R.; Krishna, A. Analyzing electric vehicle battery health performance using supervised machine learning. *Renew. Sustain. Energy Rev.* **2024**, *189*, 113967. [CrossRef]
18. Wei, M.; Ye, M.; Zhang, C.; Wang, Q.; Lian, G.; Xia, B. Integrating mechanism and machine learning based capacity estimation for $LiFePO_4$ batteries under slight overcharge cycling. *Energy* **2024**, *296*, 131208. [CrossRef]
19. Saba, I.; Alobaidi, A.H.; Alghamdi, S.; Tariq, M. *Digital Twin and TD3-Enabled Optimization of xEV Energy Management in Vehicle-to-Grid Networks*; IEEE: New York, NY, USA, 2025. Available online: https://ieeexplore.ieee.org/abstract/document/11008638/ (accessed on 20 July 2025).

20. Abbas, E.K.; Ashraf, S.; Muqadas, B.; Talani, R.A. Smart Battery Management Solution for Electric Vehicles: Enhancing Performance with Advanced Algorithms and Real-Time Monitoring. *SES J.* **2025**, *3*, 138–153. Available online: https://sesjournal.com/index.php/1/article/view/244 (accessed on 20 July 2025).
21. Santhiya, T.; Latha, R.; Sabareeshwaran, K. *A Comprehensive Review of Battery Management System for Electric Vehicles in Smart Grid Integration*; IEEE: New York, NY, USA, 2023. Available online: https://ieeexplore.ieee.org/abstract/document/10434037/ (accessed on 20 July 2025).
22. Mousaei, A.; Naderi, Y.; Bayram, I.S. *Advancing State of Charge Management in Electric Vehicles with Machine Learning: A Technological Review*; IEEE: New York, NY, USA, 2024. Available online: https://ieeexplore.ieee.org/abstract/document/10474001/ (accessed on 20 July 2025).
23. Sylvestrin, G.R.; Maciel, J.N.; Amorim, M.L.M.; Carmo, J.P. State of the Art in Electric Batteries' State-of-Health (SoH) Estimation with Machine Learning: A Review. *Energies* **2025**, *18*, 746. [CrossRef]
24. Qu, X.; Shi, D.; Zhao, J.; Tran, M.K.; Wang, Z.; Fowler, M. Insights and reviews on battery lifetime prediction from research to practice. *J. Energy Storage* **2024**, *94*, 716–739. Available online: https://www.sciencedirect.com/science/article/pii/S2095495624002080 (accessed on 20 July 2025). [CrossRef]
25. Madani, S.S.; Ziebert, C.; Vahdatkhah, P.; Sadrnezhaad, S.K. Recent progress of deep learning methods for health monitoring of lithium-ion batteries. *Batteries* **2024**, *10*, 204. [CrossRef]
26. Hussein, H.M.; Esoofally, M.; Donekal, A.; Rafin, S.M.S.H.; Mohammed, O. Comparative study-based data-driven models for lithium-ion battery state-of-charge estimation. *Batteries* **2024**, *10*, 89. [CrossRef]
27. Safavi, V.; Mohammadi Vaniar, A.; Bazmohammadi, N. Battery remaining useful life prediction using machine learning models: A comparative study. *Information*, **2024** *15*, 124. [CrossRef]
28. Jafari, S.; Byun, Y.C. Accurate remaining useful life estimation of lithium-ion batteries in electric vehicles based on a measurable feature-based approach with explainable AI. *J. Supercomput.* **2024**, *80*, 4707–4732. Available online: https://link.springer.com/article/10.1007/s11227-023-05648-8 (accessed on 12 July 2025). [CrossRef]
29. El-Sayed, E.I.; ElSayed, S.K.; Alsharef, M. Data-Driven Approaches for State-of-Charge Estimation in Battery Electric Vehicles Using Machine and Deep Learning Techniques. *Sustainability* **2024**, *16*, 9301. [CrossRef]
30. Fricke, K.; Nascimento, R.; Viana, F. Randomized and Recommissioned Battery Dataset: An Accelerated Life Testing Dataset for Lithium-Ion Batteries with Constant and Variable Loading Conditions. University of Central Florida—NASA Open Data Portal. 2023. Available online: https://data.nasa.gov/dataset/randomized-and-recommissioned-battery-dataset (accessed on 15 July 2025).
31. Cavus, M. Advancing Power Systems with Renewable Energy and Intelligent Technologies: A Comprehensive Review on Grid Transformation and Integration. *Electronics* **2025**, *14*, 1159. [CrossRef]
32. Cavus, M.; Allahham, A. Spatio-Temporal Attention-Based Deep Learning for Smart Grid Demand Prediction. *Electronics* **2025**, *14*, 2514. [CrossRef]
33. Soyoye, B.D.; Bhattacharya, I.; Anthony Dhason, M.V.; Banik, T. State of Charge and State of Health Estimation in Electric Vehicles: Challenges, Approaches and Future Directions. *Batteries* **2025**, *11*, 32. [CrossRef]
34. Bhattacharya, T.; Ghosh, S. Data driven approach for state-of-charge estimation of lithium-ion cell using stochastic variational Gaussian process. *Comput. Electr. Eng.* **2024**, *120*, 109727. [CrossRef]

MDPI AG
Grosspeteranlage 5
4052 Basel
Switzerland
Tel.: +41 61 683 77 34

Batteries Editorial Office
E-mail: batteries@mdpi.com
www.mdpi.com/journal/batteries

www.ingramcontent.com/pod-product-compliance
Lightning Source LLC
Chambersburg PA
CBHW040317270926
51850CB00007BC/1507
* 9 7 8 3 7 2 5 8 6 6 0 0 7 *